Lecture Notes in Physics

Founding Editors

Wolf Beiglböck

Jürgen Ehlers

Klaus Hepp

Hans-Arwed Weidenmüller

Volume 1047

Series Editors

Roberta Citro, Salerno, Italy

Peter Hänggi, Augsburg, Germany

Betti Hartmann, London, UK

Morten Hjorth-Jensen, Oslo, Norway

Maciej Lewenstein, Barcelona, Spain

Satya N. Majumdar, Orsay, France

Luciano Rezzolla, Frankfurt am Main, Germany

Angel Rubio, Hamburg, Germany

Wolfgang Schleich, Ulm, Germany

Stefan Theisen, Potsdam, Germany

James D. Wells, Ann Arbor, MI, USA

Gary P. Zank, Huntsville, AL, USA

The series Lecture Notes in Physics (LNP), founded in 1969, reports new developments in physics research and teaching - quickly and informally, but with a high quality and the explicit aim to summarize and communicate current knowledge in an accessible way. Books published in this series are conceived as bridging material between advanced graduate textbooks and the forefront of research and to serve three purposes:

- to be a compact and modern up-to-date source of reference on a well-defined topic;
- to serve as an accessible introduction to the field to postgraduate students and non-specialist researchers from related areas;
- to be a source of advanced teaching material for specialized seminars, courses and schools.

Both monographs and multi-author volumes will be considered for publication. Edited volumes should however consist of a very limited number of contributions only. Proceedings will not be considered for LNP.

Volumes published in LNP are disseminated both in print and in electronic formats, the electronic archive being available at springerlink.com. The series content is indexed, abstracted and referenced by many abstracting and information services, bibliographic networks, subscription agencies, library networks, and consortia.

Proposals should be sent to a member of the Editorial Board, or directly to the responsible editor at Springer:

Dr Lisa Scalone
lisa.scalone@springernature.com

Mikko Laine · Simona Procacci

From Inflation to Hot Big Bang

A Tutorial on Cosmological Perturbations

 Springer

Mikko Laine
AEC, Institute for Theoretical Physics
University of Bern
Bern, Switzerland

Simona Procacci
Department of Theoretical Physics
University of Geneva
Geneva, Switzerland

ISSN 0075-8450   ISSN 1616-6361   (electronic)
Lecture Notes in Physics
ISBN 978-3-032-09892-4   ISBN 978-3-032-09893-1   (eBook)
https://doi.org/10.1007/978-3-032-09893-1

This eBook was published Open Access with funding support from the Sponsoring Consortium for Open Access Publishing in Particle Physics (SCOAP3).

This Springer imprint is published by the registered company Springer Nature Switzerland AG
The registered company address is: Gewerbestrasse 11, 6330 Cham, Switzerland

If disposing of this product, please recycle the paper.

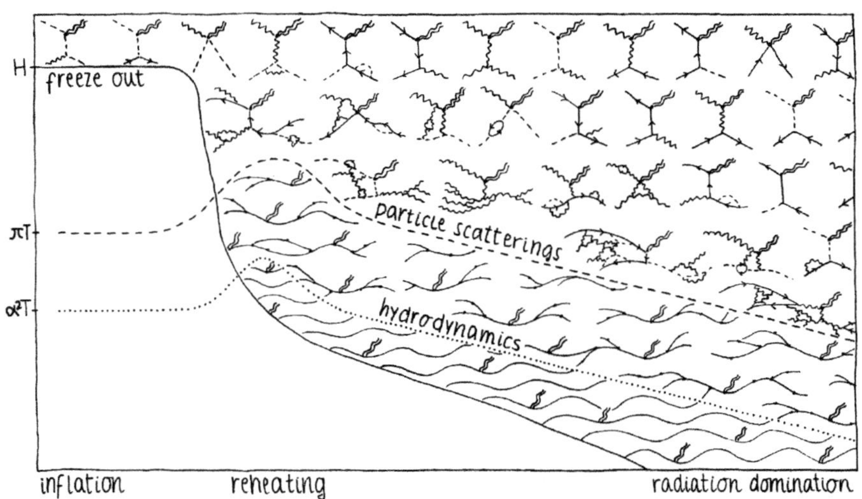

H
freeze out
πT
α²T
particle scatterings
hydrodynamics
inflation
reheating
radiation domination

Preface

These lecture notes grew out of the PhD thesis of one of the authors [1]. The latter was, in turn, influenced by unpublished but wonderfully detailed lecture notes by Hannu Kurki-Suonio [2]. Over the years, we have also benefitted from many standard books and review articles on cosmological perturbations, such as refs. [3–13]. In addition, we have had the chance to present parts of this material, notably on primordial gravitational waves, at graduate schools and small lecture series.

While preparing lectures, we developed the feeling that the basic ingredients of the inflationary paradigm do not come across in a sufficiently transparent manner from existing research monographs, despite their otherwise superb quality. They typically operate, so to say, at a higher level than what appears necessary for a first exposure. We therefore decided to attempt a more elementary treatment, specifically for pedagogical purposes, with the details carefully worked out, but with the price that not every modern development is covered. That said, we do want to convey a perspective on how gravitational-wave observations and new astrophysical data may help to probe the physics of inflation and reheating in the foreseeable future.

For helpful discussions or collaboration on topics that touched upon this work, we would like to thank (in alphabetical order) Maria Berti, Simone Biondini, Matthias Blau, Dietrich Bödeker, Chiara Caprini, Sveva Castello, Ruth Durrer, Miguel Escudero, Jacopo Ghiglieri, Nastassia Grimm, Armando Hauser, Mark Hindmarsh, Greg Jackson, Keijo Kajantie, Philipp Klose, Helena Kolesova, Antonino Midiri, Germano Nardini, Sami Nurmi, Arttu Rajantie, Alica Rogelj, Stefan Sandner, Jan Schütte-Engel, Francesco Sorrenti, Enrico Speranza, Anna Tokareva, Jorinde van de Vis, Sebastian Zell, Yan Zhu, and many others.

<table>
<tr><td>Bern, Switzerland</td><td>Mikko Laine</td></tr>
<tr><td>Geneva, Switzerland</td><td>Simona Procacci</td></tr>
<tr><td>Towards late August 2025</td><td></td></tr>
</table>

Literature

1. Procacci, S.: Towards a thermally complete study of inflationary predictions. PhD thesis, University of Bern, Switzerland (2023). https://boristheses.unibe.ch/4522/
2. Kurki-Suonio, H.: Cosmological Perturbation Theory. University of Helsinki, Finland (2024). https://www.mv.helsinki.fi/home/hkurkisu/
3. Kodama, H., Sasaki, M.: Cosmological Perturbation Theory. Prog. Theor. Phys. Suppl. **78**, 1 (1984). https://doi.org/10.1143/PTPS.78.1
4. Linde, A.D.: Particle Physics and Inflationary Cosmology. Contemp. Concepts Phys. **5**, 1 (1990). hep-th/0503203
5. Mukhanov, V.F., Feldman, H.A., Brandenberger, R.H.: Theory of cosmological perturbations. Phys. Rept. **215**, 203 (1992). https://doi.org/10.1016/0370-1573(92)90044-Z
6. Liddle, A.R., Lyth, D.H.: Cosmological Inflation and Large-Scale Structure. Cambridge University Press (2000). https://doi.org/10.1017/CBO9781139175180
7. Mukhanov, V., Winitzki, S.: Introduction to Quantum Effects in Gravity. Cambridge University Press (2007). https://doi.org/10.1017/CBO9780511809149
8. Weinberg, S.: Cosmology. Oxford University Press (2008)
9. Durrer, R.: The Cosmic Microwave Background. Cambridge University Press (2020). https://doi.org/10.1017/9781316471524
10. Malik, K.A., Wands, D.: Cosmological perturbations. Phys. Rept. **475**, 1 (2009). https://doi.org/10.1016/j.physrep.2009.03.001, 0809.4944
11. Baumann, D.: Inflation. Contribution to TASI (2009). https://doi.org/10.1142/9789814327183_0010, 0907.5424
12. Gorbunov, D.S., Rubakov, V.A.: Introduction to the Theory of the Early Universe: Cosmological Perturbations and Inflationary Theory. World Scientific (2011). https://doi.org/10.1142/7873
13. Huterer, D.: A Course in Cosmology—From Theory to Practice. Cambridge University Press (2023). https://doi.org/10.1017/9781009070232

Competing Interests The authors have no competing interests to declare that are relevant to the content of this manuscript.

Units, Notation, and Hints on Software Tools

We employ units in which the speed of light, c, the reduced Planck constant, $\hbar$, and the Boltzmann constant, k_B, equal unity. The Einstein equations are parametrized by the Newton constant, G, which can conveniently be re-expressed via the Planck mass,

$$G \equiv \frac{1}{m_\mathrm{pl}^2}, \quad m_\mathrm{pl} = 1.2209 \times 10^{19}\,\mathrm{GeV}. \tag{1}$$

In the literature, the so-called reduced Planck mass, $M_\mathrm{pl} \equiv m_\mathrm{pl}/\sqrt{8\pi}$, is often employed. In astrophysics, distances are expressed in units of a parsec (pc),

$$1\mathrm{pc} = 3.0857 \times 10^{16}\,\mathrm{m} \simeq 3.2616\,\mathrm{ly}, \tag{2}$$

where a light year (ly) is the distance that light travels in a year. Temperatures are measured in energy units, obtained through a multiplication of Kelvin (K) with the Boltzmann constant,

$$k_\mathrm{B} \times (1\,\mathrm{K}) = 8.6173 \times 10^{-5}\,\mathrm{eV} \iff 1\mathrm{eV} \overset{k_\mathrm{B}=1}{=} 11\,605\,\mathrm{K}. \tag{3}$$

Indices denoted by small Greek letters describe components in the four-dimensional space-time, $\{\alpha, \beta, \mu, \nu, \rho, \sigma, ...\} \in \{0, 1, 2, 3\}$. Small Roman letters are used for spatial indices in three dimensions, $\{i, j, k, l, m, n, ...\} \in \{1, 2, 3\}$. Derivatives of a generic function Q of an arbitrary variable x and coordinates $x^\mu = (x^0, x^i)$ are mostly abbreviated as

$$Q_{,x} \equiv \partial_x Q(x, \ldots) \quad \textit{derivative with respect to the variable } x, \tag{4}$$

$$Q_{;\mu} \equiv D_\mu Q(x^\mu, \ldots) \quad \textit{covariant derivative with respect to } x^\mu. \tag{5}$$

On a manifold with metric $g_{\mu\nu}$ and inverse metric $g^{\mu\nu}$, covariant derivatives read

$$Q^{\alpha\beta\cdots\,;\nu}_{\rho\sigma\cdots} = Q^{\alpha\beta\cdots}_{\rho\sigma\cdots\,;\mu}g^{\mu\nu}\,, \tag{6}$$

$$Q^{\alpha\beta\cdots}_{\rho\sigma\cdots\,;\mu} \equiv Q^{\alpha\beta\cdots}_{\rho\sigma\cdots\,,\mu} + \Gamma^{\alpha}_{\mu\nu}Q^{\nu\beta\cdots}_{\rho\sigma\cdots} + \Gamma^{\beta}_{\mu\nu}Q^{\alpha\nu\cdots}_{\rho\sigma\cdots} - \Gamma^{\nu}_{\mu\rho}\,Q^{\alpha\beta\cdots}_{\nu\sigma\cdots} - \Gamma^{\nu}_{\mu\sigma}\,Q^{\alpha\beta\cdots}_{\rho\nu\cdots} + \cdots\,, \tag{7}$$

where $(\ldots)_{,\mu} \equiv \partial(\ldots)/\partial x^{\mu}$ is a partial derivative, and $\Gamma^{\alpha}_{\mu\nu}$ are the *Christoffel symbols*. Repeated indices are summed over, also when they are purely spatial.

When the expansion of the universe plays an important role, it is often useful to work with so-called comoving (or conformal) coordinates, and denote them by $\mathcal{X} \equiv (\tau, \mathbf{x})$. The *cosmological scale factor* is denoted by a. Physical (or local Minkowskian) coordinates are $\mathcal{R} \equiv (t, \mathbf{r})$, with $\mathbf{r} \equiv a\mathbf{x}$. For spatial momenta, $\mathbf{k}$ is comoving, $\mathbf{p} \equiv \mathbf{k}/a$ is local Minkowskian. The absolute value of the *comoving momentum* is $k \equiv |\mathbf{k}|$. The absolute value of the *physical momentum*, $p \equiv |\mathbf{p}|$, unfortunately coincides with the notation used for the pressure; given that the dimensions and physical roles of these quantities are different, it should be clear from the context which is meant. More relations between comoving and physical coordinates will be introduced in Sect. 1.2.

For Fourier analysis in spatial directions, we use the conventions

$$Q(\cdot, \mathbf{x}) = \int_{\mathbf{k}} Q(\cdot, \mathbf{k})e^{i\mathbf{k}\cdot\mathbf{x}}\,, \qquad\qquad \int_{\mathbf{k}} \equiv \int \frac{\mathrm{d}^3\mathbf{k}}{(2\pi)^3}\,, \tag{8}$$

$$Q(\cdot, \mathbf{k}) = \int_{\mathbf{x}} Q(\cdot, \mathbf{x})e^{-i\mathbf{k}\cdot\mathbf{x}}\,, \qquad\qquad \int_{\mathbf{x}} \equiv \int \mathrm{d}^3\mathbf{x}\,. \tag{9}$$

In four-dimensional, conformal and comoving space-time, denoting $\mathcal{K} \equiv (\omega, \mathbf{k})$, Fourier analysis is defined as

$$Q(\mathcal{X}) = \int_{\mathcal{K}} Q(\mathcal{K})e^{i(-\omega\tau + \mathbf{k}\cdot\mathbf{x})}\,, \qquad \int_{\mathcal{K}} \equiv \int \frac{\mathrm{d}\omega}{2\pi}\int_{\mathbf{k}}\,, \tag{10}$$

$$Q(\mathcal{K}) = \int_{\mathcal{X}} Q(\mathcal{X})e^{i(+\omega\tau - \mathbf{k}\cdot\mathbf{x})}\,, \qquad \int_{\mathcal{X}} \equiv \int \mathrm{d}\tau\int_{\mathbf{x}}\,. \tag{11}$$

If there is a danger of confusion, we denote Fourier representations as $\tilde{Q}(\mathcal{K}) \equiv Q(\mathcal{K})$. In local Minkowskian coordinates, the metric signature is taken to be $\eta_{00} = -1$ for the *Minkowski metric*, $\eta_{\mu\nu}$. The scalar product is thus $\mathcal{K}\cdot\mathcal{X} = -\omega\tau + \mathbf{k}\cdot\mathbf{x}$.

In a number of appendices, we provide `python` codes for numerical or algebraic computations. `Python` is a scripting language, meaning that the programs are not compiled, but executed line by line. The execution can happen via a graphical interface, or as scripts. To give an example, on a `linux` operating system, the basic kernel and the necessary libraries can be installed as

```
sudo apt-get install python3

sudo apt-get install python3-numpy

sudo apt-get install python3-scipy

sudo apt-get install python3-sympy
```

Subsequently, a script, if named `*.py`, can be executed from the command line with "`python3 *.py`". Documentation (tutorials and reference manuals) can be efficiently found with internet search, with `docs.python.org`, `numpy.org`, `docs.scipy.org`, and `docs.sympy.org` being the "official" sources.

We have tested our scripts with `python` release `3.12.3`, `numpy 1.26.4`, `scipy 1.11.4`, `sympy 1.12.7`. New versions can be obtained by running the `install` commands again, however the `linux` kernel needs to be up-to-date in order for the upgrades to proceed.

We end by remarking that, at the time of writing, the symbolic package `sympy` is not efficient, i.e. it becomes very slow with long expressions. However, the symbolic routines can be straightforwardly transcribed to other languages, such as `Mathematica`. All our `python` codes, as well as `Mathematica` transcriptions of the symbolic routines, are available at `https://github.com/laineproc acci/From-inflation-to-hot-big-bang/`.

General Outline

Understanding the origins of the visible universe is an age-old challenge. And yet, we may learn more about this in the next couple of decades. This is thanks to a qualitatively new tool that will become available, as several gravitational-wave interferometers are foreseen to go in operation, opening up new frequency windows for observing the distant past. In addition, the properties of the cosmic microwave background, and of the large-scale structure in galaxies and galaxy clusters, will be mapped with increasing sensitivity.

At the same time, on the theoretical side, the predictions for early universe observables are not always robust, both because they are model-dependent, and because the simplest computational tools sometimes invoke unintended approximations. So there is room for progress, calling for the attention of the next generations of curious and critical minds.

The purpose of these lectures is to offer for a detailed hands-on introduction to the earliest moments of the universe that can presumably be understood *without* having a theory of quantum gravity at our disposal. The main tool is then just general relativity, expanded to linear order in perturbations around a cosmological background solution. For certain aspects, notably initial conditions, we need to recall basic ingredients from quantum field theory as well. Statistical physics and elements of thermal field theory are employed to study the transition from a postulated vacuum-dominated, inflationary, universe to the observationally established hot epoch. We mostly do *not* delve into the recently popular and interrelated topics of (quantum) loop corrections and (classical) second-order perturbations around cosmological backgrounds. However, we make an exception with so-called scalar-induced gravitational waves, which lend us an opportunity to discuss some of the challenges that arise, when trying to improve on the accuracy of inflationary predictions beyond the linear order.

The overall structure of our presentation is as follows. In Part I (Chaps. 1–4), we develop the general model-independent formalism for how small perturbations evolve in a cosmological background, and what kind of observable signatures they lead to. In Part II (Chaps. 5–10), we turn to the inflationary paradigm, as a possible explanation for the origin of the perturbations. The dynamics of inflationary perturbations coupled with an emergent thermal plasma and the corresponding observables are then studied using the results derived in Part I.

Contents

Part I
Basics of Perturbation Dynamics in Cosmology

Einstein Equations and the Cosmological Background Solution

1

Abstract

We review the contents of the Einstein equations of general relativity. The ingredients needed for their left-hand side, the Einstein tensor, are explained. The right-hand side, energy-momentum tensor, is specified for typical systems appearing in early universe cosmology (weakly coupled scalar field, thermalized plasma, a coupled system). The concept of a homogeneous and isotropic Friedmann-Lemaître-Robertson-Walker (FLRW) universe is introduced, and prototypical 'background' solutions of the Einstein equations are displayed. We explain why a 'spatially flat' background is generally adopted as a sensible initial condition for the universe's evolution, even if in the later sections perturbations of the spatial curvature are introduced and turn out to play a key role.

Keywords

Cosmic microwave background · Blackbody spectrum · Metric tensor · Physical time · Conformal time · Christoffel symbols · Ricci tensor · Ricci scalar · Einstein tensor · Spatial curvature · Friedmann equations · Thermal plasma · Elementary scalar field · Langevin equation · De Sitter space-time · Matter domination · Radiation domination · Kination

1.1 Overview of Early-Universe Cosmology

As in all fields of science, when we attempt to understand the early universe, the method is to use empirical facts as a constraint on the type of mathematical models that we develop. Arguably, the single most important observation is the presence of a *Cosmic Microwave Background* (CMB) radiation. It is an electromagnetic signal that looks almost the same in every direction in the sky that we observe. Therefore, we assume the universe to be homogeneous and isotropic at scales as large as our observable patch. Furthermore, the CMB displays very accurately a *blackbody spec-*

© The Author(s) 2026

3

M. Laine and S. Procacci, *From Inflation to Hot Big Bang*, Lecture Notes in Physics 1047, https://doi.org/10.1007/978-3-032-09893-1_1

trum (Planck spectrum). From this we deduce that the universe was 'hot' at early times, forming a plasma-like state, which emitted the CMB photons that we observe.

We also know that the universe expands. The expansion rate is inferred from two types of measurements. The spectral lines of distant astrophysical objects show a redshift, as if they were moving away from us with a certain velocity. Their distance from us is estimated from the luminosity of other objects that are assumed to be well understood, such as certain types of supernovea. The distance and redshift show an approximately linear relationship to each other. Extrapolating backwards in time, the objects must have been closer to us and to each other at earlier times.

Combining these observations has led to the notion of a *hot big bang*. There are several other ingredients supporting the paradigm, notably successful nucleosynthesis of light elements, a credible theory for how large-scale structures could form from small initial inhomogeneities, and indirect evidence for the presence of a cosmological neutrino background, having similar thermal characteristics as the CMB.

Historically, the hot big bang was envisaged as some kind of an *initial singularity*, possibly associated with an infinite temperature. From the modern perspective, however, we would not like to extrapolate to higher temperatures than are empirically testable. The reason is that the notion of a temperature assumes the presence of interactions, so that parts of the Fock space can be repopulated as the temperature changes. Given that interactions require time, i.e. cannot take place infinitely fast, it appears more likely that the initial state of the universe was some non-thermal (quantum) state, which later on equilibrated and reached a *maximal temperature*.

Let us add some numbers to this overall picture. The hottest epochs on which we have indirect evidence are those at which *big bang nucleosynthesis* (BBN) starts. The corresponding temperatures are around nuclear physics energy scales, $T \sim 0.1\,\mathrm{MeV}$, and the times, as measured from the moment at which the temperature would diverge if we naively extrapolated backwards, are $t > 1$ s. At these energies, the behaviour of electrons, positrons and photons is described by the well-established theory of Quantum Electrodynamics (QED). Therefore we are confident to find a relativistic plasma (cf., e.g., refs. [1–3]), where photons interact with ionized matter via *Compton scattering* (sometimes we also refer to its non-relativistic limit, *Thomson scattering*). As the universe cools down to $T \sim 3000\,\mathrm{K}$, or 0.3 eV, after $\sim 300\,000$ years, free electrons combine with the positive protons to form a gas [4], in a process that we call *recombination*. The universe thus becomes transparent to photons, which at this moment *decouple*, and start travelling freely across space and time. These are the CMB photons that we observe as background electromagnetic radiation today.

Because of the expansion of the universe, resulting in a redshift, the decoupled CMB photons are observed to have a temperature $T_0 = 2.7255\,\mathrm{K}$ today [5]. The subscript $(...)_0$ refers to the current value. The temperature is almost the same in every direction, with fluctuations of the order of $\delta T/T_0 \sim 10^{-5}$ at small scales [6]. Within the visible universe, we observe many structures (galaxies, galaxy clusters, etc.), but on the average it looks the same at distances larger than $\sim 200\,\mathrm{Mpc}$, where the astronomical distance unit pc corresponds to about 3 light years, or 3.1×10^{16} m (cf. Eq. (2)). It is conceivable that the structure seen at distances smaller than $\sim 200\,\mathrm{Mpc}$ could have formed via gravitational collapse from early perturbations. Then, the early universe would have been almost homogeneous and isotropic. We return to this issue from an *a posteriori* perspective at the end of Sect. 9.6.

Given this empirical picture, we describe the early universe within a *perturbative approach*. We first determine a *background solution*, which is assumed to be exactly homogeneous and isotropic (in the language of statistical physics, this could be called the *mean-field solution*). This is the topic of the present chapter. Subsequently, we add perturbations, and see how they evolve around the background.

1.2 Choice of Coordinates for a Homogeneous and Isotropic Universe

We now proceed to a mathematical description of the background solution. In a homogeneous and isotropic universe, we expect physical quantities Q to only depend on time, and, for later purposes, we denote these as $Q(\tau, \mathbf{x}) \to \bar{Q}(\tau)$.

The basic object of Einstein's general relativity is the space-time metric, $g_{\mu\nu}$, or equivalently, the invariant separation between two space-time points, $\mathrm{d}s^2 = g_{\mu\nu}\,\mathrm{d}x^\mu\,\mathrm{d}x^\nu$, where repeated indices are summed over. The inverse metric is denoted by $g^{\mu\nu}$, and it satisfies

$$g^{\mu\rho} g_{\rho\nu} \;=\; \delta^\mu_\nu \;\equiv\; \begin{cases} 1\,, & \mu = \nu \\ 0\,, & \mu \neq \nu \end{cases} . \tag{1.1}$$

An expanding, homogeneous and isotropic universe is described by the Friedmann-Lemaître-Robertson-Walker (FLRW) metric,

$$\mathrm{d}s^2_{\mathrm{FLRW}} \;\equiv\; -\mathrm{d}t^2 + a^2(t)\,\mathrm{d}\mathbf{x}^2_\kappa\,, \qquad \mathrm{d}\mathbf{x}^2_\kappa \;\equiv\; \frac{\mathrm{d}r^2}{1 - \kappa r^2} + r^2\,\mathrm{d}\mathbf{\Omega}^2\,. \tag{1.2}$$

Here κ is a constant representing overall spatial curvature (cf. Sect. 1.5), and $\mathrm{d}\mathbf{\Omega}^2 = \mathrm{d}\theta^2 + \sin^2\theta\,\mathrm{d}\phi^2$ is the angular distance squared. The parameter $a(t)$ is the *cosmological scale factor*, describing the expansion of the universe, and t is the *physical time*. We assume a positive-definite time, $t \in [0, \infty)$, even if it is not possible to give a physical definition of the moment $t = 0$ (the basic equations are time-translation invariant).

Another useful coordinate system is defined by the *conformal time* direction, $x^0 = \tau$, such that

$$\mathrm{d}\tau \;\equiv\; \frac{\mathrm{d}t}{a(t)} \qquad \Rightarrow \qquad \mathrm{d}s^2_{\mathrm{FLRW}} \;=\; a^2(\tau)\left[-\mathrm{d}\tau^2 + \mathrm{d}\mathbf{x}^2_\kappa \right]. \tag{1.3}$$

For the conformal time, it is conventional to consider negative values as well, with the initial moment pushed to $-\infty$. The name *conformal* is motivated by the fact that, if the spatial curvature vanishes, $\kappa = 0\,\mathrm{m}^{-2}$, the metric in Eq. (1.3) is related to the Minkowski metric by a conformal transformation, i.e. an overall scaling of dimensionful quantities.

The relative change of the scale factor a in time measures how fast the universe expands, and is called the *Hubble rate*. In the two coordinate systems of Eqs. (1.2) and (1.3), it is defined as

$$H \equiv \frac{\dot{a}}{a}\,, \quad \mathcal{H} \equiv \frac{a'}{a}\,, \tag{1.4}$$

respectively. Here, to distinguish time derivatives of an arbitrary physical quantity Q, we have introduced the notation

$$\dot{Q} \equiv \partial_t Q\,, \quad Q' \equiv \partial_\tau Q\,. \tag{1.5}$$

From Eq. (1.3), we obtain relations between the physical Hubble rate and its conformal counterpart,

$$\mathcal{H} \overset{(1.3)}{\underset{(1.4)}{=}} aH\,, \qquad\qquad \mathcal{H}' = a^2(H^2 + \dot{H}) = a\ddot{a}\,, \tag{1.6}$$

$$Q' \overset{(1.3)}{\underset{(1.5)}{=}} a\dot{Q}\,, \qquad\qquad Q'' = a^2(\ddot{Q} + H\dot{Q})\,. \tag{1.7}$$

Let us now derive the Einstein tensor for a background universe described by the FLRW metric with conformal time and generic spatial curvature κ, cf. Eq. (1.3). Using reduced-circumference polar coordinates, (τ, r, θ, ϕ), the corresponding background metric is

$$\bar{g}_{\mu\nu} \overset{(1.2)}{\underset{(1.3)}{=}} a^2 \operatorname{diag}\!\left(-1,\ (1 - \kappa r^2)^{-1},\ r^2,\ r^2 \sin^2\theta\right)\,. \tag{1.8}$$

The *Christoffel symbols* are given by

$$\bar{\Gamma}^{\rho}_{\mu\nu} \equiv \frac{1}{2}\bar{g}^{\rho\sigma}(\bar{g}_{\sigma\mu,\nu} + \bar{g}_{\sigma\nu,\mu} - \bar{g}_{\mu\nu,\sigma})\,, \tag{1.9}$$

where $(\dots)_{,\mu} \equiv \partial(\dots)/\partial x^\mu$. Noting the symmetry $\bar\Gamma^\rho_{\nu\mu} = \bar\Gamma^\rho_{\mu\nu}$, there are 4×10 independent ones. Carrying out an explicit computation, the non-vanishing among them are

$$\bar\Gamma^\tau_{\tau\tau} = \bar\Gamma^r_{\tau r} = \bar\Gamma^\theta_{\tau\theta} = \bar\Gamma^\phi_{\tau\phi} = \frac{1}{2}a^{-2}(a^2)' = \mathcal{H}\,, \tag{1.10}$$

$$\bar\Gamma^\tau_{rr} = \frac{1}{2}\underbrace{(-a^{-2})}_{\bar g^{\tau\tau}}\underbrace{\left(-\frac{a^2}{1-\kappa r^2}\right)'}_{-\bar g_{rr}} = \frac{\mathcal{H}}{1-\kappa r^2}\,, \tag{1.11}$$

$$\bar\Gamma^\tau_{\theta\theta} = \frac{1}{2}(-a^{-2})\underbrace{(-a^2 r^2)'}_{-\bar g_{\theta\theta}} = r^2\mathcal{H}\,, \tag{1.12}$$

$$\bar\Gamma^\tau_{\phi\phi} = \frac{1}{2}(-a^{-2})\underbrace{(-a^2 r^2\sin^2\theta)'}_{-\bar g_{\phi\phi}} = r^2\sin^2\theta\,\mathcal{H}\,, \tag{1.13}$$

$$\bar\Gamma^r_{rr} = \frac{1}{2}\underbrace{a^{-2}(1-\kappa r^2)}_{\bar g^{rr}}\,\partial_r\left(\frac{a^2}{1-\kappa r^2}\right) = \frac{\kappa r}{1-\kappa r^2}\,, \tag{1.14}$$

$$\bar\Gamma^r_{\theta\theta} = \frac{1}{2}a^{-2}(1-\kappa r^2)\partial_r(-a^2 r^2) = -r\,(1-\kappa r^2)\,, \tag{1.15}$$

$$\bar\Gamma^r_{\phi\phi} = \frac{1}{2}a^{-2}(1-\kappa r^2)\partial_r(-a^2 r^2\sin^2\theta) = -r\sin^2\theta\,(1-\kappa r^2)\,, \tag{1.16}$$

$$\bar\Gamma^\theta_{r\theta} = \bar\Gamma^\phi_{r\phi} = \frac{1}{2}r^{-2}\partial_r(r^2) = r^{-1}\,, \tag{1.17}$$

$$\bar\Gamma^\theta_{\phi\phi} = \frac{1}{2}\underbrace{a^{-2}r^{-2}}_{\bar g^{\theta\theta}}\,\partial_\theta(-a^2 r^2\sin^2\theta) = -\sin\theta\cos\theta\,, \tag{1.18}$$

$$\bar\Gamma^\phi_{\theta\phi} = \frac{1}{2}\underbrace{a^{-2}r^{-2}\sin^{-2}\theta}_{\bar g^{\phi\phi}}\,\partial_\theta(a^2 r^2\sin^2\theta) = \frac{\cos\theta}{\sin\theta}\,. \tag{1.19}$$

Given the Christoffel symbols, the *Ricci tensor* is defined as

$$\bar R_{\mu\nu} \equiv \bar\Gamma^\alpha_{\mu\nu,\alpha} - \bar\Gamma^\alpha_{\mu\alpha,\nu} + \bar\Gamma^\beta_{\mu\nu}\bar\Gamma^\alpha_{\beta\alpha} - \bar\Gamma^\beta_{\mu\alpha}\bar\Gamma^\alpha_{\nu\beta}\,. \tag{1.20}$$

Inserting Eqs. (1.10)–(1.19), the result is diagonal after cancellations, with the entries

$$\bar R_{\tau\tau} = \underbrace{\mathcal{H}'}_{\bar\Gamma^\alpha_{\tau\tau,\alpha}} - \underbrace{4\mathcal{H}'}_{\bar\Gamma^\alpha_{\tau\alpha,\tau}} + \underbrace{\mathcal{H}(4\mathcal{H})}_{\bar\Gamma^\beta_{\tau\tau}\bar\Gamma^\alpha_{\beta\alpha}} - \underbrace{4\mathcal{H}^2}_{\bar\Gamma^\beta_{\tau\alpha}\bar\Gamma^\alpha_{\tau\beta}} = -3\mathcal{H}'\,, \tag{1.21}$$

$$\bar{R}_{rr} = \underbrace{\frac{1}{1-\kappa r^2}\left(\mathcal{H}' + \cancel{\kappa} + \cancel{\frac{2\kappa^2 r^2}{1-\kappa r^2}}\right)}_{\bar\Gamma^\alpha_{rr,\alpha}} - \underbrace{\left[\frac{1}{1-\kappa r^2}\left(\cancel{\kappa} + \cancel{\frac{2\kappa^2 r^2}{1-\kappa r^2}}\right) - \cancel{\frac{2}{r^2}}\right]}_{\bar\Gamma^\alpha_{r\alpha,r}}$$

$$+ \underbrace{\frac{1}{1-\kappa r^2}\left(4\mathcal{H}^2 + 2\kappa + \cancel{\frac{\kappa^2 r^2}{1-\kappa r^2}}\right)}_{\bar\Gamma^\beta_{rr}\bar\Gamma^\alpha_{\beta\alpha}} - \underbrace{\left[\frac{1}{1-\kappa r^2}\left(2\mathcal{H}^2 + \cancel{\frac{\kappa^2 r^2}{1-\kappa r^2}}\right) + \cancel{\frac{2}{r^2}}\right]}_{\bar\Gamma^\beta_{r\alpha}\bar\Gamma^\alpha_{r\beta}}$$

$$= \frac{\mathcal{H}' + 2\mathcal{H}^2 + 2\kappa}{1-\kappa r^2}, \tag{1.22}$$

$$\bar{R}_{\theta\theta} = \underbrace{r^2(\mathcal{H}'+2\kappa) - (\cancel{1}-\cancel{\kappa r^2})}_{\bar\Gamma^\alpha_{\theta\theta,\alpha}} + \underbrace{\cancel{1} + \cancel{\frac{\cos^2\theta}{\sin^2\theta}}}_{-\bar\Gamma^\alpha_{\theta\alpha,\theta}}$$

$$+ \underbrace{r^2(4\mathcal{H}^2 - \cancel{\kappa}) - 2(1-\cancel{\kappa r^2})}_{\bar\Gamma^\beta_{\theta\theta}\bar\Gamma^\alpha_{\beta\alpha}} - \underbrace{\left[2r^2\mathcal{H}^2 - 2(1-\cancel{\kappa r^2}) + \frac{\cos^2\theta}{\sin^2\theta}\right]}_{\bar\Gamma^\beta_{\theta\alpha}\bar\Gamma^\alpha_{\theta\beta}}$$

$$= r^2(\mathcal{H}' + 2\mathcal{H}^2 + 2\kappa), \tag{1.23}$$

$$\bar{R}_{\phi\phi} = \underbrace{r^2\sin^2\theta\,(\mathcal{H}'+2\kappa) - \sin^2\theta(\cancel{1}-\cancel{\kappa r^2}) - \cancel{\partial_\theta[\sin\theta\cos\theta]}}_{\bar\Gamma^\alpha_{\phi\phi,\alpha}} - \underbrace{0}_{\bar\Gamma^\alpha_{\phi\alpha,\phi}}$$

$$+ \underbrace{r^2\sin^2\theta\,(4\mathcal{H}^2-\cancel{\kappa}) - \cancel{2\sin^2\theta(1-\kappa r^2)} - \cancel{\sin\theta\cos\theta\,\frac{\cos\theta}{\sin\theta}}}_{\bar\Gamma^\beta_{\phi\phi}\bar\Gamma^\alpha_{\beta\alpha}}$$

$$- \underbrace{\left[r^2\sin^2\theta\,(2\mathcal{H}^2) - \cancel{2\sin^2\theta(1-\kappa r^2)} - \cancel{2\sin\theta\cos\theta\,\frac{\cos\theta}{\sin\theta}}\right]}_{\bar\Gamma^\beta_{\phi\alpha}\bar\Gamma^\alpha_{\phi\beta}}$$

$$= r^2\sin^2\theta\,(\mathcal{H}' + 2\mathcal{H}^2 + 2\kappa). \tag{1.24}$$

Raising one index,

$$\bar{R}^{\mu}{}_{\nu} = \bar{g}^{\mu\rho}\,\bar{R}_{\rho\nu}\,, \tag{1.25}$$

the components of the Ricci tensor simplify to

$$\bar{R}^{\tau}{}_{\tau} \overset{(1.21)}{\underset{(1.8)}{=}} 3a^{-2}\mathcal{H}'\,, \tag{1.26}$$

$$\bar{R}^{r}{}_{r} \overset{(1.22)}{\underset{(1.8)}{=}} \bar{R}^{\theta}{}_{\theta} \overset{(1.23)}{\underset{(1.8)}{=}} \bar{R}^{\phi}{}_{\phi} \overset{(1.24)}{\underset{(1.8)}{=}} a^{-2}(\mathcal{H}' + 2\mathcal{H}^2 + 2\kappa)\,. \tag{1.27}$$

The *Ricci scalar* (sometimes also referred to as *curvature*) is therefore

$$\bar{R} \equiv \bar{R}^{\mu}{}_{\mu} \overset{(1.26)}{\underset{(1.27)}{=}} 6a^{-2}(\mathcal{H}' + \mathcal{H}^2 + \kappa)\,. \tag{1.28}$$

This yields the *Einstein tensor*, for which we consider two versions,

$$\bar{G}^{\mu}{}_{\nu} \equiv \bar{R}^{\mu}{}_{\nu} - \frac{1}{2}\delta^{\mu}_{\nu}\bar{R}\,, \quad \bar{G}_{\mu\nu} = \bar{R}_{\mu\nu} - \frac{1}{2}\bar{g}_{\mu\nu}\bar{R}\,. \tag{1.29}$$

The former index placement leads to the components

$$\bar{G}^{\tau}{}_{\tau} \overset{(1.26)}{\underset{(1.28)}{=}} -3a^{-2}(\mathcal{H}^2 + \kappa)\,, \tag{1.30}$$

$$\bar{G}^{r}{}_{r} = \bar{G}^{\theta}{}_{\theta} = \bar{G}^{\phi}{}_{\phi} \overset{(1.27)}{\underset{(1.28)}{=}} -a^{-2}(2\mathcal{H}' + \mathcal{H}^2 + \kappa)\,. \tag{1.31}$$

Therefore, evaluating the components of the Einstein equations,

$$\boxed{\bar{G}^{\mu}{}_{\nu} = 8\pi G\,\bar{T}^{\mu}{}_{\nu}\,,} \tag{1.32}$$

the off-diagonals vanish. We are left with two identities, from the 00 and ij-components.

   In the literature, the index placement in Eq. (1.32) is probably the one most frequently adopted. Its technical benefit is that then $g^{\mu}{}_{\nu} = \delta^{\mu}_{\nu}$ is trivial to all orders, simplifying the derivation of the Einstein tensor, particularly when we go to perturbations. However, there is also a drawback, namely that with the mixed index ordering the Ricci and Einstein tensors are in general not symmetric. If we instead consider

$$\boxed{\bar{G}_{\mu\nu} = 8\pi G\,\bar{T}_{\mu\nu}\,,} \tag{1.33}$$

then the tensors are manifestly symmetric. This simplifies their decomposition into scalar, vector and tensor parts, on which we rely when we study perturbations (cf. Sect. 3.1).

1.3   Einstein Equations for Different Matter Contents

In order to solve Eq. (1.32) or Eq. (1.33), we need to specify the energy-momentum tensor appearing on the right-hand side. To achieve this, physical input is needed. The simplest system is a thermalized one, because in its rest frame its state is fully characterized by few quantities, a local temperature and possibly chemical potentials, irrespective of the microscopic properties of the particles that it is composed of. If the system does not equilibrate, specifying its energy-momentum tensor requires more information. For instance, we need to know the spin of the constituents; the simplest example are spin-0 particles, described by a scalar field. For understanding realistic situations, we need to consider multicomponent systems, with some thermalized degrees of freedom (e.g. electrons and photons) and other non-equilibrium ones (e.g. neutrinos and dark matter). In the present section, we define the energy-momentum tensor and background evolution equations for three prototype systems: a thermal plasma, an elementary scalar field, and a two-component system made of an elementary scalar field interacting with a plasma.

An ideal thermal plasma (or fluid)

In the local rest frame of a thermalized plasma, the component $-T^0{}_0$ of the energy-momentum tensor equals the *energy density* ($\equiv e$), and the components $T^i{}_j$, with $i = j$, equal the *pressure* ($\equiv p$). Denoting by u^μ the *plasma flow velocity*, and by $\bar{p}$, $\bar{e}$, $\bar{u}^\mu$ the background values of the given quantities, and assuming homogeneity so that no spatial derivatives appear, a covariant form of the energy-momentum tensor reads

$$\bar{T}^\mu{}_\nu = (\bar{e} + \bar{p})\,\bar{u}^\mu \bar{u}_\nu + \bar{p}\,\delta^\mu_\nu\,, \qquad \bar{T}_{\mu\nu} = (\bar{e} + \bar{p})\,\bar{u}_\mu \bar{u}_\nu + \bar{p}\,\bar{g}_{\mu\nu}\,. \tag{1.34}$$

If e and p are parametrized by a single quantity, for instance temperature or chemical potential (but not both at the same time), their background values are related to each other by an *equation of state*,

$$\bar{p} = \bar{p}(\bar{e})\,, \qquad c_s^2 \equiv \frac{\partial \bar{p}}{\partial \bar{e}}\,, \tag{1.35}$$

where c_s is the *speed of sound*. Because of isotropy, we may assume the background fluid to be at rest, so that $\bar{u}^i = 0$. Then the background velocity in conformal coordinates is

$$\bar{u}^\mu \equiv \left(\frac{d\tau}{dt}, \mathbf{0}\right) \overset{(1.3)}{=} a^{-1}(1, \mathbf{0})\,, \qquad \bar{u}_\mu \overset{\bar{u}^\mu \bar{u}_\mu = -1}{=} a(-1, \mathbf{0})\,. \tag{1.36}$$

Inserting Eqs. (1.34) and (1.36) into (1.30) and (1.31), the two identities resulting from the Einstein equations of a homogeneous and isotropic universe are

$$3a^{-2}(\mathcal{H}^2 + \kappa) \overset{(1.30)}{\underset{(1.34),\,(1.36)}{=}} 8\pi G\,\bar{e}\,, \tag{1.37}$$

$$-a^{-2}(2\mathcal{H}' + \mathcal{H}^2 + \kappa)\delta^i_j \overset{(1.31)}{\underset{(1.34),\,(1.36)}{=}} 8\pi G\,\bar{p}\,\delta^i_j\,. \tag{1.38}$$

These can be manipulated into

$$\mathcal{H}^2 + \kappa \overset{(1.37)}{=} \frac{8\pi G}{3}\bar{e}a^2\,, \tag{1.39}$$

$$\mathcal{H}' \overset{(1.37)}{\underset{(1.38)}{=}} -\frac{4\pi G}{3}(\bar{e} + 3\bar{p})\,a^2\,. \tag{1.40}$$

Another piece of information that is frequently helpful is *energy-momentum conservation*. The Einstein tensor has by construction a vanishing four-divergence, and correspondingly

$$T^\mu{}_{\nu;\mu} \overset{(7)}{=} T^\mu{}_{\nu,\mu} + \Gamma^\mu_{\mu\alpha}T^\alpha{}_\nu - \Gamma^\alpha_{\mu\nu}T^\mu{}_\alpha = 0\,. \tag{1.41}$$

Inserting $\nu = 0$ and making use of the Christoffel symbols in Eq. (1.10), this leads to

$$\bar{T}^\mu{}_{0;\mu} \overset{(1.41)}{\underset{(1.10)}{=}} \bar{T}^0{}_0{}' + \underbrace{4\mathcal{H}\,\bar{T}^0{}_0}_{\Gamma^\mu_{\mu\tau}} - \underbrace{\mathcal{H}\,\bar{T}^0{}_0}_{\Gamma^\tau_{\tau\tau}} - \underbrace{\mathcal{H}\delta^i_j\,\bar{T}^j{}_i}_{\Gamma^i_{j\tau}} = 0 \quad \Rightarrow \quad \bar{e}' = -3\mathcal{H}(\bar{e} + \bar{p})\,.$$

$$\tag{1.42}$$

The three equations obtained, Eqs. (1.39), (1.40) and (1.42), are not independent of each other. For instance, Eq. (1.40) follows by differentiating Eq. (1.39) with respect to conformal time and using the *continuity equation* (1.42). To describe the evolution of the background, it is convenient to choose Eqs. (1.39) and (1.42) as the basic relations. Converting them to physical time, we thus obtain

$$\mathcal{H}^2 + \kappa \overset{(1.39)}{=} \frac{8\pi G}{3}\bar{e}a^2 \underset{(1.6)}{\overset{\tau \leftrightarrow t}{\Longleftrightarrow}} H^2 + \frac{\kappa}{a^2} = \frac{8\pi G}{3}\bar{e}\,, \tag{1.43}$$

$$\bar{e}' \overset{(1.42)}{=} -3\mathcal{H}(\bar{e} + \bar{p}) \underset{(1.6),\,(1.7)}{\overset{\tau \leftrightarrow t}{\Longleftrightarrow}} \dot{\bar{e}} + 3\,H(\bar{e} + \bar{p}) = 0\,. \tag{1.44}$$

We refer to this system as the *Friedmann equations*.

An elementary scalar field

As a second example, we consider a scalar field, φ, as constituting the sole matter content of the universe. The scalar field has a *self-interaction potential*, $V(\varphi)$. For simplicity we assume that φ is *'minimally coupled'* to gravity, implying that the scalar field action is postulated to contain only the terms that also appear in Minkowskian space-time. Then the corresponding *Einstein-Hilbert action* takes the form

$$S \supset \frac{1}{16\pi G} \int_{\mathcal{X}} \sqrt{-g}\, R - \int_{\mathcal{X}} \sqrt{-g} \left[\frac{1}{2} \varphi_{,\mu} \varphi^{,\mu} + V(\varphi) \right] , \tag{1.45}$$

where $g \equiv \det g_{\mu\nu}$. However, if we include all possible dimension-4 operators, then the non-minimal term $\sim \varphi^2 R$ also appears, with its associated dimensionless coupling (cf., e.g., Ref. [7] for a review). Such terms must generally be included once the theory is quantized, given that they can anyways be generated by loop effects. The only exception is, if a symmetry forbids them, for instance an invariance under a shift $\varphi \to \varphi + c$, $c \in \mathbb{R}$, like in so-called *natural inflation* [8].

With the given action, we can make use of the Hamilton principle, leading to the Euler-Lagrange equations of motion. Specifically, we can vary the action with respect to $g_{\mu\nu}$ and φ. For the metric variation, we write $g = \det(\bar{g}_{\mu\nu} + \delta g_{\mu\nu}) = \det\{(\bar{g}_{\mu\nu})[\mathbb{1} + (\bar{g}_{\mu\nu})^{-1}(\delta g_{\mu\nu})]\} \approx \bar{g}\,\{1 + \mathrm{tr}[(\bar{g}_{\mu\nu})^{-1}(\delta g_{\mu\nu})]\} = \bar{g}\,(1 + \bar{g}^{\mu\nu}\delta g_{\mu\nu})$, which implies that $\delta g / \delta g_{\mu\nu} = \bar{g}\,\bar{g}^{\mu\nu}$, and subsequently $\delta\sqrt{-g}/\delta g_{\mu\nu} = \frac{1}{2}\sqrt{-\bar{g}}\,\bar{g}^{\mu\nu}$. Furthermore, expanding the identity in Eq. (1.1) at first order in $\delta g_{\mu\nu}$, $(\bar{g}^{\mu\rho} + \delta g^{\mu\rho})(\bar{g}_{\rho\nu} + \delta g_{\rho\nu}) = \delta^{\mu}_{\nu}$, leads to $\delta g_{\mu\nu} = -\bar{g}_{\mu\rho}\bar{g}_{\nu\sigma}\delta g^{\rho\sigma}$. Combining these ingredients, we find that

$$\frac{\delta\sqrt{-g}}{\delta g^{\mu\nu}} = -\frac{1}{2}\sqrt{-g}\, g_{\mu\nu} . \tag{1.46}$$

The variation ($0 = \delta S/\delta g^{\mu\nu}$) then yields the Einstein equations $G_{\mu\nu} = 8\pi G T_{\mu\nu}$, where

$$T_{\mu\nu} \equiv \frac{-2}{\sqrt{-g}} \frac{\delta S|_{\varphi}}{\delta g^{\mu\nu}} \overset{(1.45)}{\underset{(1.46)}{=}} \varphi_{,\mu}\, \varphi_{,\nu} - g_{\mu\nu} \left(\frac{1}{2}\varphi_{,\alpha}\varphi^{,\alpha} + V \right) . \tag{1.47}$$

At the same time, varying with respect to φ ($0 = \delta S/\delta\varphi$) leads to the scalar field equation

$$(\sqrt{-g}\varphi^{,\mu})_{,\mu} \overset{(1.45)}{=} \sqrt{-g}\, V_{,\varphi} , \qquad V_{,xy\ldots} \equiv \partial_x \partial_y \cdots V . \tag{1.48}$$

For a background solution, we assume the scalar field to be spatially homogeneous, and denote the corresponding value by $\varphi(\tau, \boldsymbol{x}) \to \bar{\varphi}(\tau)$. With the metric determinant

$$\sqrt{-\bar{g}(\tau)} \overset{(1.8)}{=} \frac{a^4(\tau)\, r^2 \sin\theta}{\sqrt{1 - \kappa r^2}} \overset{\tau \leftrightarrow t}{\underset{(1.3)}{\Longleftrightarrow}} \sqrt{-\bar{g}(t)} = \frac{a^3(t)\, r^2 \sin\theta}{\sqrt{1 - \kappa r^2}} , \tag{1.49}$$

and noting that

$$\bar{\varphi}^{,\mu} = \bar{g}^{\mu\nu}\bar{\varphi}_{,\nu} = \overbrace{\bar{g}^{\mu\tau}}^{-a^{-2}\delta^{\mu}_{\tau}}\bar{\varphi}' = \overbrace{\bar{g}^{\mu t}}^{-\delta^{\mu}_{t}}\dot{\bar{\varphi}}\,, \tag{1.50}$$

where $\bar{g}^{\mu\nu}$ is diagonal, only the index choice $\mu = \tau$ or $\mu = t$ gives a contribution. We find

$$\bar{g}^{\tau\tau}\sqrt{-\bar{g}} = -a^2\frac{r^2\sin\theta}{\sqrt{1-\kappa r^2}} \quad\Rightarrow\quad (\bar{g}^{\tau\tau}\sqrt{-\bar{g}}\,\bar{\varphi}')' = -a^{-2}\sqrt{-\bar{g}}\,(\bar{\varphi}'' + 2\mathcal{H}\bar{\varphi}')\,, \tag{1.51}$$

$$\bar{g}^{tt}\sqrt{-\bar{g}} = -a^3\frac{r^2\sin\theta}{\sqrt{1-\kappa r^2}} \quad\Rightarrow\quad (\bar{g}^{tt}\sqrt{-\bar{g}}\,\dot{\bar{\varphi}})^{\cdot} = -\sqrt{-\bar{g}}\,(\ddot{\bar{\varphi}} + 3\,H\dot{\bar{\varphi}})\,. \tag{1.52}$$

With these, the evolution equation that follows from Eq. (1.48) is independent of κ,

$$\boxed{\bar{\varphi}'' + 2\mathcal{H}\bar{\varphi}' + a^2 V_{,\varphi}(\bar{\varphi}) \overset{(1.48)}{\underset{(1.51)}{=}} 0 \quad\overset{\tau\leftrightarrow t}{\underset{(1.7)}{\Longleftrightarrow}}\quad \ddot{\bar{\varphi}} + 3H\dot{\bar{\varphi}} + V_{,\varphi}(\bar{\varphi}) \overset{(1.48)}{\underset{(1.52)}{=}} 0\,.} \tag{1.53}$$

As far as the energy-momentum tensor from Eq. (1.47) goes, it is diagonal on the background level, with

$$-\bar{T}^0{}_0 \overset{(1.47)}{\underset{(1.50)}{=}} \frac{(\bar{\varphi}')^2}{2a^2} + V(\bar{\varphi}) \overset{(1.6)}{=} \frac{\dot{\bar{\varphi}}^2}{2} + V(\bar{\varphi})\,, \tag{1.54}$$

$$\bar{T}^i{}_j \overset{(1.47)}{\underset{(1.50)}{=}} \delta^i_j\left[\frac{(\bar{\varphi}')^2}{2a^2} - V(\bar{\varphi})\right] \overset{(1.6)}{=} \delta^i_j\left[\frac{\dot{\bar{\varphi}}^2}{2} - V(\bar{\varphi})\right]\,. \tag{1.55}$$

Using now Eq. (1.54) in connection with Eq. (1.30) yields

$$\mathcal{H}^2 + \kappa \overset{(1.54)}{\underset{(1.30)}{=}} \frac{8\pi G}{3}\left[\frac{(\bar{\varphi}')^2}{2} + a^2 V(\bar{\varphi})\right] \tag{1.56}$$

$$\overset{(1.6)}{\underset{(1.7)}{\Longleftrightarrow}} H^2 + \frac{\kappa}{a^2} = \frac{8\pi G}{3}\left[\frac{\dot{\bar{\varphi}}^2}{2} + V(\bar{\varphi})\right]\,. \tag{1.57}$$

From Eqs. (1.55) and (1.31), we similarly get

$$-(2\mathcal{H}' + \mathcal{H}^2 + \kappa) \overset{(1.55)}{\underset{(1.31)}{=}} 8\pi G\left[\frac{(\bar{\varphi}')^2}{2} - a^2 V(\bar{\varphi})\right] \tag{1.58}$$

$$\overset{(1.6)}{\underset{(1.7)}{\Longleftrightarrow}} -\left(2\dot{H} + 3\,H^2 + \frac{\kappa}{a^2}\right) = 8\pi G\left[\frac{\dot{\bar{\varphi}}^2}{2} - V(\bar{\varphi})\right]\,. \tag{1.59}$$

Eliminating $\mathcal{H}^2$ with the help of Eq. (1.56), we find κ-independent evolution equations,

$$\mathcal{H}' \overset{(1.56)}{\underset{(1.58)}{=}} -\frac{8\pi G}{3}\left[(\bar{\varphi}')^2 - a^2\, V(\bar{\varphi})\right] \tag{1.60}$$

$$\overset{(1.6)}{\underset{(1.7)}{\Longleftrightarrow}} \dot{H} + H^2 = -\frac{8\pi G}{3}\left[\dot{\bar{\varphi}}^2 - V(\bar{\varphi})\right]. \tag{1.61}$$

Equation (1.60) can also be obtained by taking ∂_τ on Eq. (1.56) and inserting then Eq. (1.53).

A scalar field interacting with a plasma

Finally we consider a system which has simultaneously an elementary scalar field, which has not equilibrated, and a thermal plasma. If the scalar field and the plasma interact with each other, such a system is necessarily *dissipative* in nature. This means that energy transfer can take place between the two components: the scalar field feels friction, and thereby loses energy to the plasma. At the same time, it also experiences thermal fluctuations, gaining energy from the plasma. These processes have a relationship to each other, known as the *fluctuation-dissipation relation*. However, they are different kind of processes: the loss can be associated with a specific differential operator, whereas the gain is the result of stochastic dynamics, void of any 'structure'.

In the past decades, a framework of *dissipative effective theories* has been under development, with the goal of writing down an action for a system of the type described, generalizing upon Eq. (1.45). However, the formalism is quite complicated, requiring amongst others a duplication of degrees of freedom. For our purposes, it is more straightforward to modify directly the evolution equations. Let us stress that the evolution equations to be introduced should be thought of as an *effective theory for the slow variables*, after 'integrating out' the fast variables. As is the case for all effective theories, this implies that the description has a limited range of validity. We return to this in Chap. 7, where a derivation of the effective description is presented.

Let us start from the scalar field evolution equation, previously in Eq. (1.48). The presence of a plasma permits for the inclusion of two further structures. First of all, the existence of a four-velocity, u^μ, determining the plasma rest frame, allows us to define a new first-order differential operator, $u^\mu \partial_\mu$. Second, the plasma induces a stochastic noise, adding an inhomogeneous term to the scalar field evolution equation. Denoting a covariant derivative by $(\ldots)_{;\mu}$ (cf. Eq. (5)), the resulting evolution equation, often referred to as the *Langevin equation*, thus reads

$$\varphi^{;\mu}{}_{;\mu} - \Upsilon\, u^\mu \varphi_{,\mu} - V_{,\varphi} + \varrho = 0. \tag{1.62}$$

Here the dissipative coefficient Υ is a 'hydrodynamic' function, similar to the previous energy density and pressure. Like V, its form depends on the underlying microscopic degrees of freedom and their interactions with φ. The noise term ϱ, on the other hand, describes the random kicks that transfer energy from the plasma to

φ. As it contains no information (or 'no memory'), we assume that it correlates as *white noise*,

$$\langle \varrho(\mathcal{X})\varrho(\mathcal{Y})\rangle \approx \frac{\Omega\,\delta^{(4)}(\mathcal{X}-\mathcal{Y})}{\sqrt{-g}}\,, \qquad \mathcal{X} \equiv (x^0, \mathbf{x})\,, \tag{1.63}$$

with the *noise autocorrelator* Ω depending on the hydrodynamic variables (notably T). The division in Eq. (1.63) makes the Dirac-δ covariant, so that $\int_{\mathcal{X}}\sqrt{-g}\,[\delta^{(4)}(\mathcal{X})/\sqrt{-g}] = 1$. The symbol '$\approx$' anticipates that the Dirac-δ may need to be 'regularized', as discussed after Eq. (7.44) and around the end of Sect. 7.7.

As for the energy-momentum tensor of the coupled system, its physics is simpler than that of the evolution equation in Eq. (1.62), given that a transfer of energy from one component to another does *not affect* the overall energy. Therefore, the coefficient Υ and the noise ϱ should be absent from $T_{\mu\nu}$. We do remark, however, that hydrodynamics is also an effective theory, representing an expansion in gradients. Dissipative coefficients, such as viscosities, as well as the associated stochastic noise terms, do make an appearance at higher orders in gradients, and we return to this in Sect. 3.7.

Aiming now to combine Eqs. (1.34) and (1.47), we write

$$T_{\mu\nu} \underset{(1.47)}{\overset{(1.34)}{\equiv}} \varphi_{,\mu}\varphi_{,\nu} - \frac{g_{\mu\nu}\,\varphi_{,\alpha}\varphi^{,\alpha}}{2} + (e+p)\,u_\mu u_\nu + p\,g_{\mu\nu}\,. \tag{1.64}$$

According to Eq. (1.47), $p \supset -V$, whereas $e+p$ contains no V. However, thermodynamically, the energy density is a Legendre transform of the pressure. Denoting by T the temperature of the plasma, we then write $e = Ts - p \equiv Tp_{,T} - p$ (cf. Sect. 2.4), where s is the *entropy density*. Therefore, if V is part of an effective rather than a fundamental theory, and therefore depends on the temperature (representing the *free energy density*), then

$$p \equiv p_r - V\,, \quad e \equiv e_r + V - TV_{,T}\,, \quad e+p = Ts = T(s_r - V_{,T})\,. \tag{1.65}$$

Here e_r, p_r and s_r denote the energy density, pressure and entropy density of the 'radiation' component, respectively, defined to be independent of φ. The average values of the combinations in Eq. (1.65) are denoted by $\bar{e}$ and $\bar{p}$. We remark that, physically, these are *not* the full energy density and pressure, since the contributions from the derivatives of φ have been kept apart in Eq. (1.64).

The background field equation, generalizing upon Eq. (1.53), now becomes

$$\bar{\varphi}'' + (2\mathcal{H} + a\Upsilon)\bar{\varphi}' + a^2\,V_{,\varphi} \underset{(1.36),\,(1.62)}{\overset{(1.53)}{=}} 0 \tag{1.66}$$

$$\underset{(1.7)}{\overset{(1.6)}{\Longleftrightarrow}} \ddot{\bar{\varphi}} + (3H + \Upsilon)\dot{\bar{\varphi}} + V_{,\varphi} = 0\,. \tag{1.67}$$

The noise term, ϱ, does not appear on the background level, because it is treated as being of the same order as local perturbations. The generalizations of Eqs. (1.56) and (1.60) are conveniently represented as linear combinations,

$$3(\mathcal{H}^2 + \kappa) \underset{(1.37),\,(1.54)}{\overset{(1.30)}{=}} 4\pi G\big[(\bar{\varphi}')^2 + 2a^2\bar{e}\big] \tag{1.68}$$

$$\underset{(1.7)}{\overset{(1.6)}{\Longleftrightarrow}} \; 3\left(H^2 + \frac{\kappa}{a^2}\right) \; = \; 4\pi G\,(\dot{\bar{\varphi}}^2 + 2\bar{e})\,, \tag{1.69}$$

$$\mathcal{H}^2 - \mathcal{H}' + \kappa \underset{(1.35),\,(1.55)}{\overset{(1.31)}{=}} 4\pi G\big[(\bar{\varphi}')^2 + a^2(\bar{e} + \bar{p})\big] \tag{1.70}$$

$$\underset{(1.7)}{\overset{(1.6)}{\Longleftrightarrow}} \; \dot{H} - \frac{\kappa}{a^2} \; = \; -4\pi G\left(\dot{\bar{\varphi}}^2 + \bar{e} + \bar{p}\right). \tag{1.71}$$

Finally, the continuity equation from Eq. (1.44) becomes

$$-(\bar{\varphi}'' + 2\mathcal{H}\bar{\varphi}')\,\bar{\varphi}' \; = \; a^2\big[\bar{e}' + 3\mathcal{H}(\bar{e} + \bar{p})\big] \tag{1.72}$$

$$\underset{(1.7)}{\overset{(1.6)}{\Longleftrightarrow}} \; -(\ddot{\bar{\varphi}} + 3\,H\dot{\bar{\varphi}})\,\dot{\bar{\varphi}} \; = \; \dot{\bar{e}} + 3\,H(\bar{e} + \bar{p})\,. \tag{1.73}$$

It is not independent, but can be obtained by taking a time derivative of Eq. (1.68), and employing then Eq. (1.70) to eliminate $\mathcal{H}'$ and Eq. (1.68) to eliminate $3(\mathcal{H}^2 + \kappa)$.

1.4   Examples of Analytic Background Solutions

Simplified equations of state

The evolution equations that we have found in Sect. 1.3 are coupled first or second-order differential equations. However, they are in general non-linear, and not analytically solvable. They also contain model-dependent functions, like the equation of state $\bar{p} = \bar{p}(\bar{e})$ from Eq. (1.35), or the self-interaction potential $V(\bar{\varphi})$ from Eqs. (1.53), (1.56) and (1.60).

If we make additional assumptions about the model-dependent functions, we may obtain simplified equations that can be solved analytically. In particular, it is conventional to introduce an *equation of state parameter*, w, as

$$w \;\equiv\; \frac{\bar{p}}{\bar{e}}\,, \tag{1.74}$$

and then look for a solution valid for as long as w is approximately constant.

We note from Eqs. (1.69) and (1.71) that, if $\dot{\bar{\varphi}}^2 \ll \bar{e}$, then these equations take the same form as if the system were an ideal plasma, described by Eqs. (1.43) and (1.44). According to Eq. (1.65), $\bar{e}$ and $\bar{p}$ include the potential V in this case. Specifically, if $\dot{\bar{\varphi}}^2 \ll \bar{e}$ and, in addition, there is no radiation, then Eq. (1.65) implies that $w \overset{\text{de Sitter}}{=} -1$. If we employ this assumption in Eq. (1.44) (i.e. we ignore the left-hand side of Eq. (1.73)), then it follows that $\bar{e}$ is approximately constant. We refer to this situation as *de Sitter space-time*.

There are other frequently used equations of state in cosmology. One is a *matter-dominated universe*, resembling dust, where $\bar{p} = 0$ and $w \stackrel{\text{matter}}{=} 0$. A *radiation-dominated universe* is instead approximated with $w \stackrel{\text{radiation}}{=} 1/3$, which is strictly speaking only correct for non-interacting blackbody radiation (cf. Eq. (7.55)). More exotic equations of state also play a role in some models, for instance *kination*, with $w \stackrel{\text{kination}}{=} 1$ [9,10].

Once the energy density $\bar{e}$ and the pressure $\bar{p}$ have been related, Eqs. (1.43) and (1.44) constitute two equations for two unknown variables: a and $\bar{e}$. Even before solving these, an important qualitative statement can be made. From Eq. (1.6), we learn that if $\mathcal{H}' < 0$, then the expansion of the universe is decelerating, $\ddot{a} < 0$, whereas $\mathcal{H}' > 0$ corresponds to an accelerating expansion, $\ddot{a} > 0$. By Eq. (1.40), after inserting $\bar{p} = w\bar{e}$, $w < -1/3$ then leads to an accelerating expansion, $w > -1/3$ to a decelerating one.

De Sitter space-time as an explanation for spatial flatness

As a first step towards a solution of the equations, let us present an argument for why the parameter κ appearing in them could be neglected. This is referred to as *spatial flatness*: as shown in Sect. 1.5, the *geometric curvature* of a constant-τ manifold is proportional to κ. Spatial flatness is considered to be in excellent agreement with empirical observations, which indicate that, at the current time, $\Omega_K \equiv |\kappa|/(a_0^2 H_0^2) < 10^{-3}$ [6]. Moreover, going to earlier times, $1/(a^2 H^2) \propto m_{\text{pl}}^2/T^2$ decreases rapidly (cf. Sect. 2.4), implying that the dimensionless ratio $|\kappa|/(a^2 H^2)$ is extremely small.

Let us consider a scenario in which the energy density remains unchanged at some initial value $(\bar{e}_i)$ for a long time, $\bar{e} \equiv \bar{e}_i \gg \dot{\bar{\varphi}}^2$. Physically, we characterize this situation by saying that the energy density is dominated by a *cosmological constant*. Then the Friedmann equation from Eq. (1.43) takes the form

$$\frac{\dot{a}^2}{a^2} + \frac{\kappa}{a^2} \stackrel{\text{(1.43)}}{=} \frac{8\pi G \bar{e}_i}{3} \equiv H_i^2 \approx \text{constant} . \tag{1.75}$$

Here we have introduced the notation H_i for a quantity which would agree with the Hubble rate if $\kappa = 0$. Our goal is to show that even if $\kappa \neq 0$, the actual Hubble rate soon approaches the value H_i, so that effectively κ plays no role.

Mathematically, Eq. (1.75) is a non-linear first-order differential equation. It is also symmetric in time reversal, $t \to -t$, so the solution should reflect this symmetry: if there is an exponentially growing solution, there must be an exponentially shrinking solution.

In order to find a solution, we may first multiply Eq. (1.75) with a^2, obtaining

$$\dot{a}^2 - H_i^2 a^2 = -\kappa . \tag{1.76}$$

We now take a time derivative of Eq. (1.76). Given that the right-hand side is a constant, this yields

$$2\dot{a}\left(\ddot{a} - H_i^2 a\right) = 0 \,. \tag{1.77}$$

Assuming that $\dot{a} \neq 0$, yields a linear differential equation of second order, which is immediately solved,

$$\ddot{a} - H_i^2 a = 0 \;\Rightarrow\; a = a_+ \, e^{H_i t} + a_- \, e^{-H_i t} \,, \quad \dot{a} = H_i \left(a_+ \, e^{H_i t} - a_- \, e^{-H_i t}\right) \,. \tag{1.78}$$

Here a_+ and a_- are integration constants. However, the original Eq. (1.76) only requires one integration constant, so we should substitute Eq. (1.78) back into Eq. (1.76) to obtain the information that we are missing. On the left-hand side, most terms cancel, yielding

$$-4H_i^2 a_+ a_- \overset{(1.76)}{\underset{(1.78)}{=}} -\kappa \;\Leftrightarrow\; a_- = \frac{\kappa}{4H_i^2 a_+} \,. \tag{1.79}$$

Inserting this into a and $\dot{a}$ from Eq. (1.78) finally gives

$$\frac{\dot{a}}{a} \overset{(1.78)}{\underset{(1.79)}{=}} H_i \, \frac{4a_+^2 H_i^2 - \kappa \, e^{-2H_i t}}{4a_+^2 H_i^2 + \kappa \, e^{-2H_i t}} \,. \tag{1.80}$$

We see from Eq. (1.80) that, if $a_+ \neq 0$ and $|\kappa|/(a_+^2 H_i^2) < 4$, the Hubble rate, $\dot{a}/a$, agrees with H_i up to exponentially small corrections, as soon as $t \gg H_i^{-1}$. This is the reason why κ can be omitted, if it is assumed that the universe underwent a period with $\bar{e}_i \approx$ constant. The resulting exponential expansion is referred to as *inflation* [11–14]. We will omit κ in the rest of this book. However, at early times, a non-zero value of κ can play an important role. In particular, it is mathematically possible that $\dot{a}/a$ crosses zero, transferring from a shrinking to an expanding universe, either smoothly (if $\kappa > 0$) or through a singularity (if $\kappa < 0$). We will not elaborate on these possibilities, however they have been discussed in the literature (cf., e.g., Ref. [15] for a review).

A prototype solution for a complete history

Let us now patch together a complete solution of a possible cosmological history. At early times, assuming $a_+ \neq 0$ in Eq. (1.80), setting $\kappa = 0$, and solving the equation, we find

$$a(t) \overset{(1.80)}{\underset{\kappa \to 0}{=}} a_i \, e^{H_i (t - t_i)} \,, \tag{1.81}$$

where we denote the values of various variables at an *initial moment* by a subscript $(...)_i$. In principle t could be defined on the real axis, but it is conventional to restrict it to positive values, $t \geq t_i > 0$. As the form of the solution shows, it would be 'natural' to express time in units of t_i, and furthermore to choose $t_i \equiv H_i^{-1}$, because then the variation of the solution is of $\mathcal{O}(1)$ when the variation of t/t_i is of $\mathcal{O}(1)$.

We assume the solution in Eq. (1.81) to hold until a time $t = t_e$, where the subscript stands for the *end of inflation*. The task then is to determine how the solution looks like at $t > t_e$. The transition from t_e to the onset of radiation domination is called *reheating* (cf. Sect. 7.1). For a simple sketch, we treat the transition *instantaneously*, assuming that at $t > t_e$, the inflaton field suddenly does not play any role any more (actually, this assumption can be relaxed, cf. the discussion following Eq. (1.107)).

Assuming that at $t \geq t_e$, the scalar field kinetic energy $\dot{\bar{\varphi}}^2$ can be omitted, and inserting $\bar{p} = w\bar{e}$ in Eq. (1.44), we consider the background dynamics predicted by

$$
0 \overset{(1.44)}{\underset{(1.74)}{=}} \dot{\bar{e}} + 3\,H\bar{e}(1+w) = \frac{\partial_t[\bar{e}\,a^{3(1+w)}]}{a^{3(1+w)}} \;\Rightarrow\; \bar{e}\,a^{3(1+w)} = \text{constant} . \tag{1.82}
$$

It follows from Eq. (1.82) that

$$
\bar{e} \overset{(1.82)}{=} \frac{\bar{e}_e\,a_e^{3(1+w)}}{a^{3(1+w)}} , \qquad t \geq t_e . \tag{1.83}
$$

Inserting into Eq. (1.43) [with $\kappa = 0$], and making use of the fact that H and $\bar{e}$ are constant during the previous de Sitter period, we find

$$
H^2 \overset{(1.43)}{\underset{(1.83)}{=}} \frac{8\pi\,G\overbrace{\bar{e}_e}^{=H_i^2}}{3}\,\frac{a_e^{3(1+w)}}{a^{3(1+w)}} \;\Leftrightarrow\; \dot{a} = H_i\,a_e^{\frac{3}{2}(1+w)}\,a^{\frac{1}{2}(-1-3w)}
$$

$$
\Leftrightarrow\; \mathrm{d}a\,a^{\frac{1}{2}(1+3w)} = H_i\,a_e^{\frac{3}{2}(1+w)}\,\mathrm{d}t . \tag{1.84}
$$

For $w \neq -1$, this can be integrated into

$$
a^{\frac{3}{2}(1+w)} - a_e^{\frac{3}{2}(1+w)} \overset{(1.84)}{\underset{w \neq -1}{=}} a_e^{\frac{3}{2}(1+w)}\,\frac{3(1+w)H_i(t-t_e)}{2}
$$

$$
\Leftrightarrow\; a = a_e\left[1 + \frac{3(1+w)H_i(t-t_e)}{2}\right]^{\frac{2}{3(1+w)}} . \tag{1.85}
$$

For $w = -1$, we go back to Eq. (1.81).

The Hubble rate corresponding to Eq. (1.85), $H = \dot{a}/a$, reads

$$
H \overset{(1.85)}{=} \frac{2\,H_i}{2 + 3(1+w)H_i(t-t_e)} , \qquad t \geq t_e . \tag{1.86}
$$

This illustrates the prototypical cosmological history in physical time: during inflation, the expansion rate, H, stays constant at H_i ($w \approx -1$), and afterwards, it

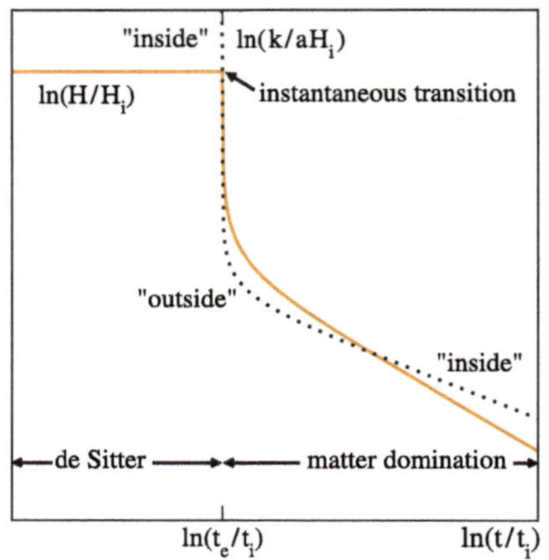

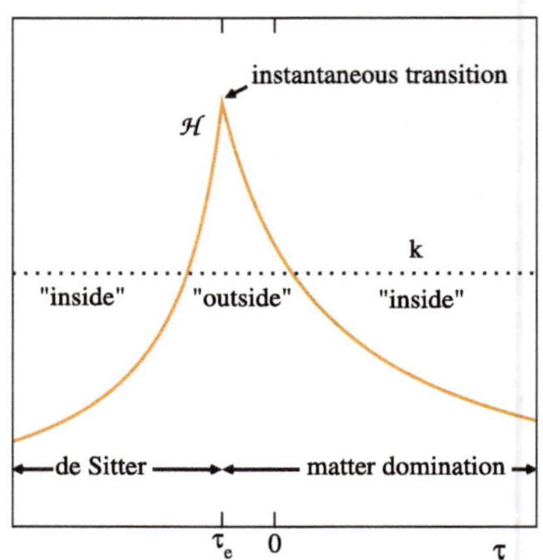

Fig. 1.1 Sketch of the Hubble rate, H, as a function of physical time (left, cf. Eq. (1.86)) and conformal time (right, cf. Eqs. (1.96) and (1.98)). By t_i we denote an initial time, at which the Hubble rate takes the constant value $H_i \equiv H(t_i)$, and by t_e the end of inflation, at which H starts to evolve. Inflation is assumed to end instantaneously, and we have set $w = 0$ afterwards (cf. Eq. (1.74)), which corresponds to matter domination. The dotted line illustrates a comoving momentum, k/a (left) or k (right). The left panel can be compared with a more realistic smoothly evolving numerical solution, shown in Fig. 1.2 in Sect. 1.7 (the axes are logarithmic in both plots)

decreases ($w \geq 0$). To understand the solution qualitatively, and anticipate a key concept of the following chapters, we compare H from Eq. (1.86) with the evolution of a physical momentum, $p \equiv k/a$ (cf. discussion between Eqs. (7) and (8)), with k being constant and a as given by Eq. (1.81) or (1.85). During de Sitter expansion, k/a decreases exponentially, and afterwards as a power law. Whenever a given momentum k satisfies $k \gg aH$, we say that it is *inside the Hubble horizon*, meaning that local processes dominate over the dilution caused by the expansion. If instead $k \ll aH$, the expansion is the dominant effect, and we say that the momentum mode is *outside the Hubble horizon*.

The behaviour in physical time is sketched in Fig. 1.1 (left). The corresponding solution in conformal time is shown in Fig. 1.1 (right), with the technical details described in Sect. 1.6. In Sect. 1.7, we also present a numerical solution, which involves a smooth rather than an instantaneous transition from the inflationary period to a matter-dominated epoch. We note that in the literature, *instantaneous reheating* often denotes a sharp transition from inflation to a radiation-dominated era, however the qualitative shapes of the plots remain the same, whether $w = 1/3$ (radiation domination) or $w = 0$ (matter domination).

Apart from the relationship between time and Hubble rate, it is often interesting to know the *relationship between time and temperature*. Let us work this out for the radiation-dominated epoch ($w = 1/3$). Then the energy density scales as T^4 (cf. Eq. (2.36)), with a coefficient conventionally parametrized through an *effective number of light degrees of freedom*, g_* (cf. Eq. (7.55)). Thereby Eq. (1.86) can be expressed as

$$H \underset{t \gg t_e,\, H_i^{-1}}{\overset{(1.86)}{\approx}} \frac{2}{3(1+w)t} \overset{w=\frac{1}{3}}{=} \frac{1}{2t} \underset{(7.55)}{\overset{(2.36)}{\approx}} \sqrt{\frac{8\pi}{3} \frac{g_*\pi^2}{30} \frac{T^2}{m_{\rm pl}}}$$ (1.87)

$$\Rightarrow t \approx \frac{3\sqrt{5}}{4\sqrt{g_*\pi^3}} \frac{m_{\rm pl}}{T^2} \approx \frac{0.301}{\sqrt{g_*}} \frac{m_{\rm pl}}{T^2} \,.$$ (1.88)

When we determine the abundance of some cosmic relic, it is sometimes practical to integrate over T rather than t, and then the Jacobian following from Eq. (1.88) is needed,

$$\frac{{\rm d}t}{{\rm d}T} \approx -\frac{3\sqrt{5}}{2\sqrt{g_*\pi^3}} \frac{m_{\rm pl}}{T^3} \,.$$ (1.89)

Since we have treated g_* as constant, Eq. (1.89) is not exact; the unapproximated version reads ${\rm d}T/{\rm d}t = -3c_s^2(T)TH(T)$, where c_s^2 is the speed of sound squared (cf. Eq. (1.35)).

1.5   More on Spatial Curvature

We elaborate here on the meaning of the parameter κ, defined in Eq. (1.2), and argued in Eq. (1.80) to be insignificant at late times, if the universe underwent de Sitter expansion at early times.

To get started, it is good to think about the nature of different parameters that characterize our cosmological model. One class of them are Lagrangian parameters of elementary particles and fields (masses, coupling constants). In an ideal 'reductionist' philosophy, we should try to determine the full solution in terms of such parameters, but in practice a more modest goal needs to be set. As less fundamental parameters we may consider 'derived quantities', such as an equation of state, in terms of which we may express the right-hand side of the Einstein equations. Yet another set of parameters is given by initial conditions. Ideally, we could hope to find a solution which is independent of initial conditions (for example, a fully thermalized state has no memory of where it came from). However, often some initial conditions are needed—for instance, obtaining a universe of the type that we observe, may require choosing a patch of space-time, where a sufficiently long period of de Sitter expansion took place.

Coming back to the parameter κ, it is not a Lagrangian parameter and, as long as we have no theory of quantum gravity, also not a derived quantity. It rather characterizes the global topology of the spatial universe (open/hyperbolic for $\kappa < 0$, flat for $\kappa = 0$, closed/spherical for $\kappa > 0$). From a general relativistic point of view, it may be viewed as an initial condition.

To give κ a more specific meaning, consider a submanifold $\tau = $ constant of the metric in Eq. (1.2). Let us compute the Ricci scalar for such a manifold,

$$R_\tau \equiv g^{ij} R_{ij} \,,$$ (1.90)

where it is implicitly assumed that only spatial indices are included, also when deriving the Ricci tensor. The background value of R_τ is denoted by $\bar{R}_\tau$. At the background level, when the metric tensor is diagonal, this simply means that we eliminate the index τ and the derivatives $(\dots)_{,\tau}$ from the derivation in Eqs. (1.10)–(1.28). This implies that we leave out $\mathcal{H}$ from Eq. (1.28), yielding

$$\bar{R}_\tau \overset{(1.28)}{\underset{\mathcal{H},\mathcal{H}' \to 0}{=}} \frac{6\kappa}{a^2} \, . \tag{1.91}$$

So, indeed, κ parametrizes the Ricci scalar of the constant-τ submanifolds, and this version of the Ricci scalar is in turn referred to as *spatial curvature*.

1.6 Background Solution in Conformal Time

While the Friedmann equations (1.43) and (1.44) are most conveniently solved in physical time, conformal time is also often made use of. Here we sketch how the background equations can be solved in conformal time.

With the same assumptions as before, Eq. (1.82) gets replaced with

$$0 \overset{(1.44)}{\underset{(1.74)}{=}} \bar{e}' + 3\mathcal{H}\bar{e}(1+w) = \frac{\partial_\tau[\bar{e}\,a^{3(1+w)}]}{a^{3(1+w)}} \Rightarrow \bar{e}\,a^{3(1+w)} = \text{constant} \, . \tag{1.92}$$

Writing the solution as $\bar{e} = \bar{e}_i\, a_i^{3(1+w)}/a^{3(1+w)}$ and inserting into Eq. (1.43), we get

$$\mathcal{H}^2 \overset{(1.43)[\kappa=0]}{\underset{(1.92)}{=}} \overbrace{\frac{8\pi G\bar{e}_i}{3}}^{\equiv H_i^2} \frac{a_i^{3(1+w)}}{a^{1+3w}} \Leftrightarrow a' = H_i a_i^{\frac{3}{2}(1+w)} a^{\frac{1}{2}(1-3w)}$$

$$\Leftrightarrow \frac{\mathrm{d}a}{a^{\frac{1}{2}(1-3w)}} = H_i a_i^{\frac{3}{2}(1+w)} \mathrm{d}\tau \, , \tag{1.93}$$

which now leads to

$$a \overset{(1.93)}{=} a_i\left[1 + \frac{(1+3w)H_i a_i(\tau - \tau_i)}{2}\right]^{\frac{2}{1+3w}} \, . \tag{1.94}$$

This applies both during de Sitter expansion and matter or radiation domination, with the underlying assumption that w is to a good approximation constant.

Consider first de Sitter expansion, $w = -1$. Then Eq. (1.94) yields

$$\text{de Sitter}: \quad a \overset{(1.94)}{\underset{w=-1}{=}} \frac{a_i}{1 + H_i a_i(\tau_i - \tau)} \, . \tag{1.95}$$

According to this result, a diverges at a time $\tau_{\mathrm{div}} = \tau_i + 1/(H_i a_i)$. It is conventional to choose $\tau_{\mathrm{div}} \equiv 0$, implying $H_i a_i \tau_i = -1$. So, Eq. (1.95) can be simplified into

$$a \overset{(1.95)}{=} -\frac{1}{H_i \tau} \,, \quad \mathcal{H} = -\frac{1}{\tau} \,, \quad \tau \leq \tau_e < 0 \,, \tag{1.96}$$

where we have denoted by τ_e a moment at which inflation ends and the expansion continues under matter or radiation domination, with $w \geq 0$.

Let us subsequently work out what happens at $\tau \geq \tau_e$. At the initial moment, the scale factor is $a_e = -1/(H_i \tau_e)$ according to Eq. (1.96). Therefore, from Eq. (1.94), with $a_i \to a_e$ and $\tau_i \to \tau_e$,

$$a \overset{(1.94)}{=} -\frac{1}{H_i \tau_e}\left[\frac{3(1+w)}{2} - \frac{(1+3w)\tau}{2\tau_e}\right]^{\frac{2}{1+3w}} \,, \quad \tau \geq \tau_e \,, \tag{1.97}$$

$$\mathcal{H} \overset{(1.97)}{=} \frac{2}{(1+3w)\tau - 3(1+w)\tau_e} \,, \quad \tau \geq \tau_e \,. \tag{1.98}$$

This behaviour is sketched in Fig. 1.1 (right) in Sect. 1.4.

1.7 Numerical Background Solution

Between Eqs. (1.81) and (1.86), we presented an analytic background solution for a prototype system. However, we had to make assumptions about the equation of state or the magnitude of $\dot{\bar{\varphi}}$ in order to achieve this. It is not difficult to solve the background equations numerically without any such assumptions, at least for a finite duration of time (e.g. for interpolating between periods with different equations of state). Here we demonstrate how this can be done.

In order to consider a simple system, we restrict for the moment to a case in which only a massive scalar field, $\bar{\varphi}$, is present. In this case, the early quasi-de Sitter period goes over to a matter-dominated expansion at late times ($w = 0$), as will be explained below.

First, let us write down the equations that we are solving. The field equation reads

$$\ddot{\bar{\varphi}} + 3H\dot{\bar{\varphi}} + V_{,\varphi} \overset{(1.53)}{=} 0 \,, \tag{1.99}$$

where for $\kappa = 0$ the Hubble rate is given by

$$H \overset{\underset{(1)}{(1.57)}}{=} \sqrt{\frac{8\pi}{3}} \frac{\sqrt{\frac{1}{2}\dot{\bar{\varphi}}^2 + V}}{m_{\mathrm{pl}}} \,. \tag{1.100}$$

As a potential we assume that of so-called *natural inflation* [8],

$$V \equiv m^2 f_a^2 \left[1 - \cos\left(\frac{\bar{\varphi}}{f_a}\right)\right] \,. \tag{1.101}$$

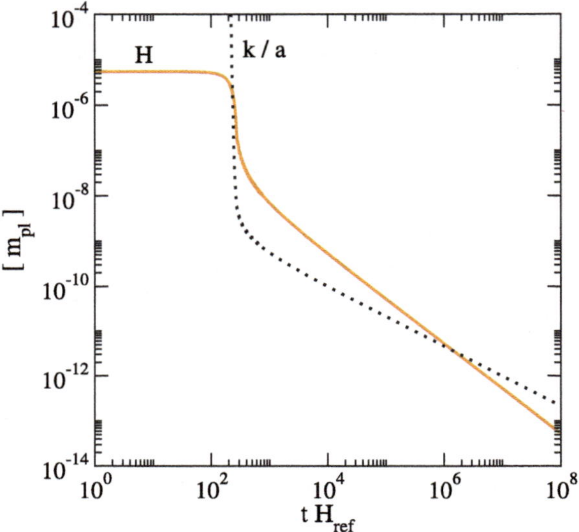

Fig. 1.2 Numerical background solution, from Eqs. (1.99)–(1.104), in physical time. For illustration, the comoving momentum mode has been chosen as $k/a(t_i) = 7.35 \times 10^{52} m_{pl}$, where the initial time $t_i \equiv H_{ref}^{-1}$ is defined according to Eq. (1.104). At late times, we have gone over from the full equation in Eq. (1.99) to the simplified version in Eq. (1.110), but the transition is smooth. This plot can be compared with the analytic expectation from Fig. 1.1 (left) in Sect. 1.4. A subsequent transition to a radiation-dominated expansion will be illustrated in Fig. 7.2 in Sect. 7.7.

Parameter values that yield semi-realistic observables (cf. Sect. 2.2) are chosen as

$$ f_a = 1.25\, m_{pl} \,, \quad m = 1.09 \times 10^{-6}\, m_{pl} \,. \tag{1.102} $$

We also need to specify initial conditions. The initial field value and its time derivative are set to

$$ \bar{\varphi}(t_i) = 3.5\, m_{pl} \,, \quad \dot{\bar{\varphi}}(t_i) = -\frac{V_{,\varphi}(\bar{\varphi}(t_i))}{3\, H_{ref}} \,, \tag{1.103} $$

where H_{ref} is defined in Eq. (1.104). The value $\bar{\varphi}(t_i)$ is close enough to the top of the potential that inflation lasts long, but not eternally. The value $\dot{\bar{\varphi}}(t_i)$ turns out to play little role (the solution is an attractor), but it has been chosen so that the attractor is reached relatively fast. The initial Hubble rate sets a 'scale' in whose units dimensionful quantities can be expressed; a particularly convenient choice for this is to define a 'reference' value of the Hubble rate, obtained as if the potential were quadratic in Eq. (1.100) $[V = \frac{1}{2} m^2 \bar{\varphi}^2]$ and $\dot{\bar{\varphi}}(t_i) = 0$,

$$ H_{ref} \equiv \sqrt{\frac{4\pi}{3}} \, \frac{m\, \bar{\varphi}(t_i)}{m_{pl}} \,. \tag{1.104} $$

The initial time is chosen as $t_i \equiv H_{\mathrm{ref}}^{-1}$.

Apart from solving for $\bar{\varphi}(t)$, we also solve for the evolution of the cosmological scale factor. The scale factor is re-parametrized in terms of *e-folds*, denoted by N, via

$$a(t) \equiv a(t_i)\, e^{N(t)} \quad \Rightarrow \quad \dot{N} = H\,, \quad N(t_i) = 0\,. \tag{1.105}$$

With a given a, we find out how *physical momenta*, defined as $k/a(t)$, redshift,

$$\frac{k}{a(t)} = \frac{k\, e^{-N(t)}}{a(t_i)}\,. \tag{1.106}$$

We now explain why the post-inflationary cosmology of this model corresponds to *early matter domination*, in the sense explained in Sect. 1.4 and sketched in Fig. 1.1. At late times, $\bar{\varphi}$ settles around its minimum at $\bar{\varphi} = 0$. Then the potential approximates to $V \approx \frac{1}{2}\, m^2 \bar{\varphi}^2$. In this regime, Eq. (1.99) becomes

$$\ddot{\bar{\varphi}} + 3H\dot{\bar{\varphi}} + m^2 \bar{\varphi} \approx 0\,. \tag{1.107}$$

By the time when $H \ll m$, this is just a damped harmonic oscillator equation of motion, with known trigonometric solutions, of periodicity $\Delta t \equiv 2\pi/m$. If we denote the energy density and pressure appearing in Eqs. (1.54) and (1.55), respectively, as

$$e_{\bar{\varphi}} \overset{(1.54)}{\equiv} \frac{\dot{\bar{\varphi}}^2}{2} + V\,, \quad p_{\bar{\varphi}} \overset{(1.55)}{\equiv} \frac{\dot{\bar{\varphi}}^2}{2} - V\,, \tag{1.108}$$

then an average over the oscillation period yields

$$\left\langle \frac{\dot{\bar{\varphi}}^2}{2} \right\rangle_{\Delta t} \overset{H \ll m}{\approx} \left\langle \frac{m^2 \bar{\varphi}^2}{2} \right\rangle_{\Delta t} \quad \Rightarrow \quad \langle e_{\bar{\varphi}} \rangle_{\Delta t} \approx \langle \dot{\bar{\varphi}}^2 \rangle_{\Delta t}\,, \quad \langle p_{\bar{\varphi}} \rangle_{\Delta t} \approx 0\,. \tag{1.109}$$

The vanishing of the (averaged) pressure implies that $w \approx 0$.

If we want to solve the evolution equations for a long time, the rapid oscillations become costly to handle. However, if we take a time derivative of Eq. (1.108) and make use of Eq. (1.99), we find

$$\dot{e}_{\bar{\varphi}} \overset{(1.108)}{\underset{(1.99)}{=}} -3H\dot{\bar{\varphi}}^2 \overset{(1.109)}{\approx} -3H e_{\bar{\varphi}}\,, \quad H \overset{(1.100)}{\underset{(1.109)}{\approx}} \sqrt{\frac{8\pi}{3}} \frac{\sqrt{e_{\bar{\varphi}}}}{m_{\mathrm{pl}}}\,. \tag{1.110}$$

This system is simple to solve for a long period of time, and contains no oscillations.

We show below a `python` script producing a benchmark solution of the set of equations (1.99), (1.100), and (1.110). The result is plotted in Fig. 1.2. We go over to Eq. (1.110) after 50 oscillations (101 crossings of zero).

```python
# numerical solution for inflation and end of inflation [numerics_bg_cold.py]
#
# import basic tools and integration routines
import numpy as np
from scipy.integrate import solve_ivp

# parameters [mpl]
fa = 1.25            # inflaton decay constant
m = 1.09e-6          # inflaton mass
koa_ref = 7.35e52    # momentum of perturbations at initial time
Pi = np.pi

# potential of inflaton (phi) [mpl^4]
def V(phi): return m*m*fa*fa*( 1 - np.cos(phi/fa) )

# derivative of inflaton potential by phi [mpl^3]
def Vd(phi): return m*m*fa*np.sin(phi/fa)

# total energy density of scalar inflaton field [mpl^4]
def etot(phi,phid): return phid*phid/2 + V(phi)

# Hubble rate [mpl]
def H(energy_density): return np.sqrt(8*Pi*np.abs(energy_density)/3)

# 2nd-order ODE as two 1st-order ODEs: solve for [dot phi, ddot phi, number of efolds]
# derivatives are taken with respect to t = Href*time
def derivatives1(t,y):
    Phi, Phid, Nfolds = y
    dphi_dt = Phid/Href
    ddphi_ddt = ( -3*H(etot(Phi, Phid))*Phid - Vd(Phi) )/Href
    dN_dt = H(etot(Phi, Phid))/Href
    dy_dt = [dphi_dt, ddphi_ddt, dN_dt]
    return dy_dt

# average over fast oscillations at late times => one 1st-order ODE
def derivatives2(t,y):
    e_phi, Nfolds = y
    dy_dt = [-3*H(e_phi)*e_phi/Href,  H(e_phi)/Href]
    return dy_dt

# initial conditions and reference values
phi_0 = 3.5                       # mpl
Href = np.sqrt(4*Pi/3)*m*phi_0    # mpl
phid_0 = - Vd(phi_0)/(3*Href)     # mpl^2

# define the event function counting the zeros of phi
def phi_crossings(t, y):  return y[0]

# integrate up to 50 oscillations (101 crossings of zero)
sol1 = solve_ivp(derivatives1, [1, 1e4], [phi_0, phid_0, 0], events=phi_crossings)
t_match = sol1.t_events[0][100]
time1 = sol1.t[sol1.t <= t_match]
n_match = len(time1)
ephi1 = etot(sol1.y[0], sol1.y[1])
efolds1 = sol1.y[2]

# initial conditions for oscillatory regime
time_0 = time1[n_match-1]
etot_0 = ephi1[n_match-1]
efolds_0 = efolds1[n_match-1]

# integrate in oscillatory regime (matter-dominated era)
sol2 = solve_ivp(derivatives2, [time_0, 1e8], [etot_0, efolds_0], method='DOP853')
```

```python
ephi2 = sol2.y[0]
efolds2 = sol2.y[1]

# append solutions
time = np.append(time1,sol2.t)
e_phi = np.append(ephi1[:n_match], ephi2)
efolds = np.append(efolds1[:n_match], efolds2)
koa = koa_ref*np.exp(-efolds)

# print to file
np.savetxt('numerics_bg_cold.dat', np.c_[time, e_phi, H(e_phi), efolds, koa],
           fmt='%.6e',newline='\n',
           header='columns: t*H_ref, e_phi/mpl**4, H/mpl, efolds, k_ref/a/mpl')
```

References

1. Weinberg, S.: Gravitation and Cosmology. Wiley, New York (1971)
2. Schwarz, D.J.: The first second of the Universe. Ann. Phys. **12**, 220 (2003). https://doi.org/10.1002/andp.200310010, astro-ph/0303574
3. Mambrini, Y.: Particles in the Dark Universe. A Student's Guide to Particle Physics and Cosmology. Springer (2021). https://doi.org/10.1007/978-3-031-66994-1
4. Rubakov, V.A., Gorbunov, D.S.: Introduction to the Theory of the Early Universe: Hot Big Bang Theory. World Scientific (2017). https://doi.org/10.1142/10447
5. Smoot, G.F., et al.: [COBE], Structure in the COBE Differential Microwave Radiometer First-Year Maps. Astrophys. J. Lett. **396**, L1 (1992). https://doi.org/10.1086/186504
6. Aghanim, N., et al.: [Planck], Planck 2018 results. I. Overview and the cosmological legacy of Planck. Astron. Astrophys. **641**, A1 (2020). https://doi.org/10.1051/0004-6361/201833880, 1807.06205
7. Hertzberg, M.P.: On inflation with non-minimal coupling. JHEP **11**, 023 (2010). https://doi.org/10.1007/JHEP11(2010)023, 1002.2995
8. Freese, K., Frieman, J.A., Olinto, A.V.: Natural inflation with pseudo Nambu-Goldstone bosons. Phys. Rev. Lett. **65**, 3233 (1990). https://doi.org/10.1103/PhysRevLett.65.3233
9. Spokoiny, B.: Deflationary Universe scenario. Phys. Lett. B **315**, 40 (1993). https://doi.org/10.1016/0370-2693(93)90155-B, gr-qc/9306008
10. Joyce, M.: Electroweak baryogenesis and the expansion rate of the Universe. Phys. Rev. D **55**, 1875 (1997). https://doi.org/10.1103/PhysRevD.55.1875, hep-ph/9606223
11. Guth, A.H.: Inflationary universe: A possible solution to the horizon and flatness problems. Phys. Rev. D **23**, 347 (1981). https://doi.org/10.1103/PhysRevD.23.347
12. Linde, A.D.: A new inflationary universe scenario: A possible solution of the horizon, flatness, homogeneity, isotropy and primordial monopole problems. Phys. Lett. B **108**, 389 (1982). https://doi.org/10.1016/0370-2693(82)91219-9
13. Albrecht, A., Steinhardt, P.J.: Cosmology for Grand Unified Theories with Radiatively Induced Symmetry Breaking. Phys. Rev. Lett. **48**, 1220 (1982). https://doi.org/10.1103/PhysRevLett.48.1220
14. Linde, A.D.: Chaotic inflation. Phys. Lett. B **129**, 177 (1983). https://doi.org/10.1016/0370-2693(83)90837-7
15. Novello, M., Bergliaffa, S.E.P.: Bouncing cosmologies. Phys. Rept. **463**, 127 (2008). https://doi.org/10.1016/j.physrep.2008.04.006, 0802.1634

Observable Signatures Beyond the Background Solution

2

Abstract

Our current universe is homogeneous and isotropic on scales that are larger than about 1% of its visible size, and spatially flat to a good precision. Its structure at smaller scales may have formed via gravitational collapse of early inhomogeneities. Reviewing experimental evidences for the universe until the emission of the Cosmic Microwave Background (CMB) radiation and the onset of structure formation, we collect the target quantities that the theoretical framework presented in the book aims to describe. The relations between present-day frequencies and wavelengths and those that play a role in the early universe are summarized. Finally, we derive basic formulae for the statistical description of small perturbations on a homogeneous and isotropic background, and compare the procedures adopted in different contexts.

Keywords

Scalar power spectrum · Amplitude A_s · Spectral tilt n_s · Tensor power spectrum · Ratio r of tensor and scalar spectra · Adiabatic initial conditions · Transfer function · Non-Gaussianity · Isocurvature perturbation · Spectral distortion · Gravitational-wave astronomy · Interferometer · Effective number of neutrino degrees of freedom

2.1　Overview

The quantitative evidence that we have about the early universe comes mostly not from the homogeneous and isotropic approximation (cf. Chap. 1), but from the fact that, on closer inspection, this approximation does not hold exactly. This can be inferred by measuring the local CMB temperatures in different directions, which show small but highly correlated fluctuations, even at large angular separations. Furthermore, going towards smaller distances, the CMB fluctuations can be mapped onto the structures seen in the roughest distribution of matter. The latter can be

© The Author(s) 2026

M. Laine and S. Procacci, *From Inflation to Hot Big Bang*, Lecture Notes in Physics 1047, https://doi.org/10.1007/978-3-032-09893-1_2

extracted with several observational techniques. The 2-point density correlation function extracted from spectroscopic galaxy redshift surveys shows a clear peak at a characteristic distance of $\sim$150 Mpc, associated with the wavelength of a single *baryon acoustic oscillation* (BAO) that had taken place by the time that baryons decoupled from photons. A 'forest' of *Lyman-α absorption lines*, superimposed on the spectra of distant quasars, reflects the presence of neutral hydrogen clouds that lie between the emitter and the observer. Finally, *gravitational lensing* (either of CMB photons or of light emitted by galaxies) can tell us about the distribution not only of matter but also of *dark matter*, which by definition is not directly visible in spectroscopic surveys or absorption lines.

Going to even smaller scales, astronomical observations reveal a lot of well-known structure (galaxies, nebulae, stars). However, it is more difficult to interpret this in terms of its primordial origin, since complicated and non-linear physics (gravitational collapse, diffusion, magnetohydrodynamics) took place in the intervening period.

In contrast, if we were able to see gravitational waves, they would be agnostic about the challenges posed by late-time astrophysical phenomena, given that gravity interacts very weakly with matter. Therefore, gravitational waves offer for a theoretically ideal window to the early universe, all the way to the inflationary period, and extending over phenomena covering an extremely broad range of distance and frequency scales. It is speculated that the first signals of primordial gravitational waves could have already been observed by Pulsar Timing Arrays.

In this chapter, we summarize the existing measurements of primordial perturbations in the CMB (technically known as scalar perturbations) (cf. Sect. 2.2), and discuss the future prospects of observing primordial gravitational waves (technically known as tensor perturbations) (cf. Sect. 2.3). To make the discussion concrete, we also review the basics of cosmological redshift and how it allows us to probe various epochs of the early universe (cf. Sect. 2.4). We end the chapter with a more foundational elaboration on what so-called power spectra are, from the mathematical and physical point of view (cf. Sect. 2.5).

2.2   The Imprints of Scalar Perturbations on the CMB

The observed temperature $T_0 = 2.7255\,\mathrm{K}$ of the CMB shows small *temperature anisotropies*. After subtracting galactic foregrounds as well as a *dipole structure* with $\delta T / T_0 \sim 10^{-3}$, originating from the motion of the Earth relative to the CMB reference frame, the remainder is of order

$$\left(\frac{\delta T}{T_0} \right)_{\mathrm{obs}} \sim 10^{-5} \simeq \left(\frac{\delta T}{T_0} \right)_{\mathrm{intr}} + \left(\frac{\delta T}{T_0} \right)_{\mathrm{jour}}. \qquad (2.1)$$

The left-hand side describes the observational result, while the right-hand side is its interpretation, composed of an intrinsic ('intr') component present at recombination, and of a journey ('jour') component originating as the photons travel to the present. In modern computations, reviewed in Sect. 9.6, no such distinction needs to be made, and the splitup is also gauge dependent (cf. Chap. 4), but it is still helpful for an intuitive picture.

Let us start with the observational side. After choosing some coordinate system, δT can be expanded in terms of spherical harmonics, $Y_{\ell m}$, from which the $\ell = 1$ component is omitted, as it corresponds to a dipole from our relative movement with respect to CMB,

$$\delta T(\theta, \phi)|_{\text{obs}} = \sum_{\ell \geq 2} \sum_{|m| \leq \ell} a_{\ell m} Y_{\ell m}(\theta, \phi) \,. \tag{2.2}$$

By definition the average value of δT vanishes, and the physical information is contained in a 2-point function of temperature fluctuations, averaged over the sky. The coefficients $a_{\ell m}$ of the spherical-harmonics expansion are independent complex variables. Averaged over the azimuthal index m, their absolute values squared are denoted by

$$C_\ell \equiv \frac{1}{2\ell + 1} \sum_{m=-\ell}^{\ell} |a_{\ell m}|^2 \,. \tag{2.3}$$

We note that a large *multipole*, ℓ, corresponds to a small angular scale $\theta_\ell \equiv \pi/\ell$. In terms of the wavelength $\lambda \sim 2\pi/k$ of the underlying density perturbations, the angular dependence originates by projecting plane waves onto spherical harmonics (this is described briefly in the paragraph below Eq. (9.64)). Then a large k corresponds to a large ℓ. The total power in the multipole ℓ is conventionally defined as

$$D_\ell \equiv \frac{\ell(\ell + 1)}{2\pi} C_\ell \,. \tag{2.4}$$

In addition to the temperature power spectrum, known as TT (so that $D_\ell \equiv D_\ell^{TT}$), photons also carry linear polarization, which is a vector. This originates from *Thomson scatterings* just before decoupling, and is not complete, but rather on the level $\sim 10\%$ of the total anisotropy [1,2]. As will be discussed in detail later on (cf. Eq. (3.7)), any vector can be decomposed into a curl-free part, which can in turn be represented with the help of a scalar potential ($\mathbf{E} \sim -\nabla A_0$), and a divergence-free part, which can be expressed through a vector potential ($\mathbf{B} \sim \nabla \times \mathbf{A}$). Even though these do *not* physically correspond to an electric or magnetic field, but are just parts of a single polarization vector, they are conventionally referred to as an E-mode and a B-mode, respectively. There is information about the cross-correlation TE and the power spectrum EE [3]. Moreover, there is an eager search for a B-mode signal (cf., e.g., Refs. [4,5]), as this does not originate directly from scalar perturbations ($\sim A_0$), and is rather believed to be influenced by gravitational waves [6,7]. That said, lensing mixes some E modes into B modes [8], and in addition there is galactic foreground emission with similar characteristics.

As sketched in Fig. 2.1, the measured D_ℓ show an oscillatory behaviour, with a global maximum at $\ell \sim 200$. The amplitude decreases as ℓ increases, starting from $\ell \sim 500$, and oscillations are damped after $\ell \sim 1000$. This shape can be explained by thinking that, just before decoupling, perturbations in the energy density propagate as *acoustic oscillations*, i.e. as a sequence of compressions and rarefactions of the

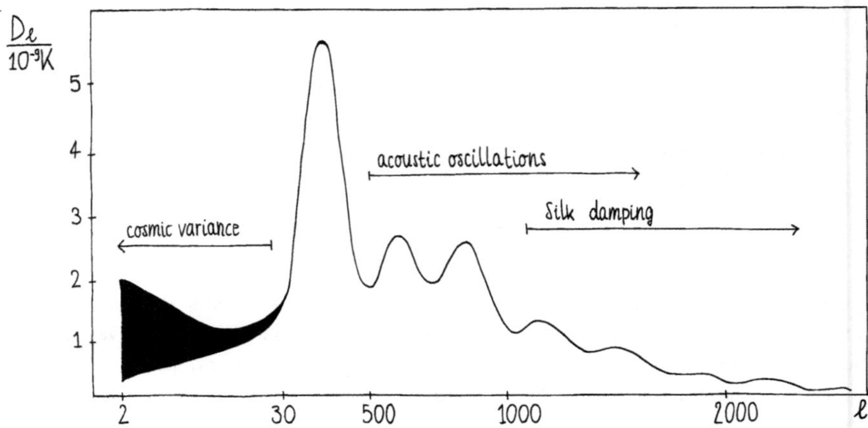

Fig. 2.1 A sketch of the angular momentum power spectrum of the CMB temperature anisotropies, D_ℓ (cf. Eq. (2.4)), at different multipole moments ℓ

medium. However, waves propagating in an interacting medium get attenuated, if their wavelength is small enough. In particular, if the wavelength is smaller than the photon mean free path, λ_{free}, some photons may escape the compressed regions before colliding with the free electrons, whereby the wave is smoothened out, and the corresponding oscillatory behaviour of D_ℓ is damped. In the CMB context this phenomenon is known as *Silk damping* [9]. We note in passing that large-scale structures such as galaxy clusters, whose size corresponding to ℓ of several thousand, start forming around inhomogeneities which did not get damped too much. This is possible if the corresponding particles feel no pressure forces, as may be the case for dark matter. After the decoupling, the apparent frequency and thus the temperature of the photons still gets modified, as they propagate in an inhomogeneous gravitational potential, through what is known as the *Sachs-Wolfe effect* [10].

Let us now rephrase the qualitative discussion above in terms of the right-hand side of Eq. (2.1). As might be easy to imagine, the physics involved is quite complicated. Apart from the hydrodynamic and collisional phenomena already alluded to, there is also the conceptual issue that, in the presence of gravitational perturbations, the definition of a temperature fluctuation is by itself not 'gauge invariant', or physical in the general relativistic sense (*gauge dependence* is defined in Chap. 4). Specifically, the division into intrinsic and journey components in Eq. (2.1) depends on the gauge chosen, whereas their sum, the observed anisotropy, is well-defined.

In order to make the problem tractable, it is conventional to *factorize* the computation into two parts (the second captures the intrinsic and journey components of Eq. (2.1)):

(i) As a first step, we consider the early universe until a moment $t_*(k)$ at which a given momentum mode k *crosses outside of the Hubble horizon*, i.e. $(k/a)(t_*) \approx H$ (see Fig. 1.1 in Sect. 1.4). We then choose a moment $t_{\mathrm{out}} > t_*$ at which the mode is well outside of the Hubble horizon. In a time window around t_{out}, there is a gauge-invariant quantity, called the curvature perturbation $\mathcal{R}_T$ (cf. Chap. 7), which is to a very good approximation constant (cf. Chap. 8). Therefore, this quantity sets the target for theoretical computations.

(ii) At a certain time after t_{out}, the mode re-enters inside the Hubble horizon (cf. Fig. 1.1 in Sect. 1.4). Then it starts to undergo oscillations, as dictated by viscous hydrodynamics (cf. Chap. 9). Viscosities are the hydrodynamic representation of the mean free path mentioned above, and lead to the damping of acoustic oscillations. The hydrodynamic description is valid as long as the photons stay equilibrated, that means until their decoupling, at $t = t_{\text{dec}}$. The decoupling period itself is usually described by a more microscopic framework than hydrodynamics, notably that of Boltzmann equations. After their decoupling, the photons 'free stream' in a curved background until the present time, when they are observed, at $t = t_{\text{obs}}$. In this period, collisions can be omitted from Boltzmann equations, and photons travel along geodesics.

If we stay at linear order in perturbations, the result of step (ii) is frequently called a *transfer function*, which is a linear mapping from initial conditions onto physical observables. Specifically, for temperature fluctuations, we talk about a scalar transfer function, $\mathcal{T}_s$. It relates the temperature anisotropies to an initial *scalar power spectrum*, $\mathcal{P}_s$, as

$$D_\ell(t) = \int_{\mathbf{k}} \mathcal{T}_s(t, t_{\text{out}}; \ell, k)\, \mathcal{P}_s(t_{\text{out}}, k)\,, \tag{2.5}$$

$$\partial_t \mathcal{T}_s(t, t_{\text{out}}; \ell, k)\big|_{t=t_{\text{out}}} = 0\,. \tag{2.6}$$

The oscillating behaviour in step (ii) above is described by a damped cosine function in $\mathcal{T}_s$, producing ultimately the features sketched in Fig. 2.1.

Assuming that the complicated physics of the transfer function can be taken care of, it is then sufficient to parametrize the power spectrum, $\mathcal{P}_s$, in terms of its amplitude and momentum dependence, around some chosen momentum, k_*, that lies in the domain of the CMB data. Furthermore, $\mathcal{P}_s$ can be identified with a 'universal' curvature power spectrum, $\mathcal{P}_s = \mathcal{P}_{\mathcal{R}_T} = \mathcal{P}_{\mathcal{R}_\varphi}$ (cf. Chap. 8). In order to simplify the notation, we normally suppress the time argument t_{out}, $\mathcal{P}_s(k) \equiv \mathcal{P}_s(t_{\text{out}}, k)$. The parametrization then reads

$$\mathcal{P}_s(k) = A_s \left(\frac{k}{k_*}\right)^{n_s - 1}\,, \tag{2.7}$$

where A_s is called the *scalar amplitude*, n_s the *spectral tilt*, and k_* the *pivot scale*. We remark that the convention for n_s in Eq. (2.7) is unfortunate, as the observed 'almost flat' spectrum corresponds to $n_s \approx 1$, so that $n_s - 1$ would be a more natural variable. The exactly flat limit, $n_s = 1$, is known as the *Harrison–Zeldovich spectrum* [11,12].

The pivot scale appearing in Eq. (2.7) is normally chosen as $k_*/a_0 \equiv (20\,\text{Mpc})^{-1}$, which lies in the middle of the CMB data domain (a frequent literature convention is to set $a_0 = 1$). The numerical values deduced from the CMB experiments are [13,14]

$$A_{\mathrm{s}}^{\mathrm{obs}} = (2.100 \pm 0.030) \times 10^{-9} \,, \tag{2.8}$$

$$n_{\mathrm{s}}^{\mathrm{obs}} = 0.9649 \pm 0.0042 \,, \tag{2.9}$$

with the precision being gradually improved upon (cf., e.g., Ref. [15]). The value $n_{\mathrm{s}}^{\mathrm{obs}} < 1$ implies that $\mathcal{P}_{\mathrm{s}}$ *decreases* slowly with increasing k.

Absence of significant non-Gaussianity

Apart from quantities that have been measured, cf. Eqs. (2.8) and (2.9), there are also quantities for which no positive signal has been found yet. This turns out to provide very important empirical information about the early universe.

A fundamental question concerns the underlying nature of the statistical ensemble. If fluctuations are distributed according to *Gaussian statistics,* then this means that all information about correlation functions of fluctuations is contained already in the 2-point correlation function (this will be discussed in more detail in Sect. 2.5). In particular, correlation functions of an odd number of fluctuations vanish, and correlation functions of an even number of fluctuations can be related to products of 2-point correlation functions, via *Isserlis' or Wick's theorem.*

The simplest probe of non-Gaussianity is therefore that we consider a 3-point correlator. Suppose that we measure temperature anisotropies, and analyse them in terms of spherical harmonics, like in Eq. (2.2). The 3-point correlator of temperature fluctuations can be projected onto the overall angular momentum mode ℓ. Recalling the rules of adding angular momenta, the result would then represent a generalization of Eq. (2.3), roughly

$$E_\ell = \sum_{\ell_i, m_i} c_\ell^{\ell_1 m_1; \ell_2 m_2; \ell_3 m_3} a_{\ell_1 m_1} a_{\ell_2 m_2} a_{\ell_3 m_3} \,. \tag{2.10}$$

In particular, taking an angular average over the full sky corresponds to $\ell(\ell_1, \ell_2, \ell_3) = 0$, and then the coefficients $c_{\ell=0}^{\cdots}$ are related to the known Wigner 3-j symbols.

We may subsequently ask if there is a statistically significant indication that $E_0 \neq 0$. If this is the case, the physical interpretation is not simple, given that non-Gaussian features can be generated both at early and at late times [16]. The ultimate goal is, however, to measure a parameter, f_{NL}, that captures the relative importance of possible non-Gaussianity. We give a schematic definition around Eq. (2.62), and for now just note that the experimental constraint can be conservatively stated as [17]

$$|f_{\mathrm{NL}}^{\mathrm{obs}}| < 100 \,. \tag{2.11}$$

Unfortunately, this is not considered to be a particularly strong constraint, because many viable inflationary models predict $|f_{\mathrm{NL}}| < 1$ (cf., e.g., Ref. [18]). Since f_{NL} is quite sensitive to momentum scales larger than the CMB ones, its constraining power is expected to grow substantially with the next generation of galaxy surveys, notably Euclid [19].

Absence of significant isocurvature perturbations

The picture described around Eq. (2.5), where the result of step (i) above Eq. (2.5) is captured by a single power spectrum, is just the simplest possibility. If it is realized, we say that the initial conditions are *adiabatic* (cf. Sect. 8.1), meaning that there is no information to distinguish between perturbations in different matter constituents, such as CMB photons and dark matter (or baryons, or neutrinos). In contrast, in some models, additional information is stored in so-called *isocurvature modes* (also known as entropy or non-adiabatic perturbations). At linear order in perturbation theory, a general scalar perturbation can then be decomposed into independent adiabatic and isocurvature components. Denoting the latter by $\mathcal{I}_i$ for the i-th independent component, different power spectra like Eq. (2.7) are needed for specifying the full initial conditions,

$$\mathcal{P}_{\mathrm{s}}(k) \longrightarrow \left\{ \mathcal{P}_{\mathcal{R}_T}(k) , \mathcal{P}_{\mathcal{I}_i}(k) \right\} , \tag{2.12}$$

and in general cross-correlations are needed as well. The evolutions at $t > t_{\mathrm{out}}$, i.e. the corresponding transfer functions, can also differ, though what exactly happens depends on the physical nature of the isocurvature mode considered.

In any case, observational evidence suggests that there is only one underlying power spectrum. Notably, the TT, TE and EE power spectra, reflecting the physics of photons, and the BAO data on large-scale structure, more sensitive to dark matter, are all consistent with each other (isocurvature modes are said to excite out-of-phase peaks and dips in CMB spectra, which are not observed). Through a non-trivial statistical analysis, this sets an upper bound on how large the amplitude of the isocurvature power spectrum can be (cf., e.g., Ref. [20]). In addition there is theoretical prejudice in favour of adiabaticity, given that simple inflationary models make this prediction (cf. Sect. 8.3), and more complicated multi-field models require tuning to avoid it (cf. Sect. 8.4). Empirically, the contribution of isocurvature modes to the CMB power spectra is constrained to [14]

$$\frac{\mathcal{P}_{\mathcal{I}_{\mathrm{dm}}}(0.04\, k_*)}{\mathcal{P}_{\mathcal{R}_T}(0.04\, k_*)} < 0.03 , \qquad \frac{0.04\, k_*}{a_0} = 0.002\,\mathrm{Mpc}^{-1} , \tag{2.13}$$

where 'dm' refers to cold dark matter. The strongest available constraint comes from the smallest momenta, like for non-Gaussianity, while the bound is weaker at $k \sim k_*$, and particularly at momenta $k \gg k_*$, which do not directly influence the CMB.

Absence of significant spectral distortions

The shape of the CMB spectrum itself may hide interesting information. Indeed there are physical processes which tend to distort it away from the blackbody form. Two main types of distortions can be distinguished: those due to non-smooth departure from thermal equilibrium, and others that occur later on, when the CMB photons free-stream through space and time. During the latter epoch, the spectrum may be influenced by lensing effects or more exotic phenomena, like energy insertions from gradually decaying particles.

In the literature, deviations from the Planck spectrum are usually parametrized via two modes, called μ-type and y-type distortions. An example for the physics of the former is that, as the primordial plasma cools down, the efficiency of photon number-changing processes decreases faster than that of *Compton scatterings*, which keep sustaining kinetic equilibrium [21]. As a result, a small photon chemical potential develops, often denoted as μ, which, in the language of text-book statistical physics, corresponds to the dimensionless ratio of chemical potential over temperature. Assuming a standard radiation-dominated era after inflation, one can estimate (cf., e.g., Ref. [22])

$$|\mu| \sim 10^{-8} \,. \tag{2.14}$$

This value is much below the bound set by the current observational results [23],

$$|\mu|^{\mathrm{obs}} < 9 \times 10^{-5} \,, \tag{2.15}$$

however, future experiments may turn spectral distortions into a useful tool.

2.3   Probes of a Primordial Gravitational-Wave Background

In many theoretical frameworks, notably within the inflationary paradigm, the same physical phenomena that produce scalar perturbations, also produce tensor perturbations. We alternatively refer to the latter as gravitational waves (the reason is explained around Eq. (3.106), and is elaborated upon in Chap. 10). Because of the similar dynamics, in order to constrain the tensor power spectrum, $\mathcal{P}_{\mathrm{t}}$, it is conventional to adopt a strategy similar to that for $\mathcal{P}_{\mathrm{s}}$. Let us assume that the evolution after Hubble horizon re-entry is well understood, and captured by a tensor transfer function, $\mathcal{T}_{\mathrm{t}}$. Following Eq. (2.7), but with a different convention, we parametrize the initial conditions with a *tensor amplitude*, A_{t}, and a *tensor tilt*, n_{t}, as

$$\mathcal{P}_{\mathrm{t}}(k) = A_{\mathrm{t}} \left(\frac{k}{k_*} \right)^{n_{\mathrm{t}}} \,. \tag{2.16}$$

Even if the primordial tensor power spectrum has not been observed at the time of writing, it is being actively searched for. Here we mention some of the avenues.

Influence on CMB polarization

When photons propagate from their decoupling until observation in the presence of tensor perturbations, they develop a B-mode polarization (cf., e.g., Refs. [24,25]).

Such a polarization signal has not been observed yet, setting an upper bound on the relative amplitude of the tensor power spectrum, or *tensor-to-scalar ratio* [14,26],

$$r_{0.002}^{\text{obs}} \equiv \frac{\mathcal{P}_{\text{t}}(0.04\,k_*)}{\mathcal{P}_{\text{s}}(0.04\,k_*)} < 0.038\,, \qquad \frac{0.04\,k_*}{a_0} = 0.002\,\text{Mpc}^{-1}\,, \tag{2.17}$$

where the momentum has been chosen like in Eq. (2.13). This turns out to be a powerful constraint, because part of the model-dependence of theoretical predictions cancels in this ratio (cf. Eqs. (10.32)–(10.34)).

Pulsar timing arrays

As gravitational waves distort time and distance measurements, we can obtain constraints on them by measuring the arrival times of signals from precisely ticking distant emitters. In particular, pulsars have been observed since several decades with this goal in mind [27–30]. However, there are plausible astrophysical sources, related for instance to supermassive black holes, which may produce background signals similar to primordial gravitational waves. It remains to be seen how the significance of the signal evolves, when more data is gathered in the next decades.

Upcoming gravitational-wave detectors

A possibility to observe gravitational waves directly is offered by large laser interferometers (in the future, atom interferometers may be tested as well, cf., e.g., Ref. [31]). The hope is that, after subtracting a 'foreground' signal generated by well-understood astrophysical processes (such as black hole mergers), a stochastic signal with a distinctive spectral shape will be left over. In the 2030s, the first orbiting interferometer, the Laser Interferometer Space Antenna (LISA), should go into operation, with its peak sensitivity in the frequency range $f_0 \sim 10^{-4}...10^{-1}$ Hz [32].

From the theoretical perspective, what LISA may measure today (t_0) is the fractional energy density carried by gravitational radiation, referred to as its *spectrum*,

$$\frac{\mathrm{d}\Omega_{\text{gw},0}}{\mathrm{d}\ln k} \equiv \frac{1}{e_{\text{crit}}} \frac{\mathrm{d}e_{\text{gw}}}{\mathrm{d}\ln k}\bigg|_{t_0} \overset{(10.141)}{=} \mathcal{T}_{\text{t}}(t_0, t_{\text{out}}, k)\,\mathcal{P}_{\text{t}}(k)\,, \qquad e_{\text{crit}} \equiv \bar{e}(t_0) \overset{(1.43)}{=} \frac{3H_0^2}{8\pi G}\,. \tag{2.18}$$

The *current value of the Hubble rate*, H_0, is parametrized in terms of the *reduced Hubble rate*, h, as $H_0 = 100\,h\,\text{km}\,\text{s}^{-1}\,\text{Mpc}^{-1}$. Instead of the comoving wave number k, the physical variable for $\Omega_{\text{gw},0}$ is the frequency, f_0, which is related to k by

$$\frac{k}{a_0} = 2\pi f_0\,. \tag{2.19}$$

Inserting h and f_0, the projected LISA sensitivity is (cf., e.g., Ref. [33, Fig. 10])

$$\frac{h^2 d\Omega_{\mathrm{gw,0}}^{\mathrm{LISA}}}{d \ln f_0} \gtrsim 10^{-13} \, , \qquad (2.20)$$

but from the cosmological point of view the actual performance depends a lot on how astrophysical foregrounds can be subtracted.

Apart from LISA, there are plans for new generations of gravitational-wave interferometers sensitive to higher frequencies, such as TianQin ($f_0 \sim 10^{-3} ... 10^0$ Hz) [34], DECIGO ($f_0 \sim 10^{-1} ... 10^1$ Hz) [35], Cosmic Explorer (CE) ($f_0 \sim 10^1 ... 10^3$ Hz) [36], Einstein Telescope (ET) ($f_0 \sim 10^1 ... 10^3$ Hz) [37], as well as Big Bang Observer (BBO), a potential successor of LISA. Proposed is also a whole class of so-called ultra-high frequency (UHF) detectors (extending possibly up to $f_0 \sim 10^{11}$ Hz or beyond) [38]. Time will tell what kind of sensitivities these concepts reach, but they would definitely offer interesting probes of the earliest moments of the universe (cf. Sect. 10.8).

Effective number of neutrino degrees of freedom

Yet another probe on the energy density that gravitational waves carry, is offered by the effective number of light degrees of freedom. This is strongly constrained, for instance, through its role in big bang nucleosynthesis (BBN), or the decoupling of CMB photons.

In BBN, fusion processes of neutrons and protons into light nuclei start at $T \sim 0.1$ MeV, when thermal motion is less energetic than the binding energy [39,40]. The efficiency depends on the relation of the expansion rate and of the fusion/fission rates. Observed abundances of light elements can therefore impose a constraint on viable expansion rates via Eq. (1.43), which, for a radiation-dominated universe ($\bar{e} \approx e_r$), takes the form

$$H^2 = \frac{8\pi G}{3} e_r \, . \qquad (2.21)$$

In this context *radiation* means more broadly light, *relativistic* particles. If we modify e_r by adding more degrees of freedom, the expansion of the universe is faster, and the abundances change. Conventionally, the energy density of additional relativistic components is parametrized by an *effective number of neutrino species*, N_{eff},

$$e_r = e_\gamma + e_\nu + \dots \, , \qquad \frac{e_\nu}{e_\gamma} \equiv \frac{7}{8}\left(\frac{4}{11}\right)^{4/3} N_{\mathrm{eff}} \, . \qquad (2.22)$$

The Standard Model prediction for N_{eff} is (cf., e.g., Refs. [41–43])

$$N_{\mathrm{eff}}^{\mathrm{SM}} = 3.044 \pm 0.001 \, . \qquad (2.23)$$

This value is larger than 3, for instance because the neutrinos obtain an extra energy injection after their decoupling, from non-equilibrium $e^+e^- \to \nu\bar{\nu}$ annihilations.

As an additional relativistic degree of freedom, primordial gravitational waves also influence the overall energy density. This contribution is commonly parametrized as

$$\frac{e_{\rm gw}}{e_\gamma} \subset \frac{\Delta e_r}{e_\gamma} \equiv \frac{7}{8}\left(\frac{4}{11}\right)^{4/3} \Delta N_{\rm eff} . \tag{2.24}$$

For a specific gravitational-wave source, $e_{\rm gw}$ can be estimated by integrating $d\Omega_{\rm gw}/d\ln k$ in Eq. (2.18) over all k. From the joint analysis of CMB measurements and BBN light element abundances we have [13,26]

$$N_{\rm eff}^{\rm obs} = 2.89 \pm 0.11 . \tag{2.25}$$

The difference between Eqs. (2.25) and (2.23) defines the experimental estimate for $\Delta N_{\rm eff}$, consistent with zero at the 2σ level, and with the upper bound

$$\Delta N_{\rm eff} < 0.2 . \tag{2.26}$$

The accuracy is expected to improve in the future, promoting $\Delta N_{\rm eff}$ to a non-trivial constraint on the energy density carried by primordial gravitational waves. We will convert this bound to units of $\Omega_{\rm gw,0}$ around Eq. (10.168).

2.4 Cosmological Redshift of Wavelengths and Frequencies

When discussing observations, *current-day frequencies*, f_0, are regularly expressed in units of Hz, and *current-day wavelengths*, λ_0, in units of parsec (pc). In theoretical considerations, we rather employ comoving momenta, k, or physical momenta, $p_0 = k/a_0 = 2\pi/\lambda_0$. We recall here the relations between these quantities, and also how they are related to the Hubble rate in the early universe. As for the current Hubble rate, we write it as

$$H_0 = h \frac{100\,{\rm km}}{{\rm s\,Mpc}} \overset{(2)}{\Rightarrow} \frac{H_0}{h} = 3.241 \times 10^{-18}\,{\rm Hz} \tag{2.27}$$

$$= h \frac{10^5 c}{2.998 \times 10^8\,{\rm Mpc}} \overset{c=1}{\Rightarrow} \frac{h}{H_0} = 2.998\,{\rm Gpc} \overset{(2)}{\simeq} 9.778 \times 10^9\,{\rm ly} . \tag{2.28}$$

Interpreted as years, Eq. (2.28) reflects the age of the universe (and is close to it when divided by $h \approx 0.7$).

In order to relate present and past scales, it is helpful to first determine how temperature evolves in the early universe (we already discussed this around Eq. (1.88), but now add the missing ingredients). So, let us assume that we can describe the energy-momentum tensor via an ideal fluid. Then we can take Eq. (1.44) as a starting point,

$$\dot{\bar{e}} + 3\,H(\bar{e} + \bar{p}) \overset{(1.44)}{=} 0 \ .$$
(2.29)

Next, we need to recall some basic thermodynamics. The first law reads

$$\mathrm{d}E = T\,\mathrm{d}S - p\,\mathrm{d}V + \mu\,\mathrm{d}N \ ,$$
(2.30)

and the overall energy can be written as

$$E = TS - pV + \mu N \ .$$
(2.31)

In cosmology, chemical potentials are usually very small, so we set $\mu \to 0$ for simplicity (that said, Eq. (2.33) also applies with $\mu \neq 0$, however in that situation it must be supplemented by a conservation equation for the corresponding comoving particle number, N). Writing $E = e\,V$ and $S = s\,V$ in Eq. (2.30), where e is the *energy density* and s is the *entropy density*, and dividing Eq. (2.31) by volume, we get

$$\begin{cases} V\,\mathrm{d}e + e\,\mathrm{d}V \overset{(2.30)}{\underset{\mu \to 0}{=}} TV\,\mathrm{d}s + Ts\,\mathrm{d}V - p\,\mathrm{d}V \\[2ex] e \overset{(2.31)}{\underset{\mu \to 0}{=}} Ts - p \end{cases} \quad \Rightarrow \quad \mathrm{d}e = T\,\mathrm{d}s \ .$$
(2.32)

We may now divide Eq. (2.32) by $\mathrm{d}t$, and apply the result to the background quantities, as they appear in Eq. (2.29). This gives

$$0 \overset{(2.29)}{\underset{(2.32)}{=}} T\dot{\bar{s}} + 3\frac{\dot{a}}{a}T\bar{s} = \frac{T}{a^3}\frac{\mathrm{d}(\bar{s}a^3)}{\mathrm{d}t} \ ,$$
(2.33)

which is known as the law of *entropy conservation*. The entropy density $\bar{s}$ is in turn proportional to T^3, with a slowly evolving coefficient (this will be discussed around Eq. (7.55)). Therefore the ratio of scale factors today (a_0) and at the end of inflation (a_e), which we call the *redshift factor*, can be written as

$$\frac{a_0}{a_e} \overset{(2.33)}{=} \frac{T_e}{T_0} \underbrace{\left(\frac{\bar{s}_e/T_e^3}{\bar{s}_0/T_0^3}\right)^{1/3}}_{\text{slowly evolving}} \ .$$
(2.34)

An important point is that the derivation of Eq. (2.34) relied on the assumption of thermal equilibrium. However, the same relation is also used when this is *not* the case. Notably, *neutrinos decouple* from the electromagnetic plasma at $T \sim 2$ MeV. Below this temperature, they still carry energy and momentum density, and therefore affect the overall expansion, but their phase space distribution is non-thermal (often it is modelled by saying that neutrinos have a different temperature than the electromagnetic plasma; sometimes chemical potentials are added for further structure). It requires a theoretical computation to determine how the decoupling affects the scale factor. Afterwards, we may continue to express the result like in Eq. (2.34), with T denoting the temperature of the electromagnetic plasma. In other words, we *redefine the meaning of the entropy density*, so that it incorporates the contribution from the decoupled neutrinos to the universe expansion (unfortunately, this definition of a non-equilibrium entropy is not unique). At the current temperature, identified with that of the CMB background, $T_0 \approx 2.7255$ K, the newly defined *present-day entropy density*, $\bar{s}_0$, can be expressed in terms of the parameter h_* from Eq. (7.55), with a value $h_{*,0} \approx 3.930$ ([44], Table 1).

From Eq. (2.34), we can estimate how many *e-folds* took place after inflation. As an example, let us choose a high temperature after inflation, $T_e = 10^{15}$ GeV. Assuming that the Standard Model equation of state [45] is the dominant one up to T_e, we obtain

$$\ln\left(\frac{a_0}{a_e}\right)\Bigg|_{T_e = 10^{15}\,\text{GeV}} \overset{(2.34)}{\underset{[45]}{=}} 64.7 \,. \tag{2.35}$$

Requiring that the current observable universe was causally connected before inflation, it is therefore likely that inflation lasted at least 65 *e*-folds. In inflationary model building, it is customary to posit that the pivot scale (cf. Eq. (2.7)) exited the Hubble horizon 50–60 *e*-folds before inflation ended, and display the corresponding variation as a systematic error band related to an unknown reheating history. A way to avoid this uncertainty and determine the correct amount of *e*-folds is explained in the context of Fig. 7.2 in Sect. 7.7.

Next, let us estimate the Hubble rate, from Eq. (1.43) (setting $\kappa \to 0$). We write it as

$$H \overset{(1.43)}{=} \sqrt{\frac{8\pi(\bar{e}/T^4)}{3} \frac{T^2}{m_{\text{pl}}}} \,, \tag{2.36}$$

where the ratio $\bar{e}/T^4$ is also slowly evolving, and (during radiation domination) often parametrized through an effective number of light degrees of freedom, via Eq. (7.55).

With these ingredients, we can estimate when a given comoving momentum mode entered inside the Hubble horizon. We write the relevant ratio, evaluated at the end of inflation, as

$$\frac{k}{a_e H_e} \overset{k = a_0 p_0}{=} \frac{p_0}{T_0} \frac{a_0}{a_e} \frac{T_0}{H_e} \overset{(2.34)}{\underset{(2.36)}{=}} \frac{p_0}{T_0} \left(\frac{\bar{s}_e/T_e^3}{\bar{s}_0/T_0^3}\right)^{1/3} \sqrt{\frac{3}{8\pi}} \frac{m_{\text{pl}}/T_e}{\sqrt{\bar{e}/T_e^4}} \,. \tag{2.37}$$

As examples of numerical values, we obtain

$$\left(\frac{\bar{s}_e/T_e^3}{\bar{s}_0/T_0^3}\right)^{1/3} \sqrt{\frac{3}{8\pi}} \frac{m_{\text{pl}}/T_e}{\sqrt{\bar{e}/T_e^4}} \overset{[45]}{\approx} \begin{cases} 2.147 \times 10^3 \,, & T_e = 10^{15}\,\text{GeV} \\ 3.145 \times 10^{21} \,, & T_e = 1\,\text{MeV} \end{cases} \,. \tag{2.38}$$

Next, we need to express p_0/T_0 in terms of f_0 or λ_0. In terms of frequency,

$$\frac{p_0}{T_0} \overset{p_0 = 2\pi f_0}{\underset{\text{Hz} = 1/s}{=}} \frac{2\pi}{s\,T_0} \frac{f_0}{\text{Hz}} \approx 1.761 \times 10^{-11} \frac{f_0}{\text{Hz}} \,. \tag{2.39}$$

Combining this with Eq. (2.38), we can find at which moment today's frequencies entered inside the Hubble horizon. For instance,

$$\frac{k}{a_e H_e}\bigg|_{T_e = 10^{15}\,\text{GeV}} = 1 \overset{(2.37),(2.38)}{\underset{(2.39)}{\Leftrightarrow}} f_0 = 2.645 \times 10^7\,\text{Hz}\,, \tag{2.40}$$

$$\frac{k}{a_e H_e}\bigg|_{T_e = 1\,\text{MeV}} = 1 \overset{(2.37),(2.38)}{\underset{(2.39)}{\Leftrightarrow}} f_0 = 1.806 \times 10^{-11}\,\text{Hz}\,. \tag{2.41}$$

If we employ wavelengths instead, then

$$\frac{p_0}{T_0} \overset{p_0 = 2\pi/\lambda_0}{=} \frac{2\pi}{\text{pc}\,T_0}\left(\frac{\lambda_0}{\text{pc}}\right)^{-1} \overset{(2)}{=} 1.711 \times 10^{-19}\left(\frac{\lambda_0}{\text{pc}}\right)^{-1}\,. \tag{2.42}$$

For instance,

$$\frac{k}{a_e H_e}\bigg|_{T_e = 10^{15}\,\text{GeV}} = 1 \overset{(2.37),(2.38)}{\underset{(2.42)}{\Leftrightarrow}} \lambda_0 = 3.673 \times 10^{-16}\,\text{pc}\,, \tag{2.43}$$

$$\frac{k}{a_e H_e}\bigg|_{T_e = 1\,\text{MeV}} = 1 \overset{(2.37),(2.38)}{\underset{(2.42)}{\Leftrightarrow}} \lambda_0 = 538.1\,\text{pc}\,. \tag{2.44}$$

Finally, we can relate frequencies and wavelengths directly,

$$\frac{f_0}{\text{Hz}} \overset{f_0 = c/\lambda_0}{=} \frac{c\,\text{s}}{\text{pc}}\left(\frac{\lambda_0}{\text{pc}}\right)^{-1} \overset{(2)}{=} 9.716 \times 10^{-9}\left(\frac{\lambda_0}{\text{pc}}\right)^{-1}\,. \tag{2.45}$$

The scales probed by the CMB correspond to $\lambda_0 = (10 \ldots 3 \times 10^3)\,\text{Mpc}$, which then implies $f_0 = (10^{-15} \ldots 3 \times 10^{-18})\,\text{Hz}$. It should be clarified that this is the frequency of a light-like perturbation that leads to an anisotropy in temperature measurements in various directions. It is not the peak frequency of the CMB Planck spectrum, which rather lies in the microwave range, $f_{\text{CMB}} \sim 10^{11}\,\text{Hz}$, as can be deduced by setting $p_0 \sim T_0$ in Eq. (2.39).

2.5   What Do Power Spectra Mean in Different Contexts?

We now turn to somewhat technical issues, needed frequently in the later chapters. Given a set of data for the perturbations δQ of a physical quantity $Q \in \mathbb{R}$, observed or simulated, we define its power spectrum, $\mathcal{P}_Q(k)$, such as $\mathcal{P}_s$ and $\mathcal{P}_t$ mentioned above, and contrast with the determination of power spectra with theoretical methods.

We work in comoving conformal coordinates, and denote by $\mathbf{x}$ and $\mathbf{k}$ the positions and momenta, respectively. The time argument τ is often left implicit, and unless stated otherwise we assume equal times in the correlations. If we place ourselves at $\mathbf{x}_0$ and consider a light ray travelling in direction $\mathbf{n}$, then a plane wave of momentum $\mathbf{k}$ has the phase $e^{i\mathbf{k}\cdot(\mathbf{x}_0 - \Delta\tau\,\mathbf{n})}$. The angular momentum that we met above, ℓ, originates from the expansion of $e^{-i\Delta\tau\,\mathbf{k}\cdot\mathbf{n}}$ in spherical harmonics (cf. the discussion below Eq. (9.64)), however here we work with $\mathbf{x}$ and $\mathbf{k}$ rather than ℓ. To streamline the notation, we define

$$\int_{\tau_i} \equiv \int_{-\infty}^{\tau} \mathrm{d}\tau_i \,, \quad \int_{\mathbf{x}_i} \equiv \int \mathrm{d}^3\mathbf{x}_i \,, \quad \int_{\mathbf{k}_i} \equiv \int \frac{\mathrm{d}^3\mathbf{k}_i}{(2\pi)^3} \,, \quad \int_{a_i,\,b_i,\,\dots} \equiv \int_{a_i}\int_{b_i}\int_{\dots} \,.$$

(2.46)

Formal definitions and basic properties

A classical *statistical or ensemble average* is defined as the weighted sum over allowed configurations,

$$\langle \mathcal{O}[Q] \rangle \equiv \sum_i p_i\, \mathcal{O}[Q^{(i)}] \,,$$

(2.47)

with p_i being the probability to realize the ith configuration. If we assume the system to be *translationally invariant*, i.e. with no preferred location, then equal-time 2-point correlations can only depend on the relative spatial separation,

$$\langle\, \delta Q(\mathbf{x}_1)\, \delta Q(\mathbf{x}_2)\, \rangle = \langle\, \delta Q(\mathbf{x}_1 - \mathbf{x}_2)\, \delta Q(\mathbf{0})\, \rangle \,.$$

(2.48)

Inverse Fourier transforming Eq. (2.48) to comoving momentum yields

$$
\begin{aligned}
\langle\, \delta Q(\mathbf{k}_1)\, \delta Q^*(\mathbf{k}_2)\, \rangle
&\overset{(9)}{=} \int_{\mathbf{x}_1,\mathbf{x}_2} e^{-i(\mathbf{k}_1\cdot\mathbf{x}_1 - \mathbf{k}_2\cdot\mathbf{x}_2)} \langle\, \delta Q(\mathbf{x}_1)\, \delta Q(\mathbf{x}_2)\, \rangle \\
&\overset{(2.48)}{=} \int_{\mathbf{x}_1,\mathbf{x}_2} e^{-i(\mathbf{k}_1\cdot\mathbf{x}_1 - \mathbf{k}_2\cdot\mathbf{x}_2)} \langle\, \delta Q(\mathbf{x}_1 - \mathbf{x}_2)\, \delta Q(\mathbf{0})\, \rangle \\
&\overset{\mathbf{x}_1 \to \underline{\mathbf{x}_1} + \mathbf{x}_2}{=} \int_{\mathbf{x}_1,\mathbf{x}_2} e^{-i(\mathbf{k}_1 - \mathbf{k}_2)\cdot\mathbf{x}_2} e^{-i\mathbf{k}_1\cdot\mathbf{x}_1} \langle\, \delta Q(\mathbf{x}_1)\, \delta Q(\mathbf{0})\, \rangle \\
&= (2\pi)^3 \delta^{(3)}(\mathbf{k}_1 - \mathbf{k}_2) \underbrace{\int_{\mathbf{x}_1} e^{-i\mathbf{k}_1\cdot\mathbf{x}_1} \langle\, \delta Q(\mathbf{x}_1)\, \delta Q(\mathbf{0})\, \rangle}_{\text{like (9): } \equiv P_Q(\mathbf{k}_1)} \,.
\end{aligned}
$$

(2.49)

The Dirac-δ can be interpreted as momentum conservation. So translational invariance implies momentum conservation, in accordance with *Noether's theorem*.

If we go back to coordinate space, and look at equal positions, we obtain

$$\langle \delta Q^2(\mathbf{x}) \rangle \overset{(8)}{=} \int_{\mathbf{k}_1,\mathbf{k}_2} e^{i(\mathbf{k}_1-\mathbf{k}_2)\cdot\mathbf{x}} \langle \delta Q(\mathbf{k}_1)\, \delta Q^*(\mathbf{k}_2) \rangle \tag{2.50}$$

$$\overset{(2.49)}{=} \int_{\mathbf{k}_1,\mathbf{k}_2} e^{i(\mathbf{k}_1-\mathbf{k}_2)\cdot\mathbf{x}} (2\pi)^3 \delta^{(3)}(\mathbf{k}_1-\mathbf{k}_2) P_Q(\mathbf{k}_1)$$

$$= \underbrace{\int_0^\infty \frac{dk_1}{k_1}}_{\int_{-\infty}^\infty d\ln k_1} \underbrace{\frac{k_1^3}{(2\pi)^3} \int d\Omega_{\mathbf{k}_1}\, P_Q(\mathbf{k}_1)}_{\equiv \mathcal{P}_Q(k_1)} \,, \tag{2.51}$$

where $d\Omega_{\mathbf{k}} \equiv d\phi\, d\theta \sin\theta$ is the *infinitesimal solid angle* in spherical coordinates, and $\mathcal{P}_Q(k)$ is what we call the *power spectrum* of δQ. We remark that this procedure is a bit formal, as nothing guarantees that the momentum integral in Eq. (2.51) is convergent. Nevertheless, the *integrand* can be defined. We also note that the overall outcome is independent of the position $\mathbf{x}$, reflecting again translational invariance.

What if the spatial volume is finite?

If we carry out a numerical simulation, or inspect the visible universe, the spatial volume is finite. In this situation there is a maximal possible wavelength, and a minimal possible momentum. In order to maintain the property of translational invariance, it is convenient to impose periodic boundary conditions over the finite volume (this is also natural if we work with angular variables). Then momenta get quantized, $\mathbf{k} = 2\pi\mathbf{m}/L$, where $\mathbf{m} \in \mathbb{Z}^3$ and the volume has been taken to be a box of size L^3. Momentum integrals become sums, $\int_{\mathbf{k}} \to \frac{1}{L^3} \sum_{\mathbf{m}}$; Dirac-$\delta$'s become Kronecker-$\delta$'s, $(2\pi)^3\delta^{(3)}(\mathbf{k}_1-\mathbf{k}_2) \to \int_{\mathbf{x}} e^{-i\mathbf{x}\cdot(\mathbf{k}_1-\mathbf{k}_2)} = L^3\delta_{\mathbf{m}_1,\mathbf{m}_2}$; and the definitions in Eqs. (2.49) and (2.51) can be adapted accordingly.

A large volume offers for an alternative to the statistical average in Eq. (2.47), namely the *spatial average*, as we now illustrate. Let us consider a generic 2-point function. If far-away regions are uncorrelated, it satisfies

$$\langle \mathcal{O}(\mathbf{x})\, \mathcal{O}(\mathbf{y}) \rangle \overset{|\mathbf{x}-\mathbf{y}|\to\infty}{\longrightarrow} \langle \mathcal{O}(\mathbf{x}) \rangle \langle \mathcal{O}(\mathbf{y}) \rangle \,. \tag{2.52}$$

In many systems, the approach to this limit is exponentially fast, such that we can write

$$\langle \mathcal{O}(\mathbf{x})\, \mathcal{O}(\mathbf{y}) \rangle = \langle \mathcal{O}(\mathbf{x}) \rangle \langle \mathcal{O}(\mathbf{y}) \rangle + f(|\mathbf{x}-\mathbf{y}|)\, e^{-|\mathbf{x}-\mathbf{y}|/\xi} \,, \tag{2.53}$$

where ξ is a characteristic *correlation length*, and the function f grows at most as a power at large argument.

Let us now insert $\delta Q^2(\mathbf{x})$ in the role of $\mathcal{O}(\mathbf{x})$. From Eq. (2.51) we know that $\langle \mathcal{O}(\mathbf{x}) \rangle$ is independent of $\mathbf{x}$. From Eq. (2.53) we see that, on average, the values of $\delta Q^2(\mathbf{x})$ and $\delta Q^2(\mathbf{y})$ are uncorrelated, if $|\mathbf{x}-\mathbf{y}| \gg \xi$. Therefore, we can view $\delta Q^2(\mathbf{x})$

and $\delta Q^2(\mathbf{y})$ as independent realizations from the statistical ensemble that defined Eq. (2.47). In other words, the statistical average in Eq. (2.47) can be replaced by a volume average, provided that the volume is large, $L \gg \xi$,

$$\langle \delta Q^2(\mathbf{x}) \rangle \overset{L \gg \xi}{\approx} \frac{1}{L^3} \int_{\mathbf{x}} \delta Q^2(\mathbf{x}) \,. \tag{2.54}$$

The same procedure also applies to unequal-position correlators, like in Eq. (2.48).

A replacement like in Eq. (2.54) is essential for observational cosmology. We have only one universe, and cannot take an average over many statistically independent realizations. Instead, we take an angular average over observation directions, which physically corresponds to a spatial average over the last scattering surface. The result can be compared with theoretical models based on a statistical ensemble, provided that the average contains many uncorrelated domains. Instead, when we consider small values of ℓ, we are correlating large angular separations, and our single universe does not have many independent patches. This is the issue of *cosmic variance* (cf. Fig. 2.1), and it restricts the information that is available about the largest structures.

What if we take a quantum rather than a statistical average?

The definition in Eq. (2.47) refers to a *classical* statistical average. In contrast, the inflationary paradigm, discussed in part II of this book, postulates that the fluctuations that we see have a quantum-mechanical origin, and the expectation value is a (distant-past) vacuum expectation value in the corresponding Hilbert space.

In the quantum-mechanical framework, we can still formally define a correlator like in Eq. (2.51). The power spectrum is extracted from the integrand as before (though the issue of non-convergence of the integral becomes concrete, because there are vacuum fluctuations at all scales). Fourier transforms, $\delta Q(\mathbf{k}_1)$, are replaced by what we call *mode functions*, $\widehat{Q}_k$, which appear with annihilation and creation operators, $w_{\mathbf{k}}$ and $w_{\mathbf{q}}^{\dagger}$, so that

$$\langle \widehat{Q}_{\varphi}^2(\mathbf{x}) \rangle \overset{\text{quantum}}{\underset{\text{average}}{\equiv}} \langle 0 | \widehat{Q}_{\varphi}(\mathbf{x}) \widehat{Q}_{\varphi}(\mathbf{x}) | 0 \rangle$$

$$\overset{(5.31)}{=} \int \frac{d^3\mathbf{k}\, d^3\mathbf{q}}{(2\pi)^3} \langle 0 | w_{\mathbf{k}} \widehat{Q}_k e^{i\mathbf{k}\cdot\mathbf{x}} w_{\mathbf{q}}^{\dagger} \widehat{Q}_q^* e^{-i\mathbf{q}\cdot\mathbf{x}} | 0 \rangle \overset{(5.33)}{=} \int_{\mathbf{k}} |\widehat{Q}_k|^2 \,. \tag{2.55}$$

Compared with the derivation leading to Eq. (2.51), the difference is that momentum conservation, $\delta^{(3)}(\mathbf{k} - \mathbf{q})$, does not arise from translational invariance (cf. Eq. (2.49)), but from a commutation relation. We return to Eq. (2.55) around Eqs. (5.52) and (8.8).

What if we average over the initial rather than the final ensemble?

Let us now restore time dependence to the problem. In theoretical considerations, the observable $\delta Q(\tau, \mathbf{x})$ is normally obtained from some differential equation, whose (special) solution at linear order takes the form

$$\mathcal{D}\, \delta Q(\tau, \mathbf{x}) = \rho(\tau, \mathbf{x}) \quad \Rightarrow \quad \delta Q(\tau, \mathbf{x}) = \int_{\tau_1, \mathbf{y}} G_{\mathrm{R}}(\tau, \tau_1, \mathbf{x}, \mathbf{y})\, \rho(\tau_1, \mathbf{y}) \ . \tag{2.56}$$

Since the background is homogeneous and isotropic, and the coefficients of $\mathcal{D}$ are determined by the background, we assume that the coefficients of $\mathcal{D}$ do not depend on $\mathbf{x}$. Then $\mathcal{D}$ and thus also the *retarded Green's function*, $G_{\mathrm{R}} \equiv \mathcal{D}^{-1}$, are translationally and rotationally invariant, and $G_{\mathrm{R}}(\tau, \tau_1, \mathbf{x}, \mathbf{y}) = G_{\mathrm{R}}(\tau, \tau_1, |\mathbf{x} - \mathbf{y}|)$.

It is helpful to transform Eq. (2.56) to comoving momentum space,

$$\delta Q(\tau, \mathbf{k}) \overset{(9)}{\underset{(2.56)}{=}} \int_{\mathbf{x}, \tau_1, \mathbf{y}} G_{\mathrm{R}}(\tau, \tau_1, |\mathbf{x} - \mathbf{y}|)\, \rho(\tau_1, \mathbf{y})\, e^{-i\mathbf{k}\cdot\mathbf{x}}$$

$$= \int_{\tau_1} \underbrace{\int_{\mathbf{y}} \rho(\tau_1, \mathbf{y}) e^{-i\mathbf{k}\cdot\mathbf{y}}}_{\text{from (9): } \rho(\tau_1, \mathbf{k})}\, \underbrace{\int_{\mathbf{x}} G_{\mathrm{R}}(\tau, \tau_1, |\mathbf{x} - \mathbf{y}|)\, e^{-i\mathbf{k}\cdot(\mathbf{x}-\mathbf{y})}}_{\equiv\, G_{\mathrm{R}}(\tau, \tau_1, k)} \ . \tag{2.57}$$

With the assumptions made, the momentum-space Green's function is real,

$$G_{\mathrm{R}}^*(\tau, \tau_1, k) \overset{(2.57)}{=} \int_{\mathbf{x}} G_{\mathrm{R}}(\tau, \tau_1, |\mathbf{x}|) e^{+i\mathbf{k}\cdot\mathbf{x}} \overset{\mathbf{x}\to-\mathbf{x}}{=} G_{\mathrm{R}}(\tau, \tau_1, k) \ . \tag{2.58}$$

With Eq. (2.57), we can evaluate the correlator in Eq. (2.50). Given that the Green's function G_{R} is deterministic, the statistical average can be transported to the initial state. Assuming the latter to be translationally invariant, we find

$$\langle \delta Q^2(\tau, \mathbf{x}) \rangle$$

$$\overset{(2.50)}{=} \int_{\mathbf{k}_1, \mathbf{k}_2} e^{i(\mathbf{k}_1 - \mathbf{k}_2)\cdot\mathbf{x}} \langle \delta Q(\tau, \mathbf{k}_1)\, \delta Q^*(\tau, \mathbf{k}_2) \rangle$$

$$\overset{(2.57)}{\underset{(2.58)}{=}} \int_{\tau_1, \tau_2, \mathbf{k}_1, \mathbf{k}_2} e^{i(\mathbf{k}_1 - \mathbf{k}_2)\cdot\mathbf{x}}\, G_{\mathrm{R}}(\tau, \tau_1, k_1)\, G_{\mathrm{R}}(\tau, \tau_2, k_2) \underbrace{\langle \rho(\tau_1, \mathbf{k}_1)\rho^*(\tau_2, \mathbf{k}_2) \rangle}_{\text{from (2.49): } (2\pi)^3 \delta^{(3)}(\mathbf{k}_1 - \mathbf{k}_2) P_\rho(\tau_1, \tau_2, \mathbf{k}_1)}$$

$$\overset{\mathbf{k}_1 \to \mathbf{k}}{\underset{(2.60)}{=}} \boxed{\int_{-\infty}^{\infty} \mathrm{d}\ln k \int_{\tau_1, \tau_2} \underbrace{G_{\mathrm{R}}(\tau, \tau_1, k)\, G_{\mathrm{R}}(\tau, \tau_2, k)\, P_\rho(\tau_1, \tau_2, k)}_{\text{from (2.51): } \mathcal{P}_Q(\tau, k)} \ .} \tag{2.59}$$

We have here converted the Fourier transform, P_ρ, to the power spectrum, $\mathcal{P}_\rho$,

$$\mathcal{P}_\rho(\tau_1, \tau_2, k) \stackrel{(2.51)}{\equiv} \frac{k^3}{(2\pi)^3} \int d\Omega_{\mathbf{k}}\, P_\rho(\tau_1, \tau_2, \mathbf{k})\,. \tag{2.60}$$

What we learn from Eq. (2.59) is that at linear order in perturbations, the statistical properties of final-state correlators are equivalent to those of initial-state correlators, except that in general we need *unequal-time* power spectra on the initial-state side.

We remark that if the source, ρ, is stochastic in nature, then the average over the source realizations in Eq. (2.59) could be either a statistical (cf. Eq. (2.47)) or a quantum one (cf. Eq. (2.55)). In the statistical case, it can be further simplified under certain assumptions. Notably, if we consider small momenta, and long-time differences, then the source fluctuations should be uncorrelated in time, $P_\rho(\tau_1, \tau_2, k) \approx \Omega(\tau_1, k)\, \delta(\tau_1 - \tau_2)$ (cf. Eq. (1.63)), meaning that we have *white noise*. In this case, Eq. (2.59) takes a simpler form,

$$\mathcal{P}_Q(\tau, k) \stackrel[(2.59)]{k \to 0}{\approx} \int_{\tau_1} G_{\mathrm{R}}^2(\tau, \tau_1, k)\, \Omega(\tau_1, k)\,. \tag{2.61}$$

On the other hand, in the quantum case, coherence implies the presence of long-time correlations. We further elaborate on the notion of a quantum noise in Sect. 6.3.

What is the role of Gaussian statistics, and how to include non-Gaussianity?

The definitions of the power spectra above do *not* require assumptions about the Gaussian nature of the statistical ensemble; such assumptions become relevant only when we discuss 3-point or 4-point correlators and their relations to 2-point correlators. On the observational side, there is in general no reason for a Gaussian premise. On the theoretical side, we did make an additional assumption in Eq. (2.56), when we restricted to linear order in perturbations. Then we found that each Fourier component, $\delta Q(\tau, \mathbf{k})$, evolves independently, so that the initial and final states have the same statistical properties. In particular, if the initial statistics is Gaussian, the final ensemble inherits this property.

The linear order in perturbations corresponds to the leading term of a *perturbative approach*, similar to the Feynman diagrams of perturbative quantum field theory. In this approach, interactions between the various Fourier modes, which lead to non-Gaussianities, are treated order-by-order in a Taylor expansion. This can be applied both to the computation of initial-state correlators, as well as to their subsequent evolution until today.

In order to illustrate the procedure, let us denote the result for δQ, obtained from linear order in perturbation theory, by $\delta Q^{(1)}$. Now, assume that we compute δQ to second order, and then take a Fourier transform. Assuming translational invariance, we might expect

$$\delta Q^{(2)}(\tau, \mathbf{k}) \simeq \delta Q^{(1)}(\tau, \mathbf{k})$$
$$+ \int_{\mathbf{k}_1, \mathbf{k}_2} (2\pi)^3 \delta^{(3)}(\mathbf{k} - \mathbf{k}_1 - \mathbf{k}_2)\, f_{\mathrm{NL}}(\tau, \mathbf{k}_1, \mathbf{k}_2)\, \delta Q^{(1)}(\tau, \mathbf{k}_1)\, \delta Q^{(1)}(\tau, \mathbf{k}_2)\,.$$

$$(2.62)$$

This ansatz is actually simplified, as it assumes locality in time. In any case, the weight function, $f_{\mathrm{NL}}(\tau, \mathbf{k}_1, \mathbf{k}_2)$, is called a *bispectrum*. Often it is simplified further, for instance by treating it as a constant, rather than as a function, which is a drastic assumption. But even so, if it is non-zero, then a correlation function like

$$\langle\, \delta Q^{(2)}(\tau, \mathbf{x})\, \delta Q^{(2)}(\tau, \mathbf{x})\, \delta Q^{(2)}(\tau, \mathbf{x})\, \rangle$$

$$= \int_{\mathbf{k}_1, \mathbf{k}_2, \mathbf{k}_3} e^{i(\mathbf{k}_1 + \mathbf{k}_2 + \mathbf{k}_3)\cdot\mathbf{x}} \langle\, \delta Q^{(2)}(\tau, \mathbf{k}_1)\, \delta Q^{(2)}(\tau, \mathbf{k}_2)\, \delta Q^{(2)}(\tau, \mathbf{k}_3)\, \rangle \qquad (2.63)$$

is also non-zero, because the terms proportional to f_{NL} yield 4-point correlators, which can be reduced to 2-point correlators with the help of Isserlis' or Wick's theorem. The 2-point correlators have the properties that we have discussed above. Therefore, we can obtain a theoretical prediction for a 3-point function, to then be compared with Eq. (2.10).

References

1. Rees, M.J.: Polarization and Spectrum of the Primeval Radiation in an Anisotropic Universe. Astrophys. J. **153**, L1 (1968). https://doi.org/10.1086/180208
2. Bond, J.R., Efstathiou, G.: Cosmic background radiation anisotropies in universes dominated by nonbaryonic dark matter. Astrophys. J. Lett. **285**, L45 (1984). https://doi.org/10.1086/184362
3. Aghanim, N., et al. [Planck].: Planck 2018 results. I. Overview and the cosmological legacy of Planck. Astron. Astrophys. **641**, A1 (2020). https://doi.org/10.1051/0004-6361/201833880, 1807.06205
4. Ade, P.A.R., et al. [BICEP and Keck].: Improved Constraints on Primordial Gravitational Waves Using Planck, WMAP, and BICEP/Keck Observations Through the 2018 Observing Season. Phys. Rev. Lett. **127**, 151301 (2021). https://doi.org/10.1103/PhysRevLett.127.151301, 2110.00483
5. Zebrowski, J.A., et al.: [SPT-3G], Constraints on Inflationary Gravitational Waves with Two Years of SPT-3G Data. 2505.02827
6. Kamionkowski, M., Kosowsky, A., Stebbins, A.: A Probe of Primordial Gravity Waves and Vorticity. Phys. Rev. Lett. **78**, 2058 (1997). https://doi.org/10.1103/PhysRevLett.78.2058, astro-ph/9609132
7. Seljak, U., Zaldarriaga, M.: Signature of Gravity Waves in Polarization of the Microwave Background. Phys. Rev. Lett. **78**, 2054 (1997). https://doi.org/10.1103/PhysRevLett.78.2054, astro-ph/9609169
8. Lewis, A., Challinor, A.: Weak gravitational lensing of the CMB. Phys. Rept. **429**, 1 (2006). https://doi.org/10.1016/j.physrep.2006.03.002, astro-ph/0601594

9. Silk, J.: Cosmic Black-Body Radiation and Galaxy Formation. Astrophys. J. **151**, 459 (1968). https://doi.org/10.1086/149449

10. Sachs, R.K., Wolfe, A.M.: Perturbations of a cosmological model and angular variations of the microwave background. Astrophys. J. **147**, 73 (1967). https://doi.org/10.1007/s10714-007-0448-9

11. Harrison, E.R.: Fluctuations at the Threshold of Classical Cosmology. Phys. Rev. D **1**, 2726 (1970). https://doi.org/10.1103/PhysRevD.1.2726

12. Zeldovich, Ya.B.: A Hypothesis, Unifying the Structure and the Entropy of the Universe. Mon. Not. R. Astron. Soc. **160**, 1 (1972). https://doi.org/10.1093/mnras/160.1.1P

13. Aghanim, N., et al. [Planck].: Planck 2018 results. VI. Cosmological parameters. Astron. Astrophys. **641**, A6 (2020); ibid. **652**, C4 (2021) (erratum). https://doi.org/10.1051/0004-6361/201833910, 1807.06209

14. Akrami, Y., et al. [Planck].: Planck 2018 results. X. Constraints on inflation. Astron. Astrophys. **641**, A10 (2020). https://doi.org/10.1051/0004-6361/201833887, 1807.06211

15. Louis, T., et al.: The Atacama Cosmology Telescope: DR6 Power Spectra, Likelihoods and ΛCDM Parameters. 2503.14452

16. Komatsu, E., Spergel, D.N.: Acoustic signatures in the primary microwave background bispectrum. Phys. Rev. D **63**, 063002 (2001). https://doi.org/10.1103/PhysRevD.63.063002, astro-ph/0005036

17. Akrami, Y., et al. [Planck].: Planck 2018 results. IX. Constraints on primordial non-Gaussianity. Astron. Astrophys. **641**, A9 (2020). https://doi.org/10.1051/0004-6361/201935891, 1905.05697

18. Lyth, D.H., Rodríguez, Y.: Inflationary Prediction for Primordial Non-Gaussianity. Phys. Rev. Lett. **95**, 121302 (2005). https://doi.org/10.1103/PhysRevLett.95.121302, astro-ph/0504045

19. Andrews, A., et al.: [Euclid], Euclid: Field-level inference of primordial non-Gaussianity and cosmic initial conditions. 2412.11945

20. Enqvist, K., Kurki-Suonio, H.: Constraining isocurvature fluctuations with the Planck Surveyor. Phys. Rev. D **61**, 043002 (2000). https://doi.org/10.1103/PhysRevD.61.043002, astro-ph/9907221

21. Zeldovich, Ya.B., Sunyaev, R.A.: The interaction of matter and radiation in a hot-model universe. Astrophys. Space Sci. **4**, 301 (1969). https://doi.org/10.1007/BF00661821

22. Chluba, J., Sunyaev, R.A.: The evolution of CMB spectral distortions in the early Universe. Mon. Not. R. Astron. Soc. **419**, 1294 (2012). https://doi.org/10.1111/j.1365-2966.2011.19786.x, 1109.6552

23. Fixsen, D.J., Cheng, E.S., Gales, J.M., Mather, J.C., Shafer, R.A., Wright, E.L.: The Cosmic Microwave Background Spectrum from the Full COBE-FIRAS Data Set. Astrophys. J. **473**, 576 (1996). https://doi.org/10.1086/178173, astro-ph/9605054

24. Fabbri, R., Pollock, M.D.: The effect of primordially produced gravitons upon the anisotropy of the cosmological microwave background radiation. Phys. Lett. B **125**, 445 (1983). https://doi.org/10.1016/0370-2693(83)91322-9

25. Polnarev, A.G.: Polarization and Anisotropy Induced in the Microwave Background by Cosmological Gravitational Waves. Sov. Astron. **29**, 607 (1985)

26. Calabrese, E., et al.: The Atacama Cosmology Telescope: DR6 Constraints on Extended Cosmological Models. 2503.14454

27. Agazie, G., et al.: The NANOGrav 15 yr Data Set: Evidence for a Gravitational-Wave Background. Astrophys. J. Lett. **951**, L8 (2023). https://doi.org/10.3847/2041-8213/acdac6, 2306.16213

28. Antoniadis, J., et al.: The second data release from the European Pulsar Timing Array III. Search for gravitational wave signals. Astron. Astrophys. **678**, A50 (2023). https://doi.org/10.1051/0004-6361/202346844, 2306.16214

29. Reardon, D.J., et al.: Search for an Isotropic Gravitational-wave Background with the Parkes Pulsar Timing Array. Astrophys. J. Lett. **951**, L6 (2023). https://doi.org/10.3847/2041-8213/acdd02, 2306.16215

30. Xu, H., et al.: Searching for the Nano-Hertz Stochastic Gravitational Wave Background with the Chinese Pulsar Timing Array Data Release I. Res. Astron. Astrophys. **23**, 075024 (2023). https://doi.org/10.1088/1674-4527/acdfa5, 2306.16216
31. Abend, S., et al.: Terrestrial very-long-baseline atom interferometry: Workshop summary. AVS Quantum Sci. **6**, 024701 (2024). https://doi.org/10.1116/5.0185291, 2310.08183
32. Colpi, M., et al.: LISA Definition Study Report. 2402.07571
33. Babak, S., Petiteau, A., Hewitson, M.: LISA Sensitivity and SNR Calculations, 2108.01167
34. Li, E.-K., et al.: Gravitational wave astronomy with TianQin. Rept. Prog. Phys. **88**, 056901 (2025). https://doi.org/10.1088/1361-6633/adc9be, 2409.19665
35. Kawamura, S., et al.: Current status of space gravitational wave antenna DECIGO and B-DECIGO. PTEP **2021**, 05A105 (2021). https://doi.org/10.1093/ptep/ptab019, 2006.13545
36. Reitze, D., et al.: Cosmic Explorer: The U.S. Contribution to Gravitational-Wave Astronomy beyond LIGO. Bull. Am. Astron. Soc. **51**, 035 (2019). 1907.04833
37. Maggiore, M., et al.: Science case for the Einstein telescope. JCAP **03**, 050 (2020). https://doi.org/10.1088/1475-7516/2020/03/050, 1912.02622
38. Aggarwal, N., et al.: Challenges and opportunities of gravitational-wave searches at MHz to GHz frequencies. Living Rev. Rel. **24**, 4 (2021). https://doi.org/10.1007/s41114-021-00032-5, 2011.12414
39. Alpher, R.A., Bethe, H., Gamow, G.: The Origin of Chemical Elements. Phys. Rev. **73**, 803 (1948). https://doi.org/10.1103/PhysRev.73.803
40. Fields, B.D., Olive, K.A., Yeh, T.H., Young, C.: Big-Bang Nucleosynthesis after Planck. JCAP **03**, 010 (2020); *ibid.***11**, E02 (2020) (erratum). https://doi.org/10.1088/1475-7516/2020/03/010, 1912.01132
41. Akita, K., Yamaguchi, M.: A precision calculation of relic neutrino decoupling. JCAP **08**, 012 (2020). https://doi.org/10.1088/1475-7516/2020/08/012, 2005.07047
42. Froustey, J., Pitrou, C., Volpe, M.C.: Neutrino decoupling including flavour oscillations and primordial nucleosynthesis. JCAP **12**, 015 (2020). https://doi.org/10.1088/1475-7516/2020/12/015, 2008.01074
43. Bennett, J.J., Buldgen, G., De Salas, P.F., Drewes, M., Gariazzo, S., Pastor, S., Wong, Y.Y.Y.: Towards a precision calculation of N_{eff} in the Standard Model. Part II. Neutrino decoupling in the presence of flavour oscillations and finite-temperature QED. JCAP **04**, 073 (2021). https://doi.org/10.1088/1475-7516/2021/04/073, 2012.02726
44. Escudero, M., Jackson, G., Laine, M., Sandner, S.: Fast and flexible neutrino decoupling. Part I. The Standard Model. JCAP **02**, 046 (2026). https://doi.org/10.1088/1475-7516/2026/02/046, 2511.04747
45. Laine, M., Meyer, M., Schröder, Y.: Data for the Standard Model equation of state. http://laine.itp.unibe.ch/eos15/

Evolution Equations for First-Order Perturbations

3

Abstract

We describe how the left and right-hand sides of the Einstein equations can be perturbed to first order around the background solution. The perturbations are then decomposed into scalar, vector and tensor components under *helicity*, or two-dimensional rotations. The resulting Einstein equations for the different types of perturbations are displayed. We demonstrate how taking linear combinations of the Einstein equations, or their integrals (if an equation relates two total derivatives), or their derivatives (like when considering energy-momentum conservation), can lead to simpler equations. We also illustrate how anisotropic stress can be included within the framework of viscous hydrodynamics, and how the steps mentioned can be implemented with computer algebra.

Keywords

Linearization · Scalar perturbations · Vector perturbations · Tensor perturbations · Anisotropic stress · Bulk and shear viscosities · Computer algebra

3.1 Definition of Perturbations

In Chap. 1 we have defined a homogeneous and isotropic 'background' solution for an expanding universe. However, as explained in Chap. 2, the observed universe is not homogeneous at all scales, but displays a definite spectrum of perturbations. The goal of this chapter is to generalize the equations introduced in Chap. 1 up to first order in perturbations.

We assume that any physical quantity $Q(x)$ can be decomposed into the sum

$$Q(\tau, \mathbf{x}) = \bar{Q}(\tau) + \delta Q(\tau, \mathbf{x}) + \mathcal{O}(\delta^2) \,, \tag{3.1}$$

of a homogeneous and isotropic background $\bar{Q}(\tau)$, and a small (linear) perturbation $\delta Q(\tau, \mathbf{x})$. In some cases the background quantity $\bar{Q}(\tau)$ can also vanish. It is important

M. Laine and S. Procacci, *From Inflation to Hot Big Bang*, Lecture Notes in Physics 1047,
https://doi.org/10.1007/978-3-032-09893-1_3

to realize that the validity of the formalism is *not* dictated by the zeroth and first-order terms, which are both accounted for, but rather by the assumption that the second-order term, which is omitted, is small compared with the sum of the zeroth and first-order terms. Unfortunately, even if empirically well motivated (cf. Eqs. (2.11) and (2.62)), it is not easy to state the conditions under which this assumption is mathematically justified; we will return to this issue in later chapters, notably at the end of Sect. 5.3, and in Sect. 10.4.

Within the perturbative method, the laws of physics are linearized around the background. Subtracting the Einstein equations at zeroth order from the ones of the full theory we obtain the first-order equations,

$$\delta G_{\mu\nu} = 8\pi G\, \delta T_{\mu\nu}\,, \tag{3.2}$$

where G denotes Newton's gravitational constant. This set of equations describes the early evolution of primordial perturbations. In a later universe, non-linear effects become important and are essential for generating the large-scale structures that we observe today.

The Einstein equations contain different types of fields. We may classify them as scalars, vectors and tensors, but attention needs to be paid to the transformation group that is meant. Roughly speaking, three different kinds of transformations are relevant: 3+1-dimensional coordinate transformations; 3-dimensional spatial rotations; and 2-dimensional rotations in a transverse plane. In order to make this distinction clear, we discuss various types of objects in turn.

We will adopt a very practical notation (even if it may require some getting used to), namely that when we refer to spatial indices i, j, k, m, in contrast to space-time indices μ, ν, ρ, σ, and they label first-order perturbations (e.g. $h_{\mu\nu}$ in Eq. (3.34), or $\delta T_{\mu\nu}$ in Eq. (3.74)), or spatial derivatives, then they are assumed Euclidean. In other words, for a generic tensorial quantity Q we denote

$$\nabla^2 Q \equiv \partial^i \partial_i Q\,, \qquad Q^i = \delta^{im} Q_m = Q_i\,, \qquad \operatorname{tr} Q_{ij} = \operatorname{tr} Q^{ij} \equiv Q^i_i\,. \tag{3.3}$$

Scalars

A quantity is a scalar if it is invariant under all of the transformations mentioned above. The first-order decomposition of a scalar quantity $A(\tau, \mathbf{x})$ according to Eq. (3.1) gives

$$A(\tau, \mathbf{x}) = \bar{A}(\tau) + \delta A(\tau, \mathbf{x}) + \mathcal{O}(\delta^2)\,. \tag{3.4}$$

Vectors

An arbitrary 4-vector, $B^\mu = (B^0, B^i)$, is made of a temporal 3-scalar component, B^0, and a spatial 3-vector component, B^i. Applying the decomposition in Eq. (3.1), the spatial components of background quantities vanish as there is no preferred direction,

$$B^\mu(\tau, \mathbf{x}) = \bar{B}^\mu(\tau) + \delta B^\mu(\tau, \mathbf{x}) + \mathcal{O}(\delta^2)\,, \tag{3.5}$$

$$\bar{B}^\mu = (\bar{B}^0, \mathbf{0})\,, \quad \delta B^\mu = (b^0, \delta B^i)\,, \quad b^0 \equiv \delta B^0\,, \tag{3.6}$$

and $\partial_i \bar{B}^0 = 0$ for all spatial directions, $i = 1, 2, 3$. The spatial perturbation δB^i can furthermore be decomposed into a divergence-free and a curl-free part,

$$b^i \equiv \delta B^i = b^i_{\mathrm{v}} + b^i_{\mathrm{s}} \,, \qquad \partial_i b^i_{\mathrm{v}} = 0 \,, \ b^i_{\mathrm{s}} = -\partial^i b = -b^{,i} \,. \tag{3.7}$$

The decomposition in Eq. (3.7) can be justified via Helmholtz's theorem, but we present a constructive proof, this being straightforward. Let us introduce a parallel projector, which in Fourier space acts on a quantity Q as

$$\mathbb{L}^i_m \, Q(\mathbf{x}) \overset{(8)}{\equiv} \int_{\mathbf{k}} \frac{k^i k_m}{\mathbf{k}^2} \, Q(\mathbf{k}) \, e^{i\mathbf{k}\cdot\mathbf{x}} \,, \tag{3.8}$$

provided that $Q(\mathbf{k})$ is not too singular at small $|\mathbf{k}|$. In coordinate space the *parallel or longitudinal projector* can formally be expressed as

$$\mathbb{L}^i_m = \frac{\partial^i \partial_m}{\nabla^2} \,. \tag{3.9}$$

A *transverse projector*, $\mathbb{P}^i_m$, can then be defined, and has basic properties as

$$\mathbb{P}^i_m \equiv \delta^i_m - \mathbb{L}^i_m = \delta^i_m - \frac{\partial^i \partial_m}{\nabla^2} \,, \qquad \partial_i \mathbb{P}^i_m = 0 \,, \quad \mathbb{P}^i_k \, \mathbb{P}^k_m = \mathbb{P}^i_m \,. \tag{3.10}$$

Consider now a general 3-vector, b^i. It can be written as

$$b^i = \delta^i_m b^m \overset{(3.10)}{=} \underbrace{\mathbb{P}^i_m b^m}_{\equiv b^i_{\mathrm{v}}} + \underbrace{\mathbb{L}^i_m b^m}_{\equiv b^i_{\mathrm{s}}} \,. \tag{3.11}$$

Here the vector part, b^i_{v}, is divergence-free (transverse),

$$\partial_i b^i_{\mathrm{v}} \overset{(3.11)}{=} \partial_i \mathbb{P}^i_m b^m \overset{(3.10)}{=} 0 \,. \tag{3.12}$$

The scalar part, b^i_{s}, can be expressed as a gradient,

$$b^i_{\mathrm{s}} \equiv -b^{,i} \,, \quad b \overset{(3.11)}{\underset{(3.9)}{=}} -\frac{\partial_m}{\nabla^2} b^m \,, \tag{3.13}$$

where the overall minus sign is a convention. To summarize, we call b the *scalar* and b^i_{v} the *vector* parts of the spatial perturbations δB^i [1], whereby these notions now refer to transformations with respect to the spatial transverse plane. Altogether δB^μ splits into two scalars, $\{b^0, b\}$, and one divergenceless 3-vector, b^i_{v}.

Symmetric tensors

A central role in Einstein's general relativity is played by symmetric rank-2 tensors, $C_{\mu\nu} = C_{\nu\mu}$. In four dimensions, they have ten independent components. We define

$$C_{\mu\nu}(\tau, \mathbf{x}) = \bar{C}_{\mu\nu}(\tau) + \underbrace{\delta C_{\mu\nu}(\tau, \mathbf{x})} + \mathcal{O}(\delta^2) \,, \tag{3.14}$$

$$\left.\begin{cases} \delta C_{00} \equiv -2c_0 \\ \delta C_{0i} = \delta C_{i0} \equiv -c_i \\ \delta C_{ij} \equiv -2\delta_{ij}c_{\mathrm{D}} + 2\gamma_{ij} \end{cases}\right\} \Rightarrow \delta C_{\mu\nu} = \begin{pmatrix} -2c_0 & -c_j \\ -c_i & -2\delta_{ij}c_{\mathrm{D}} + 2\gamma_{ij} \end{pmatrix} \,, \tag{3.15}$$

where $\gamma_{ij} = \gamma_{ji}$ is traceless, $\gamma_i^i = 0$. The homogeneous and isotropic background satisfies

$$\bar{C}_{0i} = \bar{C}_{i0} = 0 \,, \quad \bar{C}_{ij} = \bar{C}\,\delta_{ij} \,, \quad \partial_i \bar{C} = \partial_i \bar{C}_{00} = 0 \quad \forall i = 1, 2, 3 \,. \tag{3.16}$$

The 3-vector c_i can be decomposed as discussed above,

$$c_i \stackrel{(3.7)}{=} c_i^{\mathrm{s}} + c_i^{\mathrm{v}} \,, \quad c_i^{\mathrm{s}} = -\partial_i c \,, \quad \partial^i c_i^{\mathrm{v}} = 0 \,, \tag{3.17}$$

while the traceless 3-tensor γ_{ij} can be written in terms of scalar, transverse vector and transverse tensor parts with respect to two-dimensional transformations,

$$\gamma_{ij} = \gamma_{ij}^{\mathrm{s}} + \gamma_{ij}^{\mathrm{v}} + \gamma_{ij}^{\mathrm{t}} \,, \tag{3.18}$$

$$\gamma_{ij}^{\mathrm{s}} = \left(\partial_i \partial_j - \delta_{ij}\nabla^2/3\right)\gamma \,, \tag{3.19}$$

$$\gamma_{ij}^{\mathrm{v}} = -\frac{1}{2}\left(\partial_i \gamma_j^{\mathrm{v}} + \partial_j \gamma_i^{\mathrm{v}}\right) \,, \quad \partial^i \gamma_i^{\mathrm{v}} = 0 \,, \tag{3.20}$$

$$\delta^{ij}\gamma_{ij}^{\mathrm{t}} = 0 \,, \qquad\qquad\qquad \partial^i \gamma_{ij}^{\mathrm{t}} = 0 \,. \tag{3.21}$$

Let us verify the statements of Eqs. (3.18)–(3.21) with the help of the projectors from Eqs. (3.9) and (3.10). In general we can write δC_{ij} as

$$\delta C_{ij} = \frac{1}{2}\underbrace{\left(\delta_i^m \delta_j^n + \delta_i^n \delta_j^m\right)}_{\text{symmetric in } i \leftrightarrow j} \delta C_{mn} + \frac{1}{2}\underbrace{\left(\delta_i^m \delta_j^n - \delta_i^n \delta_j^m\right)}_{\epsilon_{ijk}\epsilon^{kmn}} \delta C_{mn} \,, \tag{3.22}$$

where ϵ_{ijk} is the antisymmetric *Levi-Civita symbol*. The contraction $\epsilon^{kmn}\delta C_{mn}$ vanishes, because we have assumed δC_{mn} symmetric. Turning our attention to the symmetric tensor, we separate its trace part as

$$\delta C_{ij} = \underbrace{\frac{1}{2}\left(\delta_i^m \delta_j^n + \delta_i^n \delta_j^m - \frac{2}{3}\delta_{ij}\delta^{mn}\right)}_{\equiv \mathbb{S}_{ij}^{mn}} \delta C_{mn} + \frac{\delta_{ij}}{3}\underbrace{\delta C_m^m}_{\equiv -6c_{\mathrm{D}}} \,. \tag{3.23}$$

In the traceless symmetric tensor, $\mathbb{S}_{ij}^{mn}$, we insert the projectors from Eq. (3.10) in the form $\delta_i^m = \mathbb{P}_i^m + \mathbb{L}_i^m$. The quadratic appearances of $\mathbb{P}$ are pulled apart into the projector

$$\mathbb{T}_{ij}^{mn} \equiv \frac{1}{2}\left(\mathbb{P}_i^m\mathbb{P}_j^n + \mathbb{P}_i^n\mathbb{P}_j^m - \mathbb{P}_{ij}\mathbb{P}^{mn}\right), \tag{3.24}$$

which is *transverse and traceless,*

$$\partial^i\mathbb{T}_{ij}^{mn} = 0 = \partial_m\mathbb{T}_{ij}^{mn}\,, \quad \delta^{ij}\mathbb{T}_{ij}^{mn} = 0 = \delta_{mn}\mathbb{T}_{ij}^{mn}\,, \quad \mathbb{T}_{ij}^{mn}\mathbb{T}_{mn}^{kl} = \mathbb{T}_{ij}^{kl}\,. \tag{3.25}$$

As far as quadratic appearances of $\mathbb{L}$ go, we note that $\mathbb{L}_i^m\mathbb{L}_j^n = \mathbb{L}_i^n\mathbb{L}_j^m = \mathbb{L}_{ij}\mathbb{L}^{mn}$. Making use of this property and the definition in Eq. (3.24), we find

$$\mathbb{S}_{ij}^{mn} \overset{(3.23)}{\underset{(3.24)}{=}} \mathbb{T}_{ij}^{mn} \tag{3.26}$$

$$+ \frac{1}{2}\left(\mathbb{L}_i^m\mathbb{P}_j^n + \mathbb{L}_i^n\mathbb{P}_j^m\right) + \frac{1}{2}\left(\mathbb{L}_j^m\mathbb{P}_i^n + \mathbb{L}_j^n\mathbb{P}_i^m\right) \tag{3.27}$$

$$+ \frac{1}{6}\left(\mathbb{P}_{ij} - 2\mathbb{L}_{ij}\right)\left(\mathbb{P}^{mn} - 2\mathbb{L}^{mn}\right). \tag{3.28}$$

Once we operate with $\mathbb{S}_{ij}^{mn}$ on δC_{mn}, Eq. (3.26) yields a transverse and traceless tensor,

$$2\gamma_{ij}^{\mathrm{t}} = \mathbb{T}_{ij}^{mn}\delta C_{mn}\,. \tag{3.29}$$

Equation (3.27) yields vectors,

$$\left[\frac{1}{2}\left(\mathbb{L}_i^m\mathbb{P}_j^n + \mathbb{L}_i^n\mathbb{P}_j^m\right) + \frac{1}{2}\left(\mathbb{L}_j^m\mathbb{P}_i^n + \mathbb{L}_j^n\mathbb{P}_i^m\right)\right]\delta C_{mn} = -\left(\partial_i\gamma_j^{\mathrm{v}} + \partial_j\gamma_i^{\mathrm{v}}\right), \tag{3.30}$$

$$\gamma_j^{\mathrm{v}} \equiv -\frac{\partial^m\mathbb{P}_j^n + \partial^n\mathbb{P}_j^m}{2\nabla^2}\delta C_{mn}\,, \quad \partial^j\gamma_j^{\mathrm{v}} \overset{(3.10)}{=} 0\,. \tag{3.31}$$

Equation (3.28) yields a scalar,

$$\frac{1}{6}\left(\mathbb{P}_{ij} - 2\mathbb{L}_{ij}\right)\left(\mathbb{P}^{mn} - 2\mathbb{L}^{mn}\right)\delta C_{mn} \overset{(3.10)}{=} 2\left(\partial_i\partial_j - \frac{\delta_{ij}\nabla^2}{3}\right)\gamma\,, \tag{3.32}$$

$$\gamma \equiv -\frac{1}{4\nabla^2}\left(\delta^{mn} - \frac{3\partial^m\partial^n}{\nabla^2}\right)\delta C_{mn}\,. \tag{3.33}$$

To summarize, symmetric 4-tensors have four scalar degrees of freedom, $\{c_0, c_{\mathrm{D}}, c, \gamma\}$, two transverse vectors, $\{c_i^{\mathrm{v}}, \gamma_i^{\mathrm{v}}\}$, and one transverse tensor, γ_{ij}^{t}, which satisfies four constraints. As a crosscheck, this sums up to $4 + 2 \times 2 + (6 - 4) = 10$ free variables.

3.2   Einstein Tensor

Apart from physical quantities, the perturbative expansion is also applied to the metric tensor describing the geometry of space-time, whose zeroth-order term is given by Eq. (1.3). The indices of proper Lorentz tensors are raised and lowered using the full, perturbed metric. However, let us anticipate that we often meet 'nontensorial' objects, for which another convention is chosen (cf. the discussion around Eq. (3.35)).

The fluctuations of the spatially flat FLRW metric in conformal coordinates can be decomposed into their scalar, vector and tensor parts with the help of Eqs. (3.14)–(3.21). We define

$$g_{\mu\nu} \;=\; \bar{g}_{\mu\nu} + \delta g_{\mu\nu} \stackrel{(1.3)}{=} a^2 \left(\eta_{\mu\nu} + h_{\mu\nu} \right) , \tag{3.34}$$

where $\eta_{\mu\nu} \equiv \mathrm{diag}(-+++)$ is the *Minkowski metric* (we have set $\kappa = 0$ in Eq. (1.3)), and $h_{\mu\nu}$ is its perturbation. The scale factor a is normally defined as unperturbed.

Now, the identification of points in the two different space-times, $g_{\mu\nu}$ and $\bar{g}_{\mu\nu}$, is gauge-dependent, as will be discussed in Chap. 4. The quantity $\delta g_{\mu\nu} = a^2 h_{\mu\nu}$ is therefore not a proper tensor. When we raise indices in $h_{\mu\nu}$, this is defined to happen with $\eta^{\mu\nu} \equiv \eta_{\mu\nu}$. We note in passing that $h_\alpha{}^\beta = h_{\alpha\mu}\eta^{\mu\beta} = h_{\mu\alpha}\eta^{\beta\mu} = \eta^{\beta\mu}h_{\mu\alpha} = h^\beta{}_\alpha$, so that the mixed index ordering is symmetric. We denote it more compactly by h^β_α. With this notation, the inverse of the metric in Eq. (3.34) is given by

$$g^{\mu\nu} \;=\; \bar{g}^{\mu\nu} + \delta g^{\mu\nu} \;=\; a^{-2} \left(\eta^{\mu\nu} - h^{\mu\nu} \right) + \mathcal{O}(\delta^2) . \tag{3.35}$$

This can be verified by an explicit computation,

$$g^{\mu\nu} g_{\nu\beta} \stackrel[(3.34)]{(3.35)}{=} \left(\eta^{\mu\nu} - h^{\mu\nu} \right)\left(\eta_{\nu\beta} + h_{\nu\beta} \right) + \mathcal{O}(\delta^2)$$

$$= \delta^\mu_\beta - \cancel{h^\mu_\beta} + \cancel{h^\mu_\beta} + \mathcal{O}(\delta^2) . \tag{3.36}$$

To recover the scalar-vector-tensor decomposition, cf. Eq. (3.15), we define

$$h_{00} \equiv -2h_0 , \tag{3.37}$$

$$h_{0i} \equiv h_{i0} \equiv -h_i , \tag{3.38}$$

$$h_{ij} \equiv 2(-h_{\mathrm{D}}\, \delta_{ij} + \vartheta_{ij}) , \tag{3.39}$$

where ϑ_{ij} is traceless. We note in passing that another notation frequently used in the literature is $h_{00} = -2A$, $h_{0i} = h_{i0} = -B_i$ and $h_{ij} = 2(-D\,\delta_{ij} + E_{ij})$, or variations thereof, but we reserve capital letters for vectors and tensors. In matrix form the metric and its inverse are hence given by

$$g_{\mu\nu} = a^2(\tau) \begin{pmatrix} -1 - 2h_0 & -h_j \\ -h_i & (1 - 2h_{\rm D})\,\delta_{ij} + 2\vartheta_{ij} \end{pmatrix}, \tag{3.40}$$

$$g^{\mu\nu} = \frac{1}{a^2(\tau)} \begin{pmatrix} -1 + 2h_0 & -h_j \\ -h_i & (1 + 2h_{\rm D})\,\delta_{ij} - 2\vartheta_{ij} \end{pmatrix} + \mathcal{O}(\delta^2). \tag{3.41}$$

The $0i$ and ij-components can be further decomposed, cf. Eqs. (3.17)–(3.21),

$$h_i = -\partial_i h + h_i^{\rm v}, \qquad\qquad\qquad\qquad \partial^i h_i^{\rm v} = 0, \tag{3.42}$$

$$\vartheta_{ij} = \vartheta_{ij}^{\rm s} + \vartheta_{ij}^{\rm v} + \vartheta_{ij}^{\rm t}, \qquad \vartheta_{ij}^{\rm s} = \left(\partial_i\partial_j - \delta_{ij}\nabla^2/3 \right)\vartheta, \tag{3.43}$$

$$\vartheta_{ij}^{\rm v} = -\frac{1}{2}\left(\partial_i\vartheta_j^{\rm v} + \partial_j\vartheta_i^{\rm v} \right), \qquad \partial^i\vartheta_i^{\rm v} = 0, \tag{3.44}$$

$$\delta^{ij}\vartheta_{ij}^{\rm t} = 0, \qquad\qquad\qquad\qquad \partial^i\vartheta_{ij}^{\rm t} = 0. \tag{3.45}$$

In order to give a geometric interpretation to the metric perturbations [2], let us denote a submanifold of constant τ, called a *time slice*, by $\mathcal{M}_\tau$. In turn, spatial coordinates separate the space-time into $x^i = $ constant *threads*. At each space-time point p, we so find a covector n_μ normal to the corresponding time slice $\mathcal{M}_\tau$, $n_\mu = (n_0, \mathbf{0})$, and a vector t^μ tangent to the thread passing through p, $t^\mu = (t^0, \mathbf{0})$. They can be normalized to unit length with $g^{\mu\nu} n_\mu n_\nu = g_{\mu\nu} t^\mu t^\nu = -1$. At first order in perturbations, we find

$$n^\mu = \frac{1}{a}(-1 + h_0, -h_i) + \mathcal{O}(\delta^2), \quad n_\mu = a\,(1 + h_0, \mathbf{0}) + \mathcal{O}(\delta^2), \tag{3.46}$$

$$t^\mu = \frac{1}{a}(1 - h_0, \mathbf{0}) + \mathcal{O}(\delta^2), \qquad t_\mu = a\,(-1 - h_0, -h_i) + \mathcal{O}(\delta^2). \tag{3.47}$$

We see that because of the non-diagonal terms in the metric, n^μ and t^μ are not parallel, but differ by the *shift vector* h_i. At the same time, h_0 determines the proper time separation between time slices, taken along the thread, and is called a *lapse function*.

With the metric tensor at hand, the next task is to construct the Einstein tensor. While in principle straightforward, this task is tedious in practice, given the many indices and lengthy expressions appearing. As a simplification, one of two strategies is often followed: either we make use of a symbolic manipulation package, or we 'choose a gauge', whereby some of the metric perturbations are put to zero. The use of a symbolic manipulation package is illustrated in Sect. 3.8, whereas a gauge-fixed computation is shown in Sect. 4.4. In both cases we have furthermore set $\kappa = 0$, so that Euclidean coordinates can be employed in spatial directions. In the remainder of this section, we outline the steps and give the results in a general gauge.

As was discussed around Eq. (1.33), it has been conventional in the literature to determine the Einstein tensor $G^\mu{}_\nu$. However, it suffers from the problem of not being symmetric in general. We therefore work out $G_{\mu\nu}$ instead.

The first step is to determine the Christoffel symbols, cf. Eq. (1.9). At first order in perturbations,

$$\delta\Gamma^\rho_{\mu\nu} \stackrel{(1.9)}{=} \frac{1}{2}\bar{g}^{\rho\sigma}\left(\delta g_{\sigma\mu,\nu} + \delta g_{\sigma\nu,\mu} - \delta g_{\mu\nu,\sigma}\right) + \frac{1}{2}\delta g^{\rho\sigma}\left(\bar{g}_{\sigma\mu,\nu} + \bar{g}_{\sigma\nu,\mu} - \bar{g}_{\mu\nu,\sigma}\right).$$

(3.48)

Here the unperturbed Christoffel symbols with vanishing curvature, $\kappa = 0$, are

$$\bar{\Gamma}^0_{0i} = \bar{\Gamma}^i_{00} = \bar{\Gamma}^i_{jk} = 0 \,, \quad \bar{\Gamma}^0_{00} = \mathcal{H} \,, \quad \bar{\Gamma}^0_{ij} = \mathcal{H}\delta_{ij} \,, \quad \bar{\Gamma}^i_{0j} = \mathcal{H}\delta^i_j \,.$$

(3.49)

From the Christoffel symbols we calculate the components of the perturbed Ricci tensor,

$$\delta R_{\mu\nu} \stackrel{(1.20)}{=} \delta\Gamma^\alpha_{\mu\nu,\alpha} - \delta\Gamma^\alpha_{\mu\alpha,\nu} + \bar{\Gamma}^\beta_{\mu\nu}\delta\Gamma^\alpha_{\beta\alpha} + \bar{\Gamma}^\alpha_{\beta\alpha}\delta\Gamma^\beta_{\mu\nu} - \bar{\Gamma}^\beta_{\mu\alpha}\delta\Gamma^\alpha_{\nu\beta} - \bar{\Gamma}^\alpha_{\nu\beta}\delta\Gamma^\beta_{\mu\alpha}\,.$$

(3.50)

The unperturbed Ricci tensor follows directly from Eqs. (1.20) and (3.49),

$$\bar{R}_{00} = -3\mathcal{H}' \,, \quad \bar{R}_{0i} = 0 \,, \quad \bar{R}_{ij} = (\mathcal{H}' + 2\mathcal{H}^2)\,\delta_{ij} \,.$$

(3.51)

Raising one index of the perturbed Ricci tensor we obtain

$$\delta R^\mu{}_\nu \stackrel{(1.25)}{=} \bar{g}^{\mu\rho}\delta R_{\rho\nu} + \delta g^{\mu\rho}\bar{R}_{\rho\nu} \,.$$

(3.52)

The perturbed Ricci scalar is obtained by contracting indices,

$$\delta R \stackrel{(1.28)}{=} \delta R^\mu{}_\mu \,.$$

(3.53)

The unperturbed trace $\bar{R} = \bar{R}^\mu{}_\mu$, originating from Eq. (3.51), reads

$$\bar{R} = 6a^{-2}(\mathcal{H}' + \mathcal{H}^2) \,.$$

(3.54)

We have now all ingredients to compute the perturbed Einstein tensor,

$$\delta G_{\mu\nu} \stackrel{(1.29)}{=} \delta R_{\mu\nu} - \frac{1}{2}\left(\bar{g}_{\mu\nu}\delta R + \delta g_{\mu\nu}\bar{R}\right) .$$

(3.55)

Proceeding as outlined above and recalling that $\vartheta_{kk} = 0$, we find

$$G_{00} = 3\mathcal{H}^2 + 2\left[(\nabla^2 - 3\mathcal{H}\partial_\tau)h_{\rm D} + \mathcal{H}h_{k,k}\right] + \vartheta_{kl,kl} + \mathcal{O}(\delta^2) \,,$$

(3.56)

$$G_{0i} = 2\left(\mathcal{H}h_{0,i} + h'_{\mathrm{D},i}\right) + (2\mathcal{H}' + \mathcal{H}^2)h_i + \frac{1}{2}(h_{i,kk} - h_{k,ik}) + \vartheta'_{ki,k} + \mathcal{O}(\delta^2), \tag{3.57}$$

$$\begin{aligned}
G_{ij} = {}& -(2\mathcal{H}' + \mathcal{H}^2)\delta_{ij} + 2(2\mathcal{H}' + \mathcal{H}^2)\big[(h_0 + h_{\mathrm{D}})\delta_{ij} - \vartheta_{ij}\big] + 2\mathcal{H}h'_0\delta_{ij} \\
& + (\nabla^2\delta_{ij} - \partial_i\partial_j)(h_0 - h_{\mathrm{D}}) + (\partial_\tau^2 + 2\mathcal{H}\partial_\tau)\big(2h_{\mathrm{D}}\delta_{ij} + \vartheta_{ij}\big) \\
& + \frac{1}{2}(\partial_\tau + 2\mathcal{H})(h_{i,j} + h_{j,i} - 2\delta_{ij}h_{k,k}) \\
& + \vartheta_{ik,jk} + \vartheta_{jk,ik} - \vartheta_{ij,kk} - \delta_{ij}\vartheta_{kl,kl} + \mathcal{O}(\delta^2).
\end{aligned} \tag{3.58}$$

For future reference, we also record the perturbed Ricci scalar to the same order, which finds use for instance in the case of a 'non-minimally coupled' scalar field,

$$\begin{aligned}
\delta R = {}& -\frac{2}{a^2}\Bigg[\, 3\mathcal{H}h'_0 + 6\big(\mathcal{H}' + \mathcal{H}^2\big)h_0 + \nabla^2(h_0 - 2h_{\mathrm{D}} + \vartheta_{kk}) \\
& + (\partial_\tau + 3\mathcal{H})(3h'_{\mathrm{D}} - h_{k,k} - \vartheta'_{kk}) - \vartheta_{kl,kl} \Bigg] \tag{3.59} \\
\overset{(3.42)-(3.45)}{=} {}& -\frac{2}{a^2}\Bigg\{\, 3\mathcal{H}h'_0 + 6\big(\mathcal{H}' + \mathcal{H}^2\big)h_0 + \nabla^2\Bigg[h_0 - 2\Big(h_{\mathrm{D}} + \frac{\nabla^2\vartheta}{3}\Big)\Bigg] \\
& + (\partial_\tau + 3\mathcal{H})(3h'_{\mathrm{D}} + \nabla^2 h) \Bigg\}.
\end{aligned} \tag{3.60}$$

We see that only scalar perturbations contribute to the Ricci scalar at first order.

Finally, we note that if we only include spatial indices in the determination of the Ricci tensor, in raising and lowering indices, and in taking the trace, we get the Ricci scalar associated with a time slice $\mathcal{M}_\tau$ (cf. Eq. (1.90)). The result can be extracted from Eq. (3.60), by eliminating time derivatives ($\mathcal{H}$, $(\ldots)'$) and the metric perturbations associated with δg_{00} (h_0) and δg_{0i} (h). This yields

$$\delta R_\tau \overset{\substack{(3.60)\\ \text{see text}}}{=} \frac{4}{a^2}\nabla^2\Big(h_{\mathrm{D}} + \frac{\nabla^2\vartheta}{3}\Big). \tag{3.61}$$

3.3   Energy-Momentum Tensor

For the right-hand side of the Einstein equations, we need to specify the energy-momentum tensor. We treat the overall energy-momentum tensor as a sum over two different components, cf. Eq. (1.64). We recall that this does not exclude the fluid and the scalar field from interacting, through a friction term and thermal fluctuations, as dictated by Eq. (1.62).

Ideal and non-ideal thermal plasmas

Let us start by considering an *ideal-fluid contribution* to the energy-momentum tensor, cf. Eq. (1.34). This means that we stay at leading order in a gradient expansion. We write

$$T_{\mu\nu} = \bar{T}_{\mu\nu} + \delta T_{\mu\nu} + \mathcal{O}(\delta^2) . \tag{3.62}$$

As a symmetric tensor, $T_{\mu\nu}$ has ten degrees of freedom in the perturbations, six of which are physical, while four are gauge degrees of freedom (cf. Chap. 4).

For the background tensor, we repeat here some results from Sect. 1.3, starting with the familiar form

$$\bar{T}_{\mu\nu} \overset{(1.34)}{=} (\bar{e} + \bar{p})\,\bar{u}_\mu \bar{u}_\nu + \bar{p}\,\bar{g}_{\mu\nu} , \tag{3.63}$$

where e is the energy density, p the pressure and u^μ the 4-velocity of the fluid. If $\bar{e}$ and $\bar{p}$ are parametrized by a single quantity, for instance temperature, the background values of e and p are related by the equation of state,

$$\bar{p} \overset{(1.35)}{=} \bar{p}(\bar{e}) , \qquad c_s^2 \overset{(1.35)}{\equiv} \frac{\partial \bar{p}}{\partial \bar{e}} , \tag{3.64}$$

where c_s is the speed of sound. Perturbations of the energy density and pressure are defined as

$$p = \bar{p} + \delta p , \quad e = \bar{e} + \delta e . \tag{3.65}$$

Because of isotropy, we assume the fluid to be at rest at leading order, $\bar{u}^i = 0$. Then the velocity is defined by

$$u^\mu = \bar{u}^\mu + \delta u^\mu , \quad \bar{u}^\mu = \left(\frac{d\tau}{dt}, \mathbf{0}\right) \overset{(1.3)}{=} a^{-1}(1, \mathbf{0}) , \quad \bar{u}_\mu \overset{(1.36)}{=} a(-1, \mathbf{0}) . \tag{3.66}$$

The linear *velocity perturbation* is conveniently expressed as

$$v_i \equiv v^i \equiv a\,\delta u^i \quad \Rightarrow \quad \delta u^i = a^{-1} v^i . \tag{3.67}$$

For the components of the covector δu_ν, we find

$$\delta u_0 = \delta g_{0\mu} \bar{u}^\mu + \bar{g}_{0\mu} \delta u^\mu = -2ah_0 - a^2 \delta u^0 , \tag{3.68}$$

$$\delta u_i = \delta g_{i\mu} \bar{u}^\mu + \bar{g}_{i\mu} \delta u^\mu = -ah_i + a^2 \delta_{ij} \delta u^i = a\,(-h_i + v_i) . \tag{3.69}$$

The value of δu^0 is inferred by imposing $u_\mu u^\mu = -1$,

$$u_\mu u^\mu = \overbrace{-1}^{\bar{u}_\mu \bar{u}^\mu} + \overbrace{a^{-1}(-2ah_0 - a^2 \delta u^0)}^{\delta u_\mu \bar{u}^\mu} - \overbrace{a\,\delta u^0}^{\bar{u}_\mu \delta u^\mu} \quad \Rightarrow \quad \delta u^0 = -a^{-1} h_0 \tag{3.70}$$

$$\Rightarrow u^\mu \overset{(3.67)}{\underset{(3.70)}{=}} a^{-1}(1 - h_0, v^i) + \mathcal{O}(\delta^2) , \tag{3.71}$$

$$\Rightarrow u_\mu \overset{(3.68)-(3.70)}{=} a(-1 - h_0, v_i - h_i) + \mathcal{O}(\delta^2) . \tag{3.72}$$

Recalling the metric from Eq. (3.40), at linear order the perfect fluid ('pf') part of the energy-momentum tensor is thus

$$T_{\mu\nu}|_{\mathrm{pf}} = \bar{T}_{\mu\nu} + (\delta e + \delta p)(a^2 \delta_{\mu 0}\,\delta_{\nu 0}) + (\bar{e} + \bar{p})\left(\bar{u}_\mu\,\delta u_\nu + \bar{u}_\nu\,\delta u_\mu\right) + \delta p\,\bar{g}_{\mu\nu} + \bar{p}\,\delta g_{\mu\nu} \tag{3.73}$$

$$\stackrel{\substack{(3.40),(3.63)\\=\\(3.72)}}{} a^2 \begin{pmatrix} \bar{e} & 0 \\ 0 & \bar{p}\,\delta_{ij} \end{pmatrix}$$

$$+ a^2 \begin{pmatrix} \delta e + 2\bar{e}h_0 & -\bar{e}\,(v_j - h_j) - \bar{p}\,v_j \\ -\bar{e}\,(v_i - h_i) - \bar{p}\,v_i & (\delta p - 2\bar{p}h_{\mathrm{D}})\,\delta_{ij} + 2\bar{p}\,\vartheta_{ij} \end{pmatrix} + \mathcal{O}(\delta^2)\,. \tag{3.74}$$

Going beyond ideal fluids, i.e. including the next order in gradients, further terms are introduced in the hydrodynamic description, involving the shear and bulk viscosities. As dictated by the *fluctuation-dissipation theorem*, and exemplified by the scalar field equation in Eq. (1.62), the introduction of dissipative coefficients necessitates the introduction of the corresponding thermal noise terms. We discuss this in more detail in Sect. 3.7. For now we just denote an additional contribution to energy-momentum perturbations by an *anisotropic stress tensor*, Π_{ij},

$$\delta T_{ij}|_{\mathrm{tot}} = \delta T_{ij}|_{\mathrm{pf}} + a^2 \Pi_{ij}\,, \tag{3.75}$$

noting that, strictly speaking, we have omitted the bulk viscous part, i.e. the term $Z_{\mu\nu}$ from Eq. (3.127). If bulk viscous corrections were to be included, the background equations should be modified as well (cf. Eq. (3.135)), and a full tensor $a^2 \Pi_{\mu\nu}$ would be needed. However, such corrections are expected to be negligible in typical cosmological scenarios (cf. Eq. (3.137)). Therefore, we restrict ourselves to the shear viscous part, whereby $a^2 \Pi_{\mu\nu}$ can be assumed purely spatial and traceless (cf. Eqs. (3.128) and (3.141)).

The vectors and tensors introduced can be decomposed into their transverse scalar, vector and tensor parts as in Eqs. (3.17)–(3.21),

$$v_i = v_i^{\mathrm{s}} + v_i^{\mathrm{v}}\,, \qquad v_i^{\mathrm{s}} = -\partial_i v\,, \qquad \partial^i v_i^{\mathrm{v}} = 0\,, \tag{3.76}$$

$$\Pi_{ij} = \Pi_{ij}^{\mathrm{s}} + \Pi_{ij}^{\mathrm{v}} + \Pi_{ij}^{\mathrm{t}}\,, \qquad \Pi_{ij}^{\mathrm{s}} = (\partial_i\partial_j - \delta_{ij}\nabla^2/3)\Pi\,, \tag{3.77}$$

$$\Pi_{ij}^{\mathrm{v}} = -\frac{1}{2}(\partial_i\Pi_j^{\mathrm{v}} + \partial_j\Pi_i^{\mathrm{v}})\,, \qquad \partial^i \Pi_i^{\mathrm{v}} = 0\,, \tag{3.78}$$

$$\delta^{ij}\Pi_{ij}^{\mathrm{t}} = 0\,, \qquad \partial^i \Pi_{ij}^{\mathrm{t}} = 0\,. \tag{3.79}$$

An elementary scalar field (or inflaton)

Turning to an elementary scalar field, the energy-momentum tensor is given in Eq. (1.47). We now write

$$\varphi = \bar{\varphi} + \delta\varphi\,. \tag{3.80}$$

At the background level this leads to

$$\bar{T}_{\mu\nu} \overset{(1.47)}{=} \bar\varphi_{,\mu}\bar\varphi_{,\nu} - \bar{g}_{\mu\nu}\left(\frac{\bar{g}^{\alpha\beta}}{2}\bar\varphi_{,\alpha}\bar\varphi_{,\beta} + \bar{V}\right)$$

$$\overset{(3.34)}{\underset{(3.35)}{=}} \begin{pmatrix} \dfrac{(\bar\varphi')^2}{2} + a^2\bar{V} & 0 \\[2ex] 0 & \delta_{ij}\left[\dfrac{(\bar\varphi')^2}{2} - a^2\bar{V}\right] \end{pmatrix} . \tag{3.81}$$

Linear perturbations originate from

$$\delta T_{\mu\nu} \overset{(1.47)}{=} \bar\varphi_{,\mu}\delta\varphi_{,\nu} + \delta\varphi_{,\mu}\bar\varphi_{,\nu} - \delta g_{\mu\nu}\left(-\frac{(\bar\varphi')^2}{2a^2} + \bar{V}\right)$$

$$- \bar{g}_{\mu\nu}\left(\frac{(\bar\varphi')^2}{2}\delta g^{00} + \bar{g}^{00}\bar\varphi'\delta\varphi' + \bar{V}_{,\varphi}\delta\varphi\right) . \tag{3.82}$$

Inserting the metric perturbations from Eqs. (3.40) and (3.41), we find

$$\delta T_{00} \overset{(3.82)}{\underset{(3.40),(3.41)}{=}} 2\bar\varphi'\delta\varphi' + 2h_0\left(-\frac{(\bar\varphi')^2}{2} + a^2\bar{V}\right) + 2h_0\frac{(\bar\varphi')^2}{2} - \bar\varphi'\delta\varphi' + a^2\bar{V}_{,\varphi}\delta\varphi$$

$$= \bar\varphi'\delta\varphi' + 2h_0\,a^2\bar{V} + a^2\bar{V}_{,\varphi}\delta\varphi , \tag{3.83}$$

$$\delta T_{0i} \overset{(3.82)}{\underset{(3.40),(3.41)}{=}} \bar\varphi'\delta\varphi_{,i} + h_i\left(-\frac{(\bar\varphi')^2}{2} + a^2\bar{V}\right) , \tag{3.84}$$

$$\delta T_{ij} \overset{(3.82)}{\underset{(3.40),(3.41)}{=}} 2(h_{\rm D}\delta_{ij} - \vartheta_{ij})\left(-\frac{(\bar\varphi')^2}{2} + a^2\bar{V}\right)$$

$$- \delta_{ij}\left(2h_0\frac{(\bar\varphi')^2}{2} - \bar\varphi'\delta\varphi' + a^2\bar{V}_{,\varphi}\delta\varphi\right) . \tag{3.85}$$

The appearances of $\bar{V}$ can be crosschecked against Eq. (3.74), after recalling from Eq. (1.65) that in the ideal-fluid notation,

$$\bar{e} \supset \bar{V} , \quad \delta e \supset \bar{V}_{,\varphi}\delta\varphi , \quad \bar{p} \supset -\bar{V} , \quad \delta p \supset -\bar{V}_{,\varphi}\delta\varphi . \tag{3.86}$$

In the following, to streamline the notation, we write $\bar{V} \to V$.

3.4 Einstein Equations

Having determined the perturbed Einstein and energy-momentum tensors in Sects. 3.2 and 3.3, respectively, we can now collect together the perturbed Einstein equations,

$$\delta G_{\mu\nu} = 8\pi G\, \delta T_{\mu\nu} \,. \tag{3.87}$$

We do this separately for scalar (s), vector (v) or tensor (t) perturbations. For reference, we also recall the background equations once more, from Eqs. (3.56), (3.58), (3.74) and (3.81), which help to simplify some of the Einstein equations:

$$3\mathcal{H}^2 \overset{(3.56)}{\underset{(3.74),(3.81)}{=}} 8\pi G\left[a^2\bar{e} + \frac{(\bar{\varphi}')^2}{2}\right], \quad -(2\mathcal{H}' + \mathcal{H}^2) \overset{(3.58)}{\underset{(3.74),(3.81)}{=}} 8\pi G\left[a^2\bar{p} + \frac{(\bar{\varphi}')^2}{2}\right]. \tag{3.88}$$

Here we have hidden the appearances of $\bar{V}$, by making use of Eq. (3.86).

Scalar perturbations

When we insert the tensor decomposition from Eqs. (3.42)–(3.45) into (3.56)–(3.58), we obtain the scalar parts of linear perturbations of the Einstein tensor,

$$\delta G_{00}^{\mathrm{s}} = 2\left[-\mathcal{H}\left(3h_{\mathrm{D}}' + \nabla^2 h\right) + \nabla^2\left(h_{\mathrm{D}} + \frac{\nabla^2\vartheta}{3}\right)\right], \tag{3.89}$$

$$\delta G_{0i}^{\mathrm{s}} = \partial_i\left[-(2\mathcal{H}' + \mathcal{H}^2)h + 2\mathcal{H}h_0 + 2\partial_\tau\left(h_{\mathrm{D}} + \frac{\nabla^2\vartheta}{3}\right)\right], \tag{3.90}$$

$$\delta G_{ij}^{\mathrm{s}} = \delta_{ij}\left[2(2\mathcal{H}' + \mathcal{H}^2)\left(h_0 + h_{\mathrm{D}} + \frac{\nabla^2\vartheta}{3}\right) + 2\mathcal{H}h_0' + \nabla^2\left(h_0 - h_{\mathrm{D}} - \frac{\nabla^2\vartheta}{3}\right)\right.$$
$$\left. + 2(\partial_\tau^2 + 2\mathcal{H}\partial_\tau)\left(h_{\mathrm{D}} + \frac{\nabla^2\vartheta}{3}\right) + (\partial_\tau + 2\mathcal{H})\nabla^2(h - \vartheta')\right]$$
$$- \partial_i\partial_j\left[2(2\mathcal{H}' + \mathcal{H}^2)\vartheta + h_0 + (\partial_\tau + 2\mathcal{H})(h - \vartheta') - h_{\mathrm{D}} - \frac{\nabla^2\vartheta}{3}\right]. \tag{3.91}$$

Comparing $\delta G_{00}^{\mathrm{s}}$ from Eq. (3.89) with $\delta T_{00}^{\mathrm{s}}$ from Eqs. (3.74) and (3.83); recalling Eq. (3.86); and substituting $2a^2\bar{e}$ with the help of Eq. (3.88), we find

$$\boxed{\begin{aligned} &-3\mathcal{H}^2 h_0 - \mathcal{H}\left(3h_{\mathrm{D}}' + \nabla^2 h\right) + \nabla^2\left(h_{\mathrm{D}} + \frac{\nabla^2\vartheta}{3}\right) \\ &= 4\pi G\left[\bar{\varphi}'(\delta\varphi - h_0\bar{\varphi}') + a^2\delta e\right]. \end{aligned}} \tag{3.92}$$

As far as $\delta G^{\mathrm{s}}_{0i}$ from Eq. (3.90) and $\delta T^{\mathrm{s}}_{0i}$ from Eqs. (3.74) and (3.84) go, we note that the part $-(2\mathcal{H}' + \mathcal{H}^2)h_{\,i}$ can be substituted with the background identity from Eq. (3.88). The rest of the equation sets two gradients equal. If we assume that the perturbations vanish at infinity, so that there is no integration constant, we obtain

$$\mathcal{H}h_0 + h'_{\mathrm{D}} + \frac{\nabla^2 \vartheta'}{3} = 4\pi G\big[\, a^2(\bar{e} + \bar{p})(v - h) + \bar{\varphi}' \delta\varphi \,\big]\,. \tag{3.93}$$

Turning to $\delta G^{\mathrm{s}}_{ij}$ from Eq. (3.91), with $\delta T^{\mathrm{s}}_{ij}$ from Eqs. (3.74) and (3.85), let us first consider the 'non-diagonal' terms, with $i \neq j$. There is a cancellation between terms containing $(2\mathcal{H}' + \mathcal{H}^2)\vartheta$ on the side of $\delta G^{\mathrm{s}}_{ij}$, and ϑ_{ij} in the side of $\delta T^{\mathrm{s}}_{ij}$, thanks to the background identity in Eq. (3.88). The remainder can again be integrated, assuming that the integration constants vanish at spatial infinity. After the cancellation of ϑ_{ij}, there is no source term from Eq. (3.74) or (3.85). The only contribution originates from the viscous anisotropic stress in Eq. (3.75), which we decompose according to Eq. (3.77). Thereby we get

$$h_0 + (\partial_\tau + 2\mathcal{H})(h - \vartheta') - h_{\mathrm{D}} - \frac{\nabla^2 \vartheta}{3} = -8\pi G a^2 \Pi\,. \tag{3.94}$$

Finally, in the parts proportional to δ_{ij}, there is a term that cancels thanks to the background identity in Eq. (3.88). In view of the tensor decomposition of ϑ_{ij} in Eq. (3.43), this concerns the terms $(2\mathcal{H}' + \mathcal{H}^2)(h_{\mathrm{D}} + \nabla^2\vartheta/3)$ on the side of $\delta G^{\mathrm{s}}_{ij}$, and $-[2a^2\bar{p} + (\bar{\varphi}')^2](h_{\mathrm{D}}\delta_{ij} - \vartheta_{ij})$ on the side of δT_{ij}. Recalling Eq. (3.86) as well, left over is then

$$\big[\,2(2\mathcal{H}' + \mathcal{H}^2) + 2\mathcal{H}\partial_\tau + \nabla^2\,\big]h_0 + \big[\,2(\partial_\tau^2 + 2\mathcal{H}\partial_\tau) - \nabla^2\,\big]\Big(h_{\mathrm{D}} + \frac{\nabla^2\vartheta}{3}\Big)$$

$$+ (\partial_\tau + 2\mathcal{H})\nabla^2(h - \vartheta') = 8\pi G\Big[\, a^2\Big(\delta p - \frac{\nabla^2\Pi}{3}\Big) + \bar{\varphi}'(\delta\varphi' - h_0\bar{\varphi}')\,\Big]\,. \tag{3.95}$$

We obtain what turns out to be a more helpful relation as a linear combination of several equations, defining $(3.96) \equiv [(3.95) - \nabla^2(3.94) - 2(3.92)]/2$, yielding

$$2(\mathcal{H}' + 2\mathcal{H}^2)h_0 + \mathcal{H}\,(h'_0 + 3h'_{\mathrm{D}} + \nabla^2 h) + (\partial_\tau^2 + 2\mathcal{H}\partial_\tau - \nabla^2)\Big(h_{\mathrm{D}} + \frac{\nabla^2\vartheta}{3}\Big)$$

$$= 4\pi G a^2\Big(\delta p - \delta e + \frac{2}{3}\nabla^2\Pi\Big)\,. \tag{3.96}$$

Equations (3.92), (3.93), (3.94) and (3.96) are independent of each other, and constitute the information that can be extracted from the Einstein equations. It is appropriate to remark, however, that in general they do *not* suffice to specify the dynamics of the system. Even if we managed to express δp, δe and Π in terms of two independent perturbations, for instance δT and $\delta\varphi$, there are 5 further variables, namely the scalar velocity v and the metric perturbations h_0, h_{D}, h and ϑ. As will be discussed in Chap. 4, only two linear combinations of the metric perturbations are 'physical', i.e. cannot be eliminated by gauge (or coordinate) transformations. This still leaves over 5 physical variables, so at least one further equation is needed. We return to this in Sect. 3.5.

Vector perturbations

Among Eqs. (3.56)–(3.58), the last two have vector parts once we insert Eqs. (3.42)–(3.45),

$$\delta G^{\mathrm{v}}_{00} = 0 \,, \tag{3.97}$$

$$\delta G^{\mathrm{v}}_{0i} = (2\mathcal{H}' + \mathcal{H}^2)h^{\mathrm{v}}_i + \frac{1}{2}\,\nabla^2(h^{\mathrm{v}}_i - \vartheta^{\mathrm{v}\prime}_i) \,, \tag{3.98}$$

$$\delta G^{\mathrm{v}}_{ij} = (2\mathcal{H}' + \mathcal{H}^2)(\vartheta^{\mathrm{v}}_{i,j} + \vartheta^{\mathrm{v}}_{j,i}) + \frac{1}{2}\,(\partial_\tau + 2\mathcal{H})\big(h^{\mathrm{v}}_{i,j} + h^{\mathrm{v}}_{j,i} - \vartheta^{\mathrm{v}\,\prime}_{i,j} - \vartheta^{\mathrm{v}\,\prime}_{j,i}\big) \,. \tag{3.99}$$

Comparing $\delta G^{\mathrm{v}}_{0i}$ from Eq. (3.98) and $\delta T^{\mathrm{v}}_{0i}$ from Eqs. (3.74) and (3.84), we note that the part $(2\mathcal{H}' + \mathcal{H}^2)h^{\mathrm{v}}_i$ can be simplified with the background identity from Eq. (3.88). This yields Eq. (3.100). With $\delta G^{\mathrm{v}}_{ij}$ from Eq. (3.99) and $\delta T^{\mathrm{v}}_{ij}$ from Eqs. (3.74), (3.75) and (3.85), and extracting the vector part of ϑ_{ij} according to Eq. (3.44), the part $(2\mathcal{H}' + \mathcal{H}^2)(\vartheta^{\mathrm{v}}_{i,j} + \vartheta^{\mathrm{v}}_{j,i})$ cancels. The remainder yields Eq. (3.101), whereby in total we have

$$\frac{1}{2}\,\nabla^2(h^{\mathrm{v}}_i - \vartheta^{\mathrm{v}\prime}_i) = 8\pi G a^2(\bar{e} + \bar{p})(h^{\mathrm{v}}_i - v^{\mathrm{v}}_i) \quad \forall i \,, \tag{3.100}$$

$$(\partial_\tau + 2\mathcal{H})\big(h^{\mathrm{v}}_{i,j} + h^{\mathrm{v}}_{j,i} - \vartheta^{\mathrm{v}\,\prime}_{i,j} - \vartheta^{\mathrm{v}\,\prime}_{j,i}\big) = -8\pi G a^2\,(\Pi^{\mathrm{v}}_{i,j} + \Pi^{\mathrm{v}}_{j,i}) \quad \forall i, j \,. \tag{3.101}$$

Here Π^{v}_i is a function of $v^{\mathrm{v}}_i - \vartheta^{\mathrm{v}\prime}_i$, cf. Eq. (3.134). We note that if we take the divergence ∂_j from Eq. (3.101) and make use of the transversality of vector perturbations, we obtain

$$(\partial_\tau + 2\mathcal{H})\nabla^2(h^{\mathrm{v}}_i - \vartheta^{\mathrm{v}\prime}_i) = -8\pi G a^2\nabla^2\Pi^{\mathrm{v}}_i \quad \forall i \,. \tag{3.102}$$

With Eqs. (3.100) and (3.102), we have two equations for two transverse vectors, $h^{\mathrm{v}}_i - \vartheta^{\mathrm{v}\prime}_i$ and $h^{\mathrm{v}}_i - v^{\mathrm{v}}_i$. If we insert Eq. (3.134) for Π^{v}_i, then the vector $h^{\mathrm{v}}_i - \vartheta^{\mathrm{v}\prime}_i$ can be eliminated, and we obtain one diffusive equation for the vector $(\bar{e} + \bar{p})(h^{\mathrm{v}}_i - v^{\mathrm{v}}_i)$.

Equation (3.101) is a local first-order time-evolution equation at each position $\mathbf{x}$. Without anisotropic stress the equation is easily solved, and leads to a rapid decay,

$$(h^{\mathrm{v}}_{i,j} + h^{\mathrm{v}}_{j,i} - \vartheta^{\mathrm{v}\,\prime}_{i,j} - \vartheta^{\mathrm{v}\,\prime}_{j,i})\,(\tau,\mathbf{x}) \;\overset{\Pi^{\mathrm{v}}\to 0}{\approx}\; \frac{c_{ij}(\mathbf{x})}{a^2(\tau)}\,, \qquad \mathrm{tr}\,c_{ij} = 0\,, \tag{3.103}$$

where c_{ij} is an integration constant. If Π^{v} is kept non-zero, vector perturbations are constantly generated by hydrodynamic fluctuations (cf. Sect. 3.7), but subsequently each mode decays, so that their average distribution corresponds to that in thermal equilibrium. For many considerations, where only non-equilibrium modes could carry enough energy density to have an observable effect, vector perturbations can be neglected.

Tensor perturbations

Among the components of $G_{\mu\nu}$, in Eqs. (3.56)–(3.58), only G_{ij} has a tensor part,

$$\delta G^{\mathrm{t}}_{00} = 0 = \delta G^{\mathrm{t}}_{0i}\,, \tag{3.104}$$

$$\delta G^{\mathrm{t}}_{ij} = -2(2\mathcal{H}' + \mathcal{H}^2)\vartheta^{\mathrm{t}}_{ij} + (\partial^2_\tau + 2\mathcal{H}\partial_\tau - \nabla^2)\vartheta^{\mathrm{t}}_{ij}\,. \tag{3.105}$$

Once again, the part proportional to $2\mathcal{H}' + \mathcal{H}^2$ cancels against the ϑ_{ij}-parts from Eqs. (3.74) and (3.85), on account of Eq. (3.88). Left over is a contribution from the tensor part of Eq. (3.75), leading to

$$\boxed{(\partial^2_\tau + 2\mathcal{H}\partial_\tau - \nabla^2)\vartheta^{\mathrm{t}}_{ij} \;=\; 8\pi G a^2 \Pi^{\mathrm{t}}_{ij}\,.} \tag{3.106}$$

We note that the differential operator on the left-hand side of Eq. (3.106) is nothing but the *d'Alembert operator in FLRW coordinates*. Perhaps the easiest way to see this is to derive the Euler-Lagrange equation for a minimally coupled scalar field, cf. Eq. (1.48). Denoting by $\bar{g}_{\mu\nu}$ the metric given by Eq. (1.2) or Eq. (1.3), and $\bar{g} \equiv \det \bar{g}_{\mu\nu}$, we obtain in physical and conformal time, respectively,

$$\Box_H Q \;\equiv\; \frac{1}{\sqrt{-\bar{g}}}\,\partial_\mu\left(\sqrt{-\bar{g}}\,\bar{g}^{\mu\nu}\partial_\nu Q\right)$$

$$= a^{-3}\big[\partial_t(-a^3 \dot{Q}) + \partial_i(a Q^{\cdot i})\big] \;=\; -\ddot{Q} - 3H\dot{Q} + a^{-2}\nabla^2 Q\,, \tag{3.107}$$

$$\Box_{\mathcal{H}} Q \;=\; a^{-4}\Big[\partial_\tau(-a^2 Q') + \partial_i(a^2 Q^{\cdot i})\Big] = a^{-2}\left(-Q'' - 2\mathcal{H}Q' + \nabla^2 Q\right)\,. \tag{3.108}$$

Therefore, the homogeneous version of Eq. (3.106) is a *wave equation*, $\Box_{\mathcal{H}}\vartheta^{\mathrm{t}}_{ij} = 0$, whose solutions are *gravitational waves*, propagating through the universe.

3.5   **Scalar Field Equation**

We have seen in Sect. 3.4 that, if both $\delta\varphi$ and δT play a role, the Einstein equations for scalar perturbations, Eqs. (3.92), (3.93), (3.94) and (3.96), are not sufficient for fixing the full solution. Here we derive the additional equation that is needed for this purpose. This is given by the scalar field equation. At the background level, its vacuum form is given in Eq. (1.48), and a form interacting with a fluid in Eq. (1.62).

The perturbations of the metric are given in (3.40) and (3.41). Recalling that ϑ_{ij} is traceless (cf. Eq. (3.39)), the determinant of $g_{\mu\nu}$ is

$$- g \overset{(3.40)}{=} a^8(1 + 2h_0)(1 - 2h_{\mathrm{D}})^3 + \mathcal{O}(\delta^2)$$

$$\Rightarrow \quad \sqrt{-g} = a^4(1 + h_0 - 3h_{\mathrm{D}}) + \mathcal{O}(\delta^2) \,. \tag{3.109}$$

The derivative from the left-hand side of (1.48), taken like in Eq. (3.108) but now to first order in perturbations, becomes

$$(\sqrt{-g}\,\varphi^{,\mu})_{,\mu} = [\sqrt{-g}\,(g^{00}\varphi' + g^{0i}\varphi_{,i})]' + [\sqrt{-g}\,(g^{i0}\varphi' + g^{ij}\varphi_{,j})]_{,i}$$

$$\overset{(3.109)}{\underset{(3.41)}{=}} -[a^2(1 - h_0 - 3h_{\mathrm{D}})\varphi' + a^2 h_i\,\varphi_{,i}]' + [a^2(-h_i\,\varphi' + \varphi_{,i})]_{,i} + \mathcal{O}(\delta^2)$$

$$= -a^2\Big[(1 - h_0 - 3h_{\mathrm{D}})(\varphi'' + 2\mathcal{H}\varphi') - \nabla^2\varphi - (h_0' + 3h_{\mathrm{D}}' - h_{i,i})\varphi'\Big] + \mathcal{O}(\delta^2) \,. \tag{3.110}$$

Here we made use of the fact that $\varphi_{,i}$ is of $\mathcal{O}(\delta)$. Equation (3.110) is divided by $\sqrt{-g}$ from Eq. (3.109), to get $\varphi^{;\mu}_{\;;\mu}$. We also note that according to Eq. (3.42), $-h_{i,i} = \nabla^2 h$, so that only scalar perturbations contribute. Inserting $\varphi = \bar\varphi + \delta\varphi$, this gives

$$\varphi^{;\mu}_{\;;\mu} \overset{(3.110)}{\underset{(3.109)}{=}} -\frac{1}{a^2}\Big[(1 - 2h_0)(\partial_\tau^2 + 2\mathcal{H}\partial_\tau)(\bar\varphi + \delta\varphi) - \nabla^2\delta\varphi$$

$$- (h_0' + 3h_{\mathrm{D}}' + \nabla^2 h)\,\bar\varphi'\Big] + \mathcal{O}(\delta^2) \,. \tag{3.111}$$

As far as the second term of Eq. (1.62) goes, we insert the velocity from Eq. (3.71), to get

$$u^\mu\varphi_{,\mu} \overset{(3.71)}{=} \frac{1}{a}\Big[(1 - h_0)\,\varphi' + v_i\varphi_{,i}\Big] + \mathcal{O}(\delta^2) = \frac{1}{a}\Big[(1 - h_0)\,\bar\varphi' + \delta\varphi'\Big] + \mathcal{O}(\delta^2) \,. \tag{3.112}$$

The coefficient Υ, multiplying the second term from Eq. (1.62), also needs to be perturbed, as it can be a function of the dynamical variables, and the same applies to the potential,

$$\Upsilon = \bar\Upsilon + \delta\Upsilon + \mathcal{O}(\delta^2) \,, \quad V_{,\varphi} = \bar V_{,\varphi} + \delta V_{,\varphi} + \mathcal{O}(\delta^2) \,. \tag{3.113}$$

The background equation then yields Eq. (1.67), where we had simplified the notation by substituting $\bar{\Upsilon} \to \Upsilon$ and $\bar{V}_{,\varphi} \to V_{,\varphi}$. At first order, combining Eqs. (3.111)–(3.113) according to Eq. (1.62), and treating the noise term ϱ as being of the same order as first-order perturbations (this is analogous to the discussion below Eq. (3.74)), yields

$$\underset{\substack{-a^2 \times (1.62)\\ \Rightarrow \\ (3.111)-(3.113)}}{} \delta\varphi'' + 2\mathcal{H}\delta\varphi' - 2h_0 \overbrace{(\bar{\varphi}'' + 2\mathcal{H}\bar{\varphi}')}^{(1.66):\ \text{becomes}\ -a\Upsilon\bar{\varphi}'-a^2\,V_{,\varphi}} - \nabla^2\delta\varphi - (h_0' + 3h_{\mathrm{D}}' + \nabla^2 h)\,\bar{\varphi}'$$

$$+ \, a\,\delta\Upsilon\bar{\varphi}' + a\Upsilon(-h_0\,\bar{\varphi}' + \delta\varphi') + a^2\delta V_{,\varphi} = a^2\varrho$$

$$\Leftrightarrow \boxed{\begin{aligned} &\delta\varphi'' + (2\mathcal{H} + a\Upsilon)\,\delta\varphi' - \nabla^2\delta\varphi - (h_0' + 3h_{\mathrm{D}}' + \nabla^2 h)\,\bar{\varphi}' \\ &+ a(\delta\Upsilon + h_0\Upsilon)\,\bar{\varphi}' + a^2(\delta V_{,\varphi} + 2h_0 V_{,\varphi}) = a^2\varrho\,. \end{aligned}} \qquad (3.114)$$

With this equation, we have the full information needed for determining the time evolution of physical scalar perturbations.

3.6   Energy-Momentum Conservation

The Einstein tensor has the property $G_{\mu\nu}{}^{;\mu} = 0$, consistent with energy-momentum conservation, $T_{\mu\nu}{}^{;\mu} = 0$. Even though formally the latter relations do *not* add information to the Einstein equations, $G_{\mu\nu} = 8\pi G T_{\mu\nu}$, they may still establish a helpful re-organization. It may be noted that in the limit of (non-perturbed) Minkowskian space-time, the Einstein tensor vanishes, yet the energy-momentum tensor remains non-trivial, and energy-momentum conservation yields essential information. For example, this is how the equations of hydrodynamics are obtained, when $T_{\mu\nu}$ is expressed in terms of fluid variables.

Inserting the Christoffel symbols, energy-momentum conservation reads

$$0 = T_{\mu\nu}{}^{;\mu} \overset{(6)}{=} T_{\mu\nu;\alpha}g^{\alpha\mu} \overset{(7)}{=} \left(T_{\mu\nu,\alpha} - T_{\mu\beta}\Gamma^{\beta}_{\nu\alpha} - T_{\beta\nu}\Gamma^{\beta}_{\mu\alpha}\right)g^{\mu\alpha} \qquad (3.115)$$

$$= \underbrace{\left(T_{0\nu}' - T_{0\beta}\Gamma^{\beta}_{\nu 0} - T_{\beta\nu}\Gamma^{\beta}_{00}\right)g^{00}}_{\mu\to 0,\ \alpha\to 0} + \underbrace{\left(T_{k\nu,l} - T_{k\beta}\Gamma^{\beta}_{\nu l} - T_{\beta\nu}\Gamma^{\beta}_{kl}\right)g^{kl}}_{\mu\to k,\ \alpha\to l}$$

$$+ \underbrace{\left(T_{0\nu,k} + T_{k\nu}' - T_{0\beta}\Gamma^{\beta}_{\nu k} - T_{k\beta}\Gamma^{\beta}_{\nu 0} - 2\,T_{\beta\nu}\,\Gamma^{\beta}_{0k}\right)g^{0k}}_{\mu\to 0,\ \alpha\to k\ \text{or}\ \mu\to k,\ \alpha\to 0}\,. \qquad (3.116)$$

At the background level ($g^{00} \to -1/a^2$, $g^{kl} \to \delta_{kl}/a^2$, $g^{0k} \to 0$), this leads to

$$0 \overset{a^2 \times (3.116)}{=} -\bar{T}'_{0\nu} + \bar{T}_{0\beta}\,\bar{\Gamma}^{\beta}_{\nu 0} + \bar{T}_{\beta\nu}\bar{\Gamma}^{\beta}_{00} + \bar{T}_{k\nu,k} - \bar{T}_{k\beta}\bar{\Gamma}^{\beta}_{\nu k} - \bar{T}_{\beta\nu}\bar{\Gamma}^{\beta}_{kk}$$

$$= \begin{cases} -\bar{T}'_{00} + 2\,\bar{T}_{00}\,\bar{\Gamma}^{0}_{00} - \bar{T}_{kl}\bar{\Gamma}^{l}_{0k} - \bar{T}_{00}\bar{\Gamma}^{0}_{kk} & (\nu = 0) \\ +\bar{T}_{00}\,\bar{\Gamma}^{0}_{0i} + \bar{T}_{ik}\bar{\Gamma}^{k}_{00} + \bar{T}_{ik,k} - \bar{T}_{kl}\bar{\Gamma}^{l}_{ik} - \bar{T}_{il}\bar{\Gamma}^{l}_{kk} & (\nu = i) \end{cases}$$

$$\overset{(3.49)}{=} \begin{cases} -\bar{T}'_{00} - \mathcal{H}(\bar{T}_{00} + \bar{T}_{kk}) & (\nu = 0) \\ +\bar{T}_{ik,k} & (\nu = i) \end{cases}$$

$$\overset{(3.74)}{\underset{(3.81)}{=}} \begin{cases} -a^2\big[\bar{e}' + 3\mathcal{H}(\bar{e} + \bar{p})\big] - \bar{\varphi}'(\bar{\varphi}'' + 2\mathcal{H}\bar{\varphi}') & (\nu = 0) \\ 0 & (\nu = i) \end{cases} . \tag{3.117}$$

The $\nu = 0$ part reproduces Eq. (1.73).

At first order, Eq. (3.115) implies

$$0 = \delta T_{\mu\nu}{}^{;\mu} \equiv \big(\bar{T}_{\mu\nu,\alpha} - \bar{T}_{\mu\beta}\bar{\Gamma}^{\beta}_{\nu\alpha} - \bar{T}_{\beta\nu}\bar{\Gamma}^{\beta}_{\mu\alpha}\big)\delta g^{\alpha\mu}$$

$$+ \big(\delta T_{\mu\nu,\alpha} - \bar{T}_{\mu\beta}\,\delta\Gamma^{\beta}_{\nu\alpha} - \delta T_{\mu\beta}\,\bar{\Gamma}^{\beta}_{\nu\alpha} - \bar{T}_{\beta\nu}\,\delta\Gamma^{\beta}_{\mu\alpha} - \delta T_{\beta\nu}\,\bar{\Gamma}^{\beta}_{\mu\alpha}\big)\bar{g}^{\alpha\mu}. \tag{3.118}$$

In order to simplify the expressions, we focus on the fluid energy-momentum tensor from Eqs. (3.74) and (3.75) in the following, omitting the scalar field contribution from Eqs. (3.83)–(3.85). Then, making use of symbolic manipulation, as explained in Sect. 3.8, or carrying out a gauge-fixed computation, as shown in Sect. 4.5, we find

$$0 = \delta T_{\mu 0}{}^{;\mu} \overset{(3.74)}{\underset{(3.75)}{\supset}} -\delta e' - 3\mathcal{H}(\delta e + \delta p) + (\bar{e} + \bar{p})(3h'_{\mathrm{D}} - v_{k,k} - \vartheta'_{kk}) - \mathcal{H}\,\Pi_{kk}$$

$$\overset{(3.42)\text{--}(3.45)}{=} -\delta e' - 3\mathcal{H}(\delta e + \delta p) + (\bar{e} + \bar{p})(3h'_{\mathrm{D}} + \nabla^2 v)\,, \tag{3.119}$$

$$0 = \delta T_{\mu i}{}^{;\mu} \overset{(3.74)}{\underset{(3.75)}{\supset}} +\delta p_{,i} + (\bar{e} + \bar{p})h_{0,i} + (\partial_\tau + 4\mathcal{H})[(\bar{e} + \bar{p})(v_i - h_i)] + \Pi_{ik,k}. \tag{3.120}$$

These will play an important role in Chap. 9. Equation (3.119) is an evolution equation for energy-density perturbations, δe, whereas Eq. (3.120) is an evolution equation for momentum-density perturbations, i.e. $(\bar{e} + \bar{p})(v_i - h_i)$. The full equations, including inflaton perturbations, will be needed in Sect. 7.5 (cf. Eqs. (7.83) and (7.84)), and are therefore part of the computer-algebraic derivation presented in Sect. 3.8.

We note that Eq. (3.119) involves only scalar perturbations (s). In contrast, Eq. (3.120) has both scalar and vector parts, and can be split into two independent relations at linear order, after inserting the decompositions from Eqs. (3.76)–(3.79),

$$\delta p + (\bar{e} + \bar{p})h_0 + (\partial_\tau + 4\mathcal{H})[(\bar{e} + \bar{p})(h - v)] + \frac{2}{3}\nabla^2\Pi \overset{(3.120)}{\underset{(3.77)}{=}} 0\,, \quad (3.121)$$

$$(\partial_\tau + 4\mathcal{H})[(\bar{e} + \bar{p})(v_i^{\mathrm{v}} - h_i^{\mathrm{v}})] - \frac{1}{2}\nabla^2\Pi_i^{\mathrm{v}} \overset{(3.120)}{\underset{(3.78)}{=}} 0\,. \quad (3.122)$$

As far as the vector part in Eq. (3.122) goes, it can also be obtained by operating with $\partial_\tau + 2\mathcal{H}$ on Eq. (3.100), and eliminating then the left-hand side with the help of Eq. (3.102). In analogy with Eq. (3.103), Eq. (3.122) implies a rapid decay of $(\bar{e} + \bar{p})(v_i^{\mathrm{v}} - h_i^{\mathrm{v}})$ at every spatial position, apart from the equilibrium fluctuations generated by anisotropic stress.

With these relations, we have completed the set of equations needed for determining the evolution of linear perturbations in the early universe. However, as mentioned below Eq. (3.96), at first sight there are more variables than equations. This is related to the possibility of coordinate transformations, and we will turn to this in Chap. 4.

3.7   Viscous Corrections to the Energy-Momentum Tensor

In Eq. (3.75), we have added an *anisotropic stress* part to perturbations of the energy-momentum tensor. However, given that it only appears for spatial indices, it is not covariant. In this section, we explain how anisotropic stress arises covariantly as a part of viscous corrections to ideal hydrodynamics. Let us mention in passing that it is also possible to define anisotropic stress within the framework of Boltzmann equations, which is valid beyond thermal equilibrium.

Hydrodynamics can be viewed as an effective theory, originating as we express $T_{\mu\nu}$ through *constitutive relations*, involving an expansion in gradients. The leading term, the perfect-fluid expression in Eq. (3.63), is of zeroth order in gradients. Terms of the first order in gradients are characterized by two new functions, the *shear viscosity*, η, and the *bulk viscosity*, ζ. Like in Eq. (1.62), when we introduce dissipative coefficients, the principle of detailed balance requires that we also introduce *hydrodynamic fluctuations* [3]. It is worth mentioning that very often the fluctuation part is omitted, but this is physically viable only if a macroscopic fluid motion sets the system far from equilibrium. For near-equilibrium fluids, the fluctuations play an essential role. Sometimes they can even be more interesting than dissipation, for instance when they source something which would otherwise not exist and which is not easily dissipated away, such as gravitational waves.

When we include viscous corrections, so that fluid motions become diffusive, the notion of a fluid velocity requires a precise definition. In the so-called *Landau-Lifshitz convention*, we require that $u^\mu u^\nu T_{\mu\nu} = e$, i.e., that viscous corrections are orthogonal to u^μ. Then the general energy-momentum tensor, up to first order in gradients, can be written as

$$T_{\mu\nu} = p\,g_{\mu\nu} + (e+p)\,u_\mu u_\nu + a^2 \Pi_{\mu\nu}\,, \tag{3.123}$$

$$a^2 \Pi_{\mu\nu} \equiv -\eta\,\mathbb{V}_\mu{}^\rho\,\mathbb{V}_\nu{}^\sigma \left(u_{\rho;\sigma} + u_{\sigma;\rho} - \frac{2\,g_{\rho\sigma}}{3}u^\gamma_{;\gamma} \right) - \zeta\,\mathbb{V}_{\mu\nu}u^\gamma_{;\gamma} + a^2 S_{\mu\nu}\,, \tag{3.124}$$

$$\mathbb{V}_{\mu\nu} \equiv g_{\mu\nu} + u_\mu u_\nu\,. \tag{3.125}$$

Here $\mathbb{V}_{\mu\nu}$ is a *projector onto directions orthogonal* to u^μ; a covariant velocity derivative reads $u_{\rho;\sigma} = u_{\rho,\sigma} - \Gamma^\alpha_{\rho\sigma}u_\alpha$ (cf. Eq. (7)); and $u^\gamma_{;\gamma} = g^{\rho\sigma}u_{\rho;\sigma}$. If we contract $\Pi_{\mu\nu}$ with $g^{\mu\nu}$, the term proportional to η drops out, whereby it represents the traceless part of viscous corrections. The trace is represented by the term proportional to ζ.

The term $S_{\mu\nu}$ in Eq. (3.124) is *not* a function of the four-velocity, but rather represents thermal fluctuations, similarly to ϱ in Eq. (1.62). Whereas viscosities dissipate energy from macroscopic fluid motion into overall thermal entropy, the noise returns energy into random fluid motions. In the context of hydrodynamics, we call these *hydrodynamic fluctuations*. We return to them at the end of this section.

We now expand the thermodynamic variables in small perturbations like before,

$$e \overset{(3.65)}{=} \bar{e} + \delta e\,, \quad p \overset{(3.65)}{=} \bar{p} + \delta p\,, \quad u_\mu \overset{(3.72)}{=} \bar{u}_\mu + \delta u_\mu + \mathcal{O}(\delta^2)\,. \tag{3.126}$$

The viscosities η and ζ could also be expanded, but here we replace them by their average values; for simplicity we denote these by η and ζ, rather than $\bar\eta$ and $\bar\zeta$. In fact, the part of $\Pi_{\mu\nu}$ proportional to η is of $\mathcal{O}(\delta)$, cf. Eq. (3.128), so it is consistent to evaluate the coefficient at $\mathcal{O}(\delta^0)$.

Let us divide the whole anisotropic stress into subparts as

$$\Pi_{\mu\nu} \overset{(3.124)}{\equiv} \underbrace{\Sigma_{\mu\nu}}_{\text{term with }\eta} + \underbrace{Z_{\mu\nu}}_{\text{term with }\zeta} + \underbrace{S_{\mu\nu}}_{\text{fluctuations}}\,. \tag{3.127}$$

A somewhat tedious computation shows that at $\mathcal{O}(\delta)$, only the spatial part of $\Sigma_{\mu\nu}$ is non-vanishing,

$$\Sigma_{ij} \overset{\substack{(3.124)\\(3.127)}}{=} -\frac{\eta}{a}\left[v_{i,j} + v_{j,i} + 2\vartheta'_{ij} - \frac{2}{3}\delta_{ij}\left(\nabla\cdot\mathbf{v} + \operatorname{tr}\vartheta'\right) \right]\,. \tag{3.128}$$

Actually, $\operatorname{tr}\vartheta = 0$, cf. Eqs. (3.43)–(3.45), but we have shown its would-be appearance, to make it explicit that Eq. (3.128) is traceless.

Let us carry out a tensor decomposition of Eq. (3.128). We can write

$$\Sigma_{ij} = \Sigma^{\mathrm{s}}_{ij} + \Sigma^{\mathrm{v}}_{ij} + \Sigma^{\mathrm{t}}_{ij}\,. \tag{3.129}$$

Following Eqs. (3.18)–(3.21), the scalar, vector and tensor parts can be expressed as

$$\Sigma^{\mathrm{s}}_{ij} = \left(\partial_i\,\partial_j - \frac{1}{3}\delta_{ij}\nabla^2\right)\Sigma\,, \tag{3.130}$$

$$\Sigma^{\mathrm{v}}_{ij} = -\frac{1}{2}\left(\partial_i\Sigma^{\mathrm{v}}_j + \partial_j\Sigma^{\mathrm{v}}_i\right)\,, \quad \partial^i\Sigma^{\mathrm{v}}_i = 0\,, \tag{3.131}$$

$$\delta_{ij}\Sigma^{\mathrm{t}}_{ij} = 0\,, \quad \partial^i\Sigma^{\mathrm{t}}_{ij} = 0\,, \tag{3.132}$$

similarly to the decomposition of ϑ_{ij}, cf. Eqs. (3.43)–(3.45). If we resolve the velocity to curl-free and divergence-free parts,

$$v_i \overset{(3.76)}{=} v_i^{\text{s}} + v_i^{\text{v}} \,, \quad v_i^{\text{s}} = -\partial_i v \,, \quad \partial^i v_i^{\text{v}} = 0 \,, \tag{3.133}$$

then it follows from Eq. (3.128) that

$$\Sigma = \frac{2\eta(v - \vartheta')}{a} \,, \quad \Sigma_i^{\text{v}} = \frac{2\eta(v_i^{\text{v}} - \vartheta_i^{\text{v}\prime})}{a} \,, \quad \Sigma_{ij}^{\text{t}} = -\frac{2\eta\,\vartheta_{ij}^{\text{t}\prime}}{a} \,. \tag{3.134}$$

To summarize, a non-vanishing shear viscosity induces anisotropic stress, $\Sigma_{ij} \neq 0$. However, given that viscous terms represent a derivative correction (cf. Eq. (3.128)), we would expect this to be a small contribution in a large universe. In fact, if the corrections were substantial, we should worry that the hydrodynamic expansion does not converge. We return to this issue more concretely in the context of bulk viscous corrections, in the paragraph below Eq. (3.136), and then again with shear viscous corrections, below Eq. (3.142).

There are important qualitative lessons to be learned from Eq. (3.134). If we insert the vector part, $\Pi_i^{\text{v}} \supset \Sigma_i^{\text{v}}$, into Eq. (3.102), or the tensor part, $\Pi_{ij}^{\text{t}} \supset \Sigma_{ij}^{\text{t}}$, into Eq. (3.106), we see that η induces a *friction term*, adding to the 'Hubble friction' already present. For the tensor part, this is shown explicitly in Eq. (3.142). This is similar to the friction Υ that affects scalar perturbations in Eq. (1.62). According to the *fluctuation-dissipation relation*, the presence of dissipation implies the presence of fluctuations, which can then also generate vector perturbations and gravitational waves. The fluctuation part originates from $S_{\mu\nu}$, as discussed around Eqs. (3.138) and (3.139).

Next, let us discuss the bulk viscous part from Eq. (3.127), denoted by $Z_{\mu\nu}$. Despite the simpler appearance in Eq. (3.124), the bulk viscous corrections are not simpler than the shear viscous ones. After another tedious computation, we find

$$Z_{\mu\nu} = -\frac{3\zeta}{a} \left\{ \begin{pmatrix} 0 & 0 \\ 0 & \mathcal{H}\,\delta_{ij} \end{pmatrix} \right.$$

$$\left. + \begin{pmatrix} 0 & -\mathcal{H}\,v_j \\ -\mathcal{H}\,v_i & 2\mathcal{H}\,\vartheta_{ij} - \delta_{ij}\left[\mathcal{H}h_0 + h_{\text{D}}' + 2\mathcal{H}h_{\text{D}} - \frac{\nabla\cdot\mathbf{v}}{3}\right] \end{pmatrix} \right\} \,. \tag{3.135}$$

The key point, from the first line of Eq. (3.135), is that a correction arises even at the background level. Comparing with Eq. (3.74), and going to physical time for a simpler interpretation, we see that the background equations now contain an *effective pressure*,

$$\bar{p}_{\text{eff}} \equiv \bar{p} - 3\zeta H \,. \tag{3.136}$$

In order to elaborate on the meaning of Eq. (3.136), we recall from Sect. 2.4, and for ζ from Ref. [4] that, up to slowly evolving coefficients, the three quantities approximate to

$$\bar{p} \sim T^4 \,, \quad H \sim \frac{T^2}{m_{\text{pl}}} \,, \quad \zeta \overset{\text{YM}}{\sim} \frac{\alpha^2 T^3}{\ln(\alpha^{-1})} \,. \tag{3.137}$$

Here we have adopted a bulk viscosity relevant for a weakly coupled Yang-Mills (YM) plasma, such as the one made of Standard Model particles, where $\alpha = g^2/(4\pi) < 1$ is a fine-structure constant related to a gauge group with gauge coupling g.

We thus see that the corrections from ζ to $\bar{p}_{\text{eff}}$ are small as long as $T < \alpha^{-2} \ln(\alpha^{-1}) m_{\text{pl}}$. This is well satisfied in normal inflationary scenarios; in fact, in the regime in which the inequality is violated, general relativity perhaps ceases to be valid as an effective theory. And in any case, if the correction is as large as the leading term, the hydrodynamic expansion breaks down. We thus see that under reasonable assumptions, bulk viscosity should *not* modify the background expansion significantly (nevertheless, there is literature where such a possibility is entertained, cf., e.g., Ref. [5]).

Let us then turn to the fluctuation part of $\Pi_{\mu\nu}$, denoted by $S_{\mu\nu}$. Unfortunately, because of the tensor structure of the energy-momentum tensor, it is a non-trivial task to work out the *noise autocorrelator*. Even if this is a text-book topic [3], for relativistic plasmas the general framework has been intensely investigated only relatively recently [6]. Here we just state that the result has a close analogy with Eq. (1.63), with the value of the noise autocorrelator being related to the viscous coefficients through a fluctuation-dissipation relation similar to that which we will work out in Eq. (7.44),

$$\langle S_{\mu\nu}(\mathcal{X}) \rangle = 0 \,, \tag{3.138}$$

$$\langle a^2 S_{\mu\nu}(\mathcal{X})\, a^2 S_{\rho\sigma}(\mathcal{Y}) \rangle \approx 2\,T \left[\eta \left(\mathbb{V}_{\mu\rho}\mathbb{V}_{\nu\sigma} + \mathbb{V}_{\mu\sigma}\mathbb{V}_{\nu\rho} \right) \right. $$
$$\left. + \left(\zeta - \frac{2\eta}{3} \right) \mathbb{V}_{\mu\nu}\mathbb{V}_{\rho\sigma} \right] \frac{\delta^{(4)}(\mathcal{X} - \mathcal{Y})}{\sqrt{-\det g_{\mu\nu}}} \,. \tag{3.139}$$

Furthermore, up to first order in perturbations,

$$\mathbb{V}_{\mu\nu} \overset{\substack{(3.125) \\ (3.40),\,(3.72)}}{=} a^2 \begin{pmatrix} 0 & -v_j \\ -v_i & (1 - 2h_{\text{D}})\,\delta_{ij} + 2\vartheta_{ij} \end{pmatrix} + \mathcal{O}(\delta^2) \,. \tag{3.140}$$

Given that on the average the noise is small, cf. Eq. (3.138), the noise is treated as being of similar magnitude as fluctuations, $S_{\mu\nu} \sim \mathcal{O}(\delta)$. Therefore, in the noise autocorrelator, we can employ the leading part of Eq. (3.140), $a^2 \delta_{ij}$. This then implies that only the spatial indices carry a non-vanishing autocorrelator. Going simultaneously to momentum space in spatial directions, the only non-zero components read

$$\langle\, S_{ij}(\tau_1,\mathbf{k})\, S_{kl}(\tau_2,\mathbf{q})\,\rangle \overset{\substack{(3.139),\,(3.140)\\[2pt](9),\,(3.34)}}{\approx} 2\,T\,\delta(\tau_1-\tau_2)\,(2\pi)^3\delta^{(3)}(\mathbf{k}+\mathbf{q})$$

$$\times\ \frac{1}{a^4}\left[\eta\left(\delta_{ik}\delta_{jl}+\delta_{il}\delta_{jk}\right)+\left(\zeta-\frac{2\eta}{3}\right)\delta_{ij}\delta_{kl}\right]+\mathcal{O}(\delta^3)\,.$$

$$(3.141)$$

An important observation can be made, if we contract Eq. (3.141) with the traceless transverse projector $\mathbb{T}^{mn}_{ij}$ from Eq. (3.24). Because of the tracelessness of the projector, the part proportional to $\zeta-2\eta/3$ drops out. But the terms proportional to η remain present. Therefore, we can say that the noise has a projection onto tensor modes; we may denote this projection by $S^{\mathrm{t}}_{ij}\equiv\mathbb{T}^{mn}_{ij}S_{mn}$. With this, we may convert the gravitational-wave equation, Eq. (3.106), including the information from Eq. (3.134), into

$$\boxed{\ \left[\,\partial_\tau^2+\left(2\mathcal{H}+16\pi Ga\eta\right)\partial_\tau-\nabla^2\,\right]\vartheta^{\mathrm{t}}_{ij}\ =\ 8\pi Ga^2 S^{\mathrm{t}}_{ij}\,.\ }$$

$$(3.142)$$

This shows how *gravitational waves are dissipated and produced in the hydrodynamic regime*. Equation (3.142) represents an analogue of the scalar equation (3.114), being however simpler, because tensor modes do not couple to other perturbations at linear order.

Let us estimate how important the friction in Eq. (3.142) could be. Going to physical time, like in Eq. (3.137), the magnitudes of the two frictions in Eq. (3.142) are

$$H\ \sim\ \frac{T^2}{m_{\mathrm{pl}}}\,,\qquad G\eta\ \overset{\mathrm{YM}}{\sim}\ \frac{T^3}{m_{\mathrm{pl}}^2\,\alpha^2\ln(\alpha^{-1})}\,,$$

$$(3.143)$$

where we have adopted the shear viscosity relevant for a weakly coupled Yang-Mills plasma [7]. We see that the correction from η is small when $T<\alpha^2\ln(\alpha^{-1})m_{\mathrm{pl}}$. This is more easily violated than the corresponding inequality for ζ, nevertheless in normal scenarios the viscous corrections should be small (as for α, typical values for Standard Model like theories at high temperatures are $\alpha\sim0.01$). In contrast, the anisotropic stress induced by decoupled neutrinos at low temperatures can have a substantial influence on low-momentum tensor perturbations, as will be discussed in Sect. 10.7.

Apart from the dissipation that η induces on the left-hand side, the right-hand side of Eq. (3.142) implies that η influences the production of gravitational waves, from hydrodynamic fluctuations. We return to this topic in Sect. 10.3 and around Eq. (10.164).

We end by remarking that relativistic hydrodynamics including first-order viscous corrections is said to be acausal (with the possibility of superluminal modes) and unstable (possessing exponentially growing solutions) (cf., e.g., Ref. [8]). However, these problems arise in the domain of large momenta, where the gradient expansion breaks down anyways, so there is no real reason for concern (moreover, hydrodynamic fluctuations are omitted from these considerations). Nevertheless, there are phenomenological recipes to rectify the issues by incorporating a subset of second-order corrections beyond ideal hydrodynamics, notably via the so-called *Mueller-Israel-Stewart theory* [9, 10].

3.8 Deriving First-Order Perturbations with Computer Algebra

We show here how the results presented in this chapter can be obtained with rudimentary symbolic manipulation. As examples, we evaluate the Ricci scalar from Eq. (3.59), and the energy-momentum conservation relations, from Eqs. (3.119) and (3.120), and in more complete form from Eqs. (7.83) and (7.84). The same methods may be extended to higher orders in perturbations, as will be demonstrated in Sect. 10.9.

The logic of the code is as follows. The metric and its inverse are inserted from Eqs. (3.40) and (3.41). A parameter 'eps' counts the order in δ. The Christoffel symbols are computed from Eq. (1.9), the Ricci tensor from Eq. (1.20), and the Ricci scalar from Eq. (1.28). The result is expanded to first order in 'eps'. A usual problem with symbolic manipulation is that the output does not automatically look simple to the human eye. A key step is thus the final one, where we introduce a tentative answer, 'testricciscalar'; insert there ingredients that can be identified from the output; re-evaluate the output after their subtraction, whereby it should now be more transparent; and keep on adding terms to the tentative answer, until all structures have been identified. Of course, this works nicely only if the code is fast, which is not the case with the current sympy release, however this will likely improve in the future. Otherwise, the code can straightforwardly be transcribed to other symbolic languages, such as Mathematica. All in all, the script looks as follows:

```python
# computer-algebraic determination of the Ricci scalar [symbolic_ricci_scalar.py]
#
# basic coordinates and functions appearing
from sympy import *
i, j, k, l = symbols('i j k l',cls=Idx); x = IndexedBase('x'); eps = symbols('eps')
a = Function('a')(x[0]); H = Function('H')(x[0])
h1 = Function('h1')(x[0],x[1],x[2],x[3]); h2 = Function('h2')(x[0],x[1],x[2],x[3])
h3 = Function('h3')(x[0],x[1],x[2],x[3])
h0 = Function('h0')(x[0],x[1],x[2],x[3]); hD = Function('hD')(x[0],x[1],x[2],x[3])
theta11 = Function('theta11')(x[0],x[1],x[2],x[3]); theta22 = Function('theta22')(x[0],x[1],x[2],x[3])
theta33 = Function('theta33')(x[0],x[1],x[2],x[3]); theta12 = Function('theta12')(x[0],x[1],x[2],x[3])
theta13 = Function('theta13')(x[0],x[1],x[2],x[3]); theta23 = Function('theta23')(x[0],x[1],x[2],x[3])

# partial derivative with respect to x[i]
def d(h,i): return diff(h,x[i],evaluate=True)
def dd(h,i,j): return diff(h,x[i],x[j],evaluate=True)          # doesn't symmetrize as should :(
def dds(h,i,j): return (dd(h,i,j)+dd(h,j,i))/2                 # symmetrized
def ddd(h,i,j,k): return diff(h,x[i],x[j],x[k],evaluate=True)  # doesn't symmetrize as should :(
def ddds(h,i,j,k): return (ddd(h,i,j,k)+ddd(h,i,k,j))/2        # symmetrized in last two arguments

# restrict to order(eps) for possibly more efficient execution
def trunc1(h): return (h.expand()).subs([(eps**5,0),(eps**4,0),(eps**3,0),(eps**2,0)])
```

```
# input metric with both indices down
a00 = (-1-2*eps*h0)
a01 = (-eps*h1);  a02 = (-eps*h2);  a03 = (-eps*h3)
a10 = a01; a20 = a02; a30 = a03
a11 = (1+2*eps*(theta11-hD)); a22 = (1+2*eps*(theta22-hD)); a33 = (1+2*eps*(theta33-hD));
a12 = (2*eps*theta12); a13 = (2*eps*theta13); a23 = (2*eps*theta23);
a21 = a12; a31 = a13; a32 = a23
gdown=a*a*Matrix([[a00,a01,a02,a03],[a10,a11,a12,a13],[a20,a21,a22,a23],[a30,a31,a32,a33]])

# input metric with both indices up
b00 = (-1+2*eps*h0)
b01 = (-eps*h1);  b02 = (-eps*h2);  b03 = (-eps*h3)
b10 = b01; b20 = b02; b30 = b03
b11 = (1+2*eps*(hD-theta11)); b22 = (1+2*eps*(hD-theta22)); b33 = (1+2*eps*(hD-theta33));
b12 = (-2*eps*theta12); b13 = (-2*eps*theta13); b23 = (-2*eps*theta23)
b21 = b12; b31 = b13; b32 = b23
gup=1/a/a*Matrix([[b00,b01,b02,b03],[b10,b11,b12,b13],[b20,b21,b22,b23],[b30,b31,b32,b33]])

# check inversion
test=(gdown*gup);  print(trunc1(test));  print("inversion checked")

# christoffel symbols
def dgdown(m,n,i): return d(gdown[m,n],i)
def gamma(r,m,n):
    return Rational(1/2)*Sum( gup[r,i]*(dgdown(i,m,n)+dgdown(i,n,m)-dgdown(m,n,i)),(i,0,3)).doit()

# ricci tensor with both indices down
def prericcidown(m,n):
    return ( Sum( d(gamma(k,m,n),k) - d(gamma(k,m,k),n),(k,0,3) ).doit() +
        Sum( gamma(l,m,n)*gamma(k,l,k) - gamma(l,m,k)*gamma(k,n,l),(k,0,3),(l,0,3) ).doit() )
riccidown=trunc1(Matrix(4,4,prericcidown))

# ricci tensor with indices up and down
ricciupdown = trunc1(gup*riccidown)

# ricci scalar
ricciscalar = ricciupdown.trace()

# go over to hubble rate
ricciscalar = ricciscalar.subs([[(Derivative(a,x[0]),a*H),(Derivative(a*H,x[0]),a*(d(H,0)+H**2))
                                ,(Derivative(a,x[0],2),a*(d(H,0)+H**2))]])

# adjust expression until the difference subtracts to zero
testricciscalar = a**(-2)*(6*(d(H,0)+H**2) - 2*eps*(
    +3*H*d(h0,0)+6*(d(H,0)+H**2)*h0
    +dd(h0,1,1)+dd(h0,2,2)+dd(h0,3,3)-2*(dd(hD,1,1)+dd(hD,2,2)+dd(hD,3,3))
    +dd(theta11,1,1)+dd(theta22,1,1)+dd(theta33,1,1)
    +dd(theta11,2,2)+dd(theta22,2,2)+dd(theta33,2,2)
    +dd(theta11,3,3)+dd(theta22,3,3)+dd(theta33,3,3)
    +3*dd(hD,0,0)+9*H*d(hD,0)  -(dds(h1,0,1)+dds(h2,0,2)+dds(h3,0,3))-3*H*(d(h1,1)+d(h2,2)+d(h3,3))
    -(dd(theta11,0,0)+dd(theta22,0,0)+dd(theta33,0,0))-3*H*(d(theta11,0)+d(theta22,0)+d(theta33,0))
    -(dd(theta11,1,1)+dd(theta22,2,2)+dd(theta33,3,3))
    -2*(dds(theta12,1,2)+dds(theta13,1,3)+dds(theta23,2,3))   ))
print(expand(a**2*(ricciscalar - testricciscalar)))
print("ricci scalar checked at 1st order")
```

Once all terms from Eq. (3.59) are included in `testricciscalar`, the last line evaluates to zero, indicating that we have identified correctly the zeroth and first-order perturbations. Below, we likewise illustrate a code that reproduces Eqs. (3.119) and (3.120), starting from an energy-momentum tensor as given in Eqs. (3.74) and (3.75), or Eqs. (7.83) and (7.84), if the scalar field part from Eqs. (3.81) and (3.82) is added.

```python
# computer-algebraic determination of energy-momentum conservation [symbolic_energy_mom_conservation.py]
#
# basic coordinates and functions appearing
from sympy import *
i, j, k, l, m = symbols('i j k l m',cls=Idx); x = IndexedBase('x'); eps = symbols('eps')
a = Function('a')(x[0]); H = Function('H')(x[0])
barphi = Function('barphi')(x[0]); bare = Function('bare')(x[0]); barp = Function('barp')(x[0])
h1 = Function('h1')(x[0],x[1],x[2],x[3]); h2 = Function('h2')(x[0],x[1],x[2],x[3])
h3 = Function('h3')(x[0],x[1],x[2],x[3])
h0 = Function('h0')(x[0],x[1],x[2],x[3]); hD = Function('hD')(x[0],x[1],x[2],x[3])
deltaphi = Function('deltaphi')(x[0],x[1],x[2],x[3])
deltae = Function('deltae')(x[0],x[1],x[2],x[3]); deltap = Function('deltap')(x[0],x[1],x[2],x[3]);
v1 = Function('v1')(x[0],x[1],x[2],x[3]); v2 = Function('v2')(x[0],x[1],x[2],x[3])
v3 = Function('v3')(x[0],x[1],x[2],x[3])
theta11 = Function('theta11')(x[0],x[1],x[2],x[3]); theta22 = Function('theta22')(x[0],x[1],x[2],x[3])
theta33 = Function('theta33')(x[0],x[1],x[2],x[3]); theta12 = Function('theta12')(x[0],x[1],x[2],x[3])
theta13 = Function('theta13')(x[0],x[1],x[2],x[3]); theta23 = Function('theta23')(x[0],x[1],x[2],x[3])
pi11 = Function('pi11')(x[0],x[1],x[2],x[3]); pi22 = Function('pi22')(x[0],x[1],x[2],x[3])
pi33 = Function('pi33')(x[0],x[1],x[2],x[3]); pi12 = Function('pi12')(x[0],x[1],x[2],x[3])
pi13 = Function('pi13')(x[0],x[1],x[2],x[3]); pi23 = Function('pi23')(x[0],x[1],x[2],x[3])

# partial derivative with respect to x[i]
def d(h,i): return diff(h,x[i],evaluate=True)
def dd(h,i,j): return diff(h,x[i],x[j],evaluate=True)           # doesn't symmetrize as should :(
def dds(h,i,j): return (dd(h,i,j)+dd(h,j,i))/2                  # symmetrized
def ddd(h,i,j,k): return diff(h,x[i],x[j],x[k],evaluate=True)  # doesn't symmetrize as should :(
def ddds(h,i,j,k): return (ddd(h,i,j,k)+ddd(h,i,k,j))/2        # symmetrized in last two arguments

# restrict to order(eps) for possibly more efficient execution
def trunc1(h): return (h.expand()).subs([(eps**5,0),(eps**4,0),(eps**3,0),(eps**2,0)])

# input metric with both indices down
a00 = (-1-2*eps*h0)
a01 = (-eps*h1);  a02 = (-eps*h2); a03 = (-eps*h3); a10 = a01; a20 = a02; a30 = a03
a11 = (1+2*eps*(theta11-hD)); a22 = (1+2*eps*(theta22-hD)); a33 = (1+2*eps*(theta33-hD));
a12 = (2*eps*theta12); a13 = (2*eps*theta13); a23 = (2*eps*theta23); a21 = a12; a31 = a13; a32 = a23
gdown=a*a*Matrix([[a00,a01,a02,a03],[a10,a11,a12,a13],[a20,a21,a22,a23],[a30,a31,a32,a33]])

# input metric with both indices up
b00 = (-1+2*eps*h0)
b01 = (-eps*h1);  b02 = (-eps*h2); b03 = (-eps*h3); b10 = b01; b20 = b02; b30 = b03
b11 = (1+2*eps*(hD-theta11)); b22 = (1+2*eps*(hD-theta22)); b33 = (1+2*eps*(hD-theta33));
b12 = (-2*eps*theta12); b13 = (-2*eps*theta13); b23 = (-2*eps*theta23); b21 = b12; b31 = b13; b32 = b23
gup=1/a/a*Matrix([[b00,b01,b02,b03],[b10,b11,b12,b13],[b20,b21,b22,b23],[b30,b31,b32,b33]])

# christoffel symbols
def dgdown(m,n,i): return d(gdown[m,n],i)
def gamma(r,m,n):
    return Rational(1/2)*Sum( gup[r,i]*(dgdown(i,m,n)+dgdown(i,n,m)-dgdown(m,n,i)),(i,0,3)).doit()

# energy-momentum tensor for perfect fluid plus varphi
t00 = d(barphi,0)**2/2 + eps*d(barphi,0)*d(deltaphi,0) + a*a*(bare + eps*(deltae + 2*h0*bare))
t01 = eps*( d(barphi,0)*(d(deltaphi,1) - d(barphi,0)/2*h1) ) + a*a*eps*(bare*(h1-v1)-barp*v1)
t02 = eps*( d(barphi,0)*(d(deltaphi,2) - d(barphi,0)/2*h2) ) + a*a*eps*(bare*(h2-v2)-barp*v2)
t03 = eps*( d(barphi,0)*(d(deltaphi,3) - d(barphi,0)/2*h3) ) + a*a*eps*(bare*(h3-v3)-barp*v3)
t10 = t01; t20 = t02; t30 = t03
t11 = d(barphi,0)*d(barphi,0)/2 + eps*( ((theta11-hD-h0)*d(barphi,0) + d(deltaphi,0))*d(barphi,0)
      ) + a*a*( barp + eps*(deltap + 2*(theta11-hD)*barp + pi11) )
t22 = d(barphi,0)*d(barphi,0)/2 + eps*( ((theta22-hD-h0)*d(barphi,0) + d(deltaphi,0))*d(barphi,0)
      ) + a*a*( barp + eps*(deltap + 2*(theta22-hD)*barp + pi22) )
t33 = d(barphi,0)*d(barphi,0)/2 + eps*( ((theta33-hD-h0)*d(barphi,0) + d(deltaphi,0))*d(barphi,0)
      ) + a*a*( barp + eps*(deltap + 2*(theta33-hD)*barp + pi33) )
t12 = eps*theta12*d(barphi,0)*d(barphi,0) + a*a*eps*(2*theta12*barp + pi12)
t13 = eps*theta13*d(barphi,0)*d(barphi,0) + a*a*eps*(2*theta13*barp + pi13)
t23 = eps*theta23*d(barphi,0)*d(barphi,0) + a*a*eps*(2*theta23*barp + pi23)
t21 = t12; t31 = t13; t32 = t23
tmunu=Matrix([[t00,t01,t02,t03],[t10,t11,t12,t13],[t20,t21,t22,t23],[t30,t31,t32,t33]])

# energy-momentum conservation equation
def preconserve(n,r):                        # second is a dummy variable
    return ( Sum( gup[m,k]*d(tmunu[m,n],k),(m,0,3),(k,0,3) ).doit() -
             Sum( gup[m,k]*(tmunu[m,l]*gamma(l,n,k)+
                            tmunu[n,l]*gamma(l,m,k)),(m,0,3),(k,0,3),(l,0,3)).doit() )
conserve=trunc1(Matrix(2,1,preconserve))    # fill only 1st spatial component to save time

# go over to hubble rate
conserve = conserve.subs([(Derivative(a,x[0]),a*H),(Derivative(a*H,x[0]),a*(d(H,0)+H**2))
```

```
                    ,(Derivative(a,x[0],2),a*(d(H,0)+H**2))])

# test energy-momentum conservation
# component nu=0
testconserve0 = -(d(bare,0)+3*H*(bare+barp))-a**(-2)*d(barphi,0)*(dd(barphi,0,0)+2*H*d(barphi,0)
    )+eps*( +a**(-2)*(
        + d(barphi,0)*(dd(deltaphi,1,1)+dd(deltaphi,2,2)+dd(deltaphi,3,3)-dd(deltaphi,0,0))
        - (dd(barphi,0,0) + 4*H*d(barphi,0))*d(deltaphi,0)
        + d(barphi,0)**2*(d(h0,0)+3*d(hD,0)-d(h1,1)-d(h2,2)-d(h3,3)
                          -d(theta11,0)-d(theta22,0)-d(theta33,0))
        + 2*d(barphi,0)*(dd(barphi,0,0) + 2*H*d(barphi,0))*h0
        )
    -d(deltae,0)-3*H*(deltae+deltap)
    +(bare+barp)*(3*d(hD,0)-d(v1,1)-d(v2,2)-d(v3,3)
                       -d(theta11,0)-d(theta22,0)-d(theta33,0))
    -H*(pi11+pi22+pi33)
    )
print(expand(conserve[0]-testconserve0))
print("energy-momentum conservation tested: nu=0")

# component nu=1
testconserve1 = eps*(
    -a**(-2)*(dd(barphi,0,0)+2*H*d(barphi,0))*d(deltaphi,1)
    +d(deltap,1)+(bare+barp)*(d(h0,1)+d(v1,0)-d(h1,0)+4*H*(v1-h1))
    +(d(bare,0)+d(barp,0))*(v1-h1)+d(pi11,1)+d(pi12,2)+d(pi13,3)
    #### remnant of non-symmetrized derivatives
    +a**(-2)*d(barphi,0)*(dd(deltaphi,1,0)-dd(deltaphi,0,1))
    )
print(expand(conserve[1]-testconserve1))
print("energy-momentum conservation tested: nu=1")
```

References

1. Bardeen, J.M.: Gauge-invariant cosmological perturbations. Phys. Rev. D **22**, 1882 (1980). https://doi.org/10.1103/PhysRevD.22.1882

2. Arnowitt, R.L., Deser, S., Misner, C.W.: The dynamics of general relativity. Gen. Rel. Grav. **40**, 1997 (2008). https://doi.org/10.1007/s10714-008-0661-1, gr-qc/0405109

3. Lifshitz, E.M., Pitaevskii, L.P.: Statistical Physics, Part 2, §88-89. Butterworth-Heinemann, Oxford (1980)

4. Arnold, P.B., Dogan, C., Moore, G.D.: Bulk viscosity of high-temperature QCD. Phys. Rev. D **74**, 085021 (2006). https://doi.org/10.1103/PhysRevD.74.085021, hep-ph/0608012

5. Zimdahl, W.: Bulk viscous cosmology. Phys. Rev. D **53**, 5483 (1996). https://doi.org/10.1103/PhysRevD.53.5483, astro-ph/9601189

6. Kapusta, J.I., Müller, B., Stephanov, M.: Relativistic theory of hydrodynamic fluctuations with applications to heavy-ion collisions. Phys. Rev. C **85**, 054906 (2012). https://doi.org/10.1103/PhysRevC.85.054906, 1112.6405

7. Arnold, P.B., Moore, G.D., Yaffe, L.G.: Transport coefficients in high temperature gauge theories (I): leading-log results. JHEP **11**, 001 (2000). https://doi.org/10.1088/1126-6708/2000/11/001, hep-ph/0010177

8. Hiscock, W.A., Lindblom, L.: Stability and causality in dissipative relativistic fluids. Annals Phys. **151**, 466 (1983). https://doi.org/10.1016/0003-4916(83)90288-9

9. Müller, I.: Zum Paradoxon der Wärmeleitungstheorie. Z. Phys. **198**, 329 (1967). https://doi.org/10.1007/BF01326412

10. Israel, W., Stewart, J.M.: Transient relativistic thermodynamics and kinetic theory. Annals Phys. **118**, 341 (1979). https://doi.org/10.1016/0003-4916(79)90130-1

Gauge Transformations and Different Gauges

4

Abstract

Coordinate covariance is a cornerstone of general relativity. At first order in small transformations, it introduces degeneracies in the identification of space-time points in the homogeneous and isotropic description with the corresponding locations in a perturbed universe. These degeneracies are referred to as gauge invariance. In the literature, many gauge choices are employed. We show how the equations can be written in a manifestly gauge-invariant form; how this permits for powerful crosschecks of the computations; and how popular gauge choices can be recovered from the general equations.

Keywords

Gauge transformation · Newtonian gauge · Zero-shear gauge · Longitudinal gauge · Poisson gauge · Comoving gauge · Spatially flat gauge · Uniform curvature gauge · Synchronous gauge · Gauge-invariant observables · Bardeen potentials · Curvature perturbations · Einstein tensor and Einstein equations in Newtonian gauge

4.1 Definition of Gauge Transformations

An intrinsic property of general relativity is the absence of a preferred frame. Yet, its manifestation within the perturbative approach is non-trivial. In particular, even if the universe were exactly homogeneous and isotropic in one coordinate system, we could make a small coordinate transformation to a new frame, and in the latter our variables would no longer be exactly homogeneous (i.e. independent of $\mathbf{x}$) and isotropic (i.e. independent of an observation direction $\mathbf{n}$). Physical quantities should be identified as those having an unambiguous coordinate-independent meaning. Such quantities are said to be *gauge invariant* or *gauge independent*.

We note that when considering transformations, there are two alternative points of view. If we talk about *active* transformations, the values of physical quantities are changed, but we stay at the same coordinate point. Instead, in the *passive* picture, which we adopt here, physics remains unchanged, but coordinates are transformed, leading to a relabelling. In the framework of cosmological perturbation theory, both points of view are valid [1].

To define the transformations, we envisage that a space-time point q, or 'event', is described in one coordinate system with the coordinates $x^\alpha(q)$. Then we relabel the same point with the new coordinates $\tilde{x}^\alpha(q)$, such that

$$\tilde{x}^\alpha(q) = x^\alpha(q) + \xi^\alpha(q) + \mathcal{O}(\delta^2) \,. \tag{4.1}$$

Denoting derivatives with respect to the original system by $(...)_{,\alpha} \equiv \partial(...)/\partial x^\alpha$, the transformation properties of space-time tensors are dictated by

$$\Xi^{\tilde{\mu}}_{\rho} \equiv \frac{\partial \tilde{x}^\mu}{\partial x^\rho} = \delta^\mu_\rho + \xi^\mu_{,\rho} + \mathcal{O}(\delta^2) \,, \quad \Xi^{\mu}_{\tilde{\rho}} \equiv \frac{\partial x^\mu}{\partial \tilde{x}^\rho} = \delta^\mu_\rho - \xi^\mu_{,\rho} + \mathcal{O}(\delta^2) \,. \tag{4.2}$$

We may also say that if we consider equivalent values of the coordinates, they correspond to different space-time points,

$$\tilde{x}^\alpha(\tilde{q}) = x^\alpha(q) \quad \overset{(4.1)}{\Rightarrow} \quad x^\alpha(\tilde{q}) = x^\alpha(q) - \xi^\alpha(q) + \mathcal{O}(\delta^2) \,. \tag{4.3}$$

In the language of the discussion around Eq. (3.47), ξ^0 changes the slicing, ξ^i the threading of the perturbed universe. The non-uniqueness of the coordinate choice leads to redundancies in the physical description, which can then be interpreted as gauge freedom.

Let us now consider some general function, $Q \equiv Q(q)$. We denote its value at the space-time point $\tilde{q}$ by $\tilde{Q} \equiv \tilde{Q}(\tilde{q})$. The function can be expanded to first order in perturbations, $Q = \bar{Q} + \delta Q$ and $\tilde{Q} = \bar{\tilde{Q}} + \delta\tilde{Q}$. Given that $\tilde{q} \neq q$, the values and perturbations are not the same. However, with the help of Eq. (4.3), we can relate them to each other. The relations depend on ξ^α, and since choices of ξ^α represent redundancies in our description, we call the relations gauge transformations. Below we work out these transformations for arbitrary scalar, vector and tensor quantities.

Starting with scalar quantities (cf. Eq. (3.4)), by definition they have the same value at the same point in every coordinate system, i.e. $\tilde{A}(\{\tilde{x}^\alpha(\tilde{q})\}) = A(\{x^\alpha(\tilde{q})\})$. Inserting Eq. (4.3) on the right-hand side of this equality, recalling that the background only depends on time, and equating the results order by order in perturbations, we find

$$\tilde{A} = \bar{\tilde{A}} + \delta\tilde{A} \overset{(4.3)}{=} \bar{A} - (\partial_\alpha \bar{A})\xi^\alpha + \delta A + \mathcal{O}(\delta^2) \quad \Rightarrow \quad \boxed{\delta\tilde{A} \approx \delta A - \bar{A}'\xi^0 \,.}$$

$$\tag{4.4}$$

Let us pause to stress an important point. Normally, for instance when we are talking about representations of compact groups, a 'scalar' quantity is one which is invariant under the transformation considered. But here, as shown by Eq. (4.4), scalar quantities do transform in a non-trivial way. According to Eq. (4.4), the only quantities which do not transform at all, are those whose background value vanishes or is time-independent.

Proceeding then to vector perturbations (cf. Eq. (3.5)), their transformation can be calculated as

$$\tilde{B}^\mu = \Xi^{\tilde{\mu}}_\rho \left[\bar{B}^\rho - (\partial_\alpha \bar{B}^\rho)\xi^\alpha + \delta B^\rho + \mathcal{O}(\delta^2) \right] \overset{(4.2)}{\approx} \bar{B}^\mu - (\partial_\alpha \bar{B}^\mu)\xi^\alpha + \bar{B}^\alpha \xi^\mu_{,\alpha} + \delta B^\mu \tag{4.5}$$

$$\Rightarrow \boxed{\delta\tilde{B}^\mu \approx \delta B^\mu - \bar{B}^{\mu\prime}\xi^0 + \bar{B}^0 \xi^{\mu\prime}\,.} \tag{4.6}$$

For the zeroth component it follows that

$$\tilde{b}^0 \overset{(3.6)}{=} b^0 - \bar{B}^{0\prime}\xi^0 + \bar{B}^0 \xi^{0\prime}\,. \tag{4.7}$$

By decomposing the spatial gauge parameter into its scalar and vector parts,

$$\xi^i = -\partial^i \xi + \xi^i_{\mathrm{v}}\,, \qquad \partial_i \xi^i_{\mathrm{v}} = 0\,, \tag{4.8}$$

and likewise $\delta B^i = b^i_{\mathrm{s}} + b^i_{\mathrm{v}}$ (cf. Eq. (3.7)), and recalling $\bar{B}^i = 0$ (cf. Eq. (3.6)), we find the transformations

$$\tilde{b}^i_{\mathrm{s}} = -\partial^i \left(b + \bar{B}^0 \xi' \right) \quad \Rightarrow \quad \tilde{b} = b + \bar{B}^0 \xi'\,, \tag{4.9}$$

$$\tilde{b}^i_{\mathrm{v}} = b^i_{\mathrm{v}} + \bar{B}^0 \xi^{i\prime}_{\mathrm{v}}\,. \tag{4.10}$$

The behaviour of tensor perturbations (cf. Eq. (3.14)) under small coordinate transformations is

$$\tilde{C}_{\mu\nu} = \Xi^\rho_{\tilde{\mu}}\, \Xi^\sigma_{\tilde{\nu}} \left[\bar{C}_{\rho\sigma} - (\partial_\alpha \bar{C}_{\rho\sigma})\xi^\alpha + \delta C_{\rho\sigma} + \mathcal{O}(\delta^2) \right] \tag{4.11}$$

$$\overset{(4.2)}{\approx} \bar{C}_{\mu\nu} - \xi^\rho_{,\mu}\bar{C}_{\rho\nu} - \xi^\rho_{,\nu}\bar{C}_{\mu\rho} - (\partial_\alpha \bar{C}_{\mu\nu})\,\xi^\alpha + \delta C_{\mu\nu} \tag{4.12}$$

$$\Rightarrow \boxed{\delta\tilde{C}_{\mu\nu} \approx \delta C_{\mu\nu} - \xi^\rho_{,\mu}\bar{C}_{\rho\nu} - \xi^\rho_{,\nu}\bar{C}_{\mu\rho} - \bar{C}'_{\mu\nu}\xi^0\,.} \tag{4.13}$$

Using Eqs. (3.15)–(3.21) we find how the single components transform. The 00-component of Eq. (4.13) yields

$$\tilde{c}_0 \overset{(4.13)}{\underset{(3.15)}{=}} c_0 + \bar{C}_{00}\,\xi^{0\prime} + \frac{1}{2}\bar{C}'_{00}\,\xi^0\,. \tag{4.14}$$

For the $0i$-components we apply Eqs. (4.8), (3.16) and (3.17), and obtain

$$\tilde{c}_i \overset{(4.13)}{\underset{(3.16)}{=}} c_i + \bar{C}\,\xi'_i + \bar{C}_{00}\,\xi^0_{,i} \overset{(4.8)}{\underset{(3.17)}{\Rightarrow}} \tilde{c} = c + \bar{C}\,\xi' - \bar{C}_{00}\,\xi^0\,, \tag{4.15}$$

$$\tilde{c}^{\mathrm{v}}_i = c^{\mathrm{v}}_i + \bar{C}\,\xi^{\mathrm{v}\prime}_i\,. \tag{4.16}$$

Analogously, the transformation behaviour of the ij-components is found,

$$\underbrace{\delta\tilde{C}_{ij}}_{(3.15):\,-2\delta_{ij}\tilde{c}_{\mathrm{D}}+2\tilde{\gamma}_{ij}} \overset{(4.13)}{=} \delta C_{ij} - \bar{C}\underbrace{\left(\xi_{j,i}+\xi_{i,j} \overbrace{-\frac{2}{3}\delta_{ij}\xi^{k}_{,k}}^{(4.8):\,\frac{2}{3}\delta_{ij}\nabla^2\xi}\right)}_{\mathrm{tr}=0} \underbrace{-\frac{2}{3}\bar{C}\delta_{ij}\xi^{k}_{,k}}_{(4.8):\,\frac{2}{3}\bar{C}\delta_{ij}\nabla^2\xi} -\bar{C}'\delta_{ij}\xi^0 \quad (4.17)$$

$$\Rightarrow \tilde{c}_{\mathrm{D}} = c_{\mathrm{D}} - \frac{1}{3}\bar{C}\,\nabla^2\xi + \frac{1}{2}\bar{C}'\xi^0 \,, \tag{4.18}$$

$$\tilde{\gamma}_{ij} = \gamma_{ij} - \frac{1}{2}\bar{C}\left(\xi_{j,i}+\xi_{i,j}\right) - \frac{1}{3}\bar{C}\delta_{ij}\nabla^2\xi \tag{4.19}$$

$$\overset{(3.18)}{\underset{(4.8)}{\Rightarrow}} \tilde{\gamma} = \gamma + \bar{C}\xi \,, \quad \tilde{\gamma}_i^{\mathrm{v}} = \gamma_i^{\mathrm{v}} + \bar{C}\xi_i^{\mathrm{v}} \,, \quad \tilde{\gamma}_{ij}^{\mathrm{t}} = \gamma_{ij}^{\mathrm{t}} \,. \tag{4.20}$$

Let us sum up the physical degrees of freedom. In addition to two scalar gauge parameters, ξ^0 and ξ, there are two transverse vector gauge parameters, ξ_{v}^i. So, in total, only two of the four scalar perturbations introduced in Eq. (3.40) are physical, and likewise, only two of the four vector perturbations are physical.

4.2　Examples of Common Gauge Choices

Given the gauge transformations defined in Sect. 4.1, not all the perturbations that we have considered in Chap. 3 are physical. There are two opposite philosophies that originate from this realization. One is that we can make use of the gauge freedom, in order to eliminate some variables. Various gauges are defined by which variables are eliminated, and in this section we explain three common choices (the list is not exhaustive, and a fourth choice is mentioned in passing in the caption of Table 4.1). The other philosophy is to manipulate the basic equations, so that the dynamical variables appear as gauge-invariant linear combinations. The latter philosophy is explained in Sect. 4.3 and is the one we in general endorse, because it offers for strong crosschecks of the computations.

To proceed with concrete examples, we need to apply Eqs. (4.14)–(4.20) to $\delta g_{\mu\nu}$ from Eq. (3.40). The background values are $\bar{C}_{00} = -a^2$ and $\bar{C} = a^2$, and when we subsequently factor out the overall a^2, we effectively need to replace $\bar{C}_{00} \to (-a^2)/a^2 = -1$, $\bar{C}'_{00} \to -2a'/a = -2\mathcal{H}$, $\bar{C} \to (a^2)/a^2 = 1$, and $\bar{C}' \to 2a'/a = 2\mathcal{H}$ in the transformation laws above. These lead to

$$\tilde{h}_0 \overset{(4.14)}{=} h_0 - \xi^{0\prime} - \mathcal{H}\xi^0 \,, \tag{4.21}$$

$$\tilde{h}_i \overset{(4.15)}{=} h_i + \xi_i' - \partial_i\xi^0 \Rightarrow \tilde{h} \overset{(4.15)}{=} h + \xi' + \xi^0 \,, \tag{4.22}$$

$$\tilde{h}_i^{\mathrm{v}} \overset{(4.16)}{=} h_i^{\mathrm{v}} + \xi_i^{\mathrm{v}\prime} \,, \tag{4.23}$$

$$\tilde{h}_{\mathrm{D}} \overset{(4.18)}{=} h_{\mathrm{D}} - \frac{1}{3}\nabla^2\xi + \mathcal{H}\xi^0 \,, \tag{4.24}$$

$$\tilde{\vartheta}_{ij} \overset{(4.19)}{=} \vartheta_{ij} - \frac{1}{2}(\partial_i \xi_j + \partial_j \xi_i) - \frac{1}{3}\delta_{ij}\nabla^2\xi \overset{(4.20)}{\Rightarrow} \tilde{\vartheta} = \vartheta + \xi \,, \tag{4.25}$$

$$\tilde{\vartheta}_i^{\text{v}} = \vartheta_i^{\text{v}} + \xi_i^{\text{v}} \,, \tag{4.26}$$

$$\tilde{\vartheta}_{ij}^{\text{t}} = \vartheta_{ij}^{\text{t}} \,. \tag{4.27}$$

For vector perturbations, we observe from Eqs. (4.23) and (4.26) the invariance of the combination which played a role in Eqs. (3.100) and (3.101), $\tilde{h}_i^{\text{v}} - \tilde{\vartheta}_i^{\text{v}\prime} = h_i^{\text{v}} - \vartheta_i^{\text{v}\prime}$.

Newtonian (or Zero-Shear, or Longitudinal, or Poisson) Gauge

A much-used and perhaps intuitive gauge choice is the one where the perturbed metric in Eq. (3.40) can be described by an almost diagonal matrix. The conformal Newtonian gauge satisfies this for scalar perturbations, imposing the two gauge conditions

$$\boxed{h^{\text{N}} = \vartheta^{\text{N}} = 0 \,.} \tag{4.28}$$

This gauge is also known as the zero-shear, longitudinal, or Poisson gauge.

We consider the gauge transformation from an arbitrary gauge to the Newtonian gauge. From Eqs. (4.22) and (4.25) we get

$$\xi \overset{(4.25)}{=} -\vartheta \,, \qquad \xi^0 \overset{(4.22)}{=} \vartheta' - h \,. \tag{4.29}$$

The remaining scalar quantities transform to

$$\phi \equiv h_0^{\text{N}} \overset{(4.21)}{\underset{(4.29)}{=}} h_0 + (\partial_\tau + \mathcal{H})(h - \vartheta') \,, \tag{4.30}$$

$$\psi \equiv h_{\text{D}}^{\text{N}} \overset{(4.24)}{\underset{(4.29)}{=}} h_{\text{D}} + \frac{1}{3}\nabla^2\vartheta + \mathcal{H}(\vartheta' - h) \,, \tag{4.31}$$

which are known as the *Bardeen potentials* [2]. The Bardeen potentials are gauge invariant: this can be proven by using the relation $\tilde{\vartheta}' - \tilde{h} = \vartheta' - h - \xi^0$, yielding

$$\tilde{\phi} \overset{(4.30)}{\underset{(4.21)}{=}} h_0 - \xi^{0\prime} - \mathcal{H}\xi^0 + (\partial_\tau + \mathcal{H})(h + \xi^0 - \vartheta')$$

$$= \phi - \xi^{0\prime} - \mathcal{H}\xi^0 + \xi^{0\prime} + \mathcal{H}\xi^0 = \phi \,, \tag{4.32}$$

$$\tilde{\psi} \overset{(4.31)}{\underset{(4.24),\,(4.25)}{=}} h_{\text{D}} - \frac{1}{3}\nabla^2\xi + \mathcal{H}\xi^0 + \frac{1}{3}\nabla^2\vartheta + \frac{1}{3}\nabla^2\xi + \mathcal{H}(\vartheta' - h - \xi^0) = \psi \,. \tag{4.33}$$

The two vector gauge conditions can be imposed, e.g. as

$$\xi_i^{\text{v}} = -\vartheta_i^{\text{v}} \,, \quad \partial^i \xi_i^{\text{v}} = -\partial^i \vartheta_i^{\text{v}} = 0 \quad \Rightarrow \quad \vartheta_i^{\text{N}}|_{\text{v}} = 0 \,, \quad h_i^{\text{N}}|_{\text{v}} = h_i^{\text{v}} - \vartheta_i^{\text{v}\prime} \,. \tag{4.34}$$

Therefore the scalar, vector and tensor parts of the perturbed metric appear as

$$h^{\mathrm{N}}_{\mu\nu}|_{\mathrm{s}} \underset{(4.28),\,(4.30),\,(4.31)}{\overset{(3.34),\,(3.40)}{=}} -2 \begin{pmatrix} \phi & 0 \\ 0 & \psi\,\delta_{ij} \end{pmatrix}, \qquad h^{\mathrm{N}}_{\mu\nu}|_{\mathrm{v}} \overset{(3.34),\,(3.40)}{\underset{(4.34)}{=}} \begin{pmatrix} 0 & -h^{\mathrm{N}}_j|_{\mathrm{v}} \\ -h^{\mathrm{N}}_i|_{\mathrm{v}} & 0 \end{pmatrix},$$

$$h^{\mathrm{N}}_{\mu\nu}|_{\mathrm{t}} \overset{(3.34)}{\underset{(3.40)}{=}} 2 \begin{pmatrix} 0 & 0 \\ 0 & \vartheta^{\mathrm{t}}_{ij} \end{pmatrix}. \tag{4.35}$$

This metric is simple enough that the Einstein tensor and the Einstein equations can be worked out by hand, as we show in Sects. 4.4 and 4.5, respectively.

Comoving Gauge (with Respect to Either Fluid or Scalar Field)

While the Newtonian gauge simplifies the left-hand side of the Einstein equations, the comoving gauge is geared towards simplifying the right-hand side, particularly if only one fluid (cf. Eq. (3.74)), or only a scalar degree of freedom appears (cf. Eq. (3.82)).

Let us start by investigating how the fluid perturbations δe, δp, v and Π transform under gauge transformations. Given that e and p are four-scalars, they obey Eq. (4.4),

$$\delta\tilde{e} = \delta e - \bar{e}'\xi^0, \qquad \delta\tilde{p} = \delta p - \bar{p}'\xi^0. \tag{4.36}$$

The four-velocity u^μ behaves as described in Eq. (4.6),

$$\delta\tilde{u}^0 \overset{(4.6)}{=} \delta u^0 - \bar{u}^{0\prime}\xi^0 + \bar{u}^0\xi^{0\prime} \overset{(3.66)}{=} \delta u^0 + \frac{1}{a}(\mathcal{H}\xi^0 + \xi^{0\prime}), \tag{4.37}$$

$$\delta\tilde{u}^i \overset{(4.6)}{=} \delta u^i + \bar{u}^0\xi^{i\prime} \overset{(3.66)}{=} \delta u^i + \frac{1}{a}\xi^{i\prime} \overset{(3.67)}{\underset{(3.76),\,(4.8)}{\Rightarrow}} \tilde{v} = v + \xi'. \tag{4.38}$$

The transformation of δu^0 corresponds to that of $-a^{-1}h_0$, cf. Eqs. (3.70) and (4.21). Finally, the anisotropic stress transforms according to Eq. (4.20), but since it has no background value ($\bar{C} = 0$), it does *not* get transformed.

A coordinate choice that simplifies the energy-momentum tensor corresponds to its rest frame. The *slicing* introduced below Eq. (3.45) is said to be *comoving*, if its normal vector aligns with the fluid 4-velocity,

$$- n^\mu \overset{(3.46)}{=} a^{-1}(1 - h_0, h^i) \overset{!}{=} u^\mu \overset{(3.71)}{=} a^{-1}(1 - h_0, v^i). \tag{4.39}$$

A comoving slicing does not necessarily exist in general. However, for scalar perturbations, $v^i_{\mathrm{s}} = -v_{,i}$ and $h^i_{\mathrm{s}} = -h_{,i}$, it always exists, as we can impose

$$\text{comoving slicing} \quad \Leftrightarrow \quad v = h. \tag{4.40}$$

According to Eqs. (4.22) and (4.38), this is obtained by choosing

$$\xi^0 = v - h \, . \tag{4.41}$$

This does not yet define a (scalar) gauge, as ξ is still arbitrary, i.e. the threading is left unspecified. The *threading* is said to be *comoving*, if the threads are world lines of comoving observers,

$$t^\mu \overset{(3.47)}{=} a^{-1}(1 - h_0, \mathbf{0}) \overset{!}{=} u^\mu \overset{(3.71)}{=} a^{-1}(1 - h_0, v^i) \, . \tag{4.42}$$

For scalar perturbations this means

$$\text{comoving threading} \quad \Leftrightarrow \quad v = 0 \, . \tag{4.43}$$

According to Eq. (4.38), we get to comoving threading by the gauge transformation

$$\xi' = -v \, . \tag{4.44}$$

A time-independent part of ξ remains unspecified by this condition.

The comoving gauge is defined by requiring both comoving slicing and comoving threading, such that the threading is orthogonal to the slicing,

$$- n^\mu = t^\mu = u^\mu \quad \Leftrightarrow \quad \boxed{v^{\mathrm{c}} = h^{\mathrm{c}} = 0 \, .} \tag{4.45}$$

We see from Eq. (3.74) that in this gauge, and in the absence of vector perturbations, the components δT_{0i} drop out from the fluid energy-momentum tensor.

A problem with the comoving gauge is that if multiple matter components are present simultaneously, in our case a scalar field and a fluid, then δT_{0i} no longer drops out, cf. Eq. (3.84). On the other hand, if *only* a scalar field is present, then, according to Eq. (4.4), we could eliminate it by choosing $\xi^0 = \delta\varphi/\bar\varphi'$. Subsequently, h could be eliminated via Eq. (4.22). Then we find $\delta\varphi^{\mathrm{c}} = 0$ and $\delta T_{0i}^{\mathrm{c}} = 0$. In other words, if φ evolves in an empty universe, constant-time hypersurfaces correspond to constant-φ hypersurfaces in this gauge, and φ is homogeneous. This is sometimes employed in the context of inflation.

Spatially Flat (or Uniform Curvature) Gauge

The spatially flat gauge is defined by vanishing time-slice Ricci scalar (cf. Eq. (3.61)),

$$\boxed{h_{\mathrm{D}}^{\mathrm{f}} + \frac{1}{3}\nabla^2 \vartheta^{\mathrm{f}} = 0 \, .} \tag{4.46}$$

We can transform to the spatially flat gauge with the scalar gauge parameter $\xi^0 = - \left(h_{\mathrm{D}} + \nabla^2 \vartheta / 3 \right) / \mathcal{H}$, cf. Eqs. (4.24) and (4.25). The other scalar gauge parameter ξ does not need to be specified to establish Eq. (4.46), however choosing it as $\xi = -\vartheta$, we may eliminate h_{D} and ϑ separately, cf. Eq. (4.25). Then the *curvature perturbation* associated with the inflaton field, denoted by $\mathcal{R}_\varphi$ in Eq. (4.60), becomes

$$\mathcal{R}_\varphi \overset{(4.60)}{=} - \left(h_{\mathrm{D}} + \frac{1}{3} \nabla^2 \vartheta \right) - \mathcal{H} \frac{\delta \varphi}{\bar{\varphi}'} \overset{(4.46)}{=} -\mathcal{H} \frac{\delta \varphi^{\mathrm{f}}}{\bar{\varphi}'} . \tag{4.47}$$

In the literature, the inflaton perturbation in this gauge is often called the *Sasaki or Mukhanov variable*, and we denote it by

$$\mathcal{Q}_\varphi \equiv - \frac{\bar{\varphi}'}{\mathcal{H}} \mathcal{R}_\varphi = \delta \varphi + \frac{\bar{\varphi}'}{\mathcal{H}} \left(h_{\mathrm{D}} + \frac{1}{3} \nabla^2 \vartheta \right) \overset{(4.47)}{=} \delta \varphi^{\mathrm{f}} . \tag{4.48}$$

This variable plays an important role in Chap. 5.

Summary

In order to organize a bit the various gauge choices and what they achieve, let us repeat again the transformation properties of scalar quantities,

$$\tilde{h}_0 \overset{(4.21)}{=} h_0 - \xi^{0\prime} - \mathcal{H} \xi^0 , \tag{4.49}$$

$$\tilde{h} \overset{(4.22)}{=} h + \xi^0 + \xi' , \tag{4.50}$$

$$\tilde{h}_{\mathrm{D}} \overset{(4.24)}{=} h_{\mathrm{D}} + \mathcal{H} \xi^0 - \frac{\nabla^2 \xi}{3} , \tag{4.51}$$

$$\tilde{\vartheta} \overset{(4.25)}{=} \vartheta + \xi , \tag{4.52}$$

$$\delta \tilde{\varphi} \overset{(4.4)}{=} \delta \varphi - \bar{\varphi}' \xi^0 , \tag{4.53}$$

$$\tilde{v} \overset{(4.38)}{=} v + \xi' . \tag{4.54}$$

The influence of the various gauge choices is then as tabulated in Table 4.1. We note that in each case, two scalar perturbations can be put to zero.

4.3   How to Eliminate Gauge Dependence from the Equations

Having shown in Sect. 4.2 how gauges can be chosen, we now proceed to discussing the opposite philosophy of keeping the gauge general, and how this can yield strong crosschecks of the computations.

The basic idea is to define combinations of the perturbations which are invariant in gauge transformations. Subsequently, we convert our basic equations into equations for these combinations. All gauge non-covariant terms must drop out while we are doing this, otherwise we have made an algebraic mistake.

Table 4.1 Examples of gauge choices, in the notation of Eqs. (4.49)–(4.54). In each case, two perturbations can be set to zero. By ϕ and ψ we refer to the Bardeen potentials, cf. Eqs. (4.30) and (4.31), and by $\mathcal{Q}_\varphi$ to a rescaled curvature perturbation, cf. Eq. (4.48). Yet another gauge discussed in the literature is the *synchronous* one, in which $\tilde{h}_0 = \tilde{h} = 0$ [3]. We fix to none of these gauges, but rather keep ξ^0 and ξ general, verifying that the results do not depend on the gauge choice

	Newtonian / zero-shear	comoving with fluid	spatially flat / uniform curvature
ξ^0	$\vartheta' - h$	$v - h$	$-\frac{1}{\mathcal{H}}\left(h_\mathrm{D} + \frac{\nabla^2\vartheta}{3}\right)$
ξ	$-\vartheta$	$-\int^\tau v$	$-\vartheta$
$\tilde{h}_0$	$h_0 + (\partial_\tau + \mathcal{H})(h - \vartheta') = \phi$	$h_0 + (\partial_\tau + \mathcal{H})(h - v)$	$h_0 + (\partial_\tau + \mathcal{H})\left[\frac{1}{\mathcal{H}}\left(h_\mathrm{D} + \frac{\nabla^2\vartheta}{3}\right)\right]$
$\tilde{h}$	0	0	$h - \vartheta' - \frac{1}{\mathcal{H}}\left(h_\mathrm{D} + \frac{\nabla^2\vartheta}{3}\right)$
$\tilde{h}_\mathrm{D}$	$h_\mathrm{D} + \frac{\nabla^2\vartheta}{3} - \mathcal{H}(h-\vartheta') = \psi$	$h_\mathrm{D} + \frac{\nabla^2\int^\tau v}{3} - \mathcal{H}(h - v)$	0
$\tilde{\vartheta}$	0	$\vartheta - \int^\tau v$	0
$\delta\tilde{\varphi}$	$\delta\varphi + \bar{\varphi}'(h - \vartheta')$	$\delta\varphi + \bar{\varphi}'(h - v)$	$\delta\varphi + \frac{\bar{\varphi}'}{\mathcal{H}}\left(h_\mathrm{D} + \frac{\nabla^2\vartheta}{3}\right) = \mathcal{Q}_\varphi$
$\tilde{v}$	$v - \vartheta'$	0	$v - \vartheta'$

Let us illustrate the procedure by considering the Einstein equation originating from the traceless (or non-diagonal) scalar part of G_{ij}, given in Eq. (3.94),

$$h_0 + (\partial_\tau + 2\mathcal{H})(h - \vartheta') - \left(h_\mathrm{D} + \frac{\nabla^2\vartheta}{3}\right) \overset{(3.94)}{=} -8\pi G a^2 \Pi. \qquad (4.55)$$

The goal is to turn this into an equation for the Bardeen potentials, ϕ and ψ.

First, from Eq. (4.30), we can solve for $h_0 = \phi - (\partial_\tau + \mathcal{H})(h - \vartheta')$. Second, from Eq. (4.31), we can solve for $h_\mathrm{D} + \frac{\nabla^2\vartheta}{3} = \psi + \mathcal{H}(h - \vartheta')$. Inserting these into Eq. (4.55), all appearances of $h - \vartheta'$ cancel, and we simply get

$$\phi - \psi \underset{(4.30),\,(4.31)}{\overset{(4.55)}{=}} -8\pi G a^2 \Pi. \qquad (4.56)$$

As discussed below Eq. (4.38), the anisotropic stress is gauge invariant. Therefore, this is a gauge-invariant relation between gauge-invariant quantities.

Actually, we can be a bit more precise concerning the anisotropic stress. The explicit expression for its shear viscous part, $\Pi \supset \Sigma$ where Σ is from Eq. (3.134), shows that it contains $v - \vartheta' = -(\psi + \mathcal{R}_v)/\mathcal{H}$, where $\mathcal{R}_v$ is from Eq. (4.61). Therefore Eq. (4.56) represents a relationship between the gauge-invariant ϕ, ψ and $\mathcal{R}_v$.

Let us contrast Eq. (4.56) with a direct derivation of the corresponding Einstein equation in the Newtonian gauge, cf. Eq. (4.94). While the result is the same, if we had only derived Eq. (4.94), we would have no internal crosscheck on the correctness of the computation. Instead, when we derived the same relation from Eq. (4.55), a strong crosscheck was offered by the exact cancellation of $h - \vartheta'$.

Another similar consideration concerns the perturbed Ricci scalar, δR. A general expression is given in Eq. (3.60), and a Newtonian result is derived in Eq. (4.85). If we set $h, \vartheta \to 0$ in Eq. (3.60), Eq. (4.85) is reproduced. However, Eq. (3.60) contains more information. If we substitute $h_0 = \phi - (\partial_\tau + \mathcal{H})(h - \vartheta')$ and $h_{\mathrm{D}} + \frac{\nabla^2 \vartheta}{3} = \psi + \mathcal{H}(h - \vartheta')$, then almost all appearances of $(h - \vartheta')$ cancel, however we are left over with

$$\delta R\big|_{(3.60)} = \delta R^{\mathrm{N}}\big|_{(4.85)} \underbrace{- \frac{6}{a^2}(\mathcal{H}'' - 2\mathcal{H}^3)(h - \vartheta')}_{-6(a''/a^3)' \underset{(1.28)}{=} -\bar{R}'} . \tag{4.57}$$

If we insert the gauge transformations from Eqs. (4.50) and (4.52), and make use of the gauge independence of ϕ and ψ, we obtain

$$\delta \tilde{R}\big|_{(3.60)} = \delta R^{\mathrm{N}}\big|_{(4.85)} - \bar{R}'(\tilde{h} - \tilde{\vartheta}') \overset{(4.50),\,(4.52)}{\underset{(4.57)}{=}} \delta R\big|_{(3.60)} - \bar{R}'\xi^0 . \tag{4.58}$$

So again we find a strong crosscheck, namely that the Ricci scalar transforms in gauge transformations as a scalar should, according to Eq. (4.4).

The procedure that led to Eq. (4.56) can be repeated for the three other Einstein equations, as well as the scalar field equation and energy-momentum conservation equations. However, we then need to introduce further gauge-invariant variables. A convenient set is offered by so-called *curvature perturbations*, $\mathcal{R}_Q$, defined from the Ricci scalar of a constant-τ slice as

$$\delta R_\tau \overset{(3.61)}{=} \frac{4}{a^2}\nabla^2\left(h_{\mathrm{D}} + \frac{\nabla^2\vartheta}{3}\right) \equiv -\frac{4}{a^2}\nabla^2\left(\mathcal{R}_Q + \mathcal{H}\frac{\delta Q}{\bar{Q}'}\right) , \tag{4.59}$$

where Q is one of the scalar quantities entering the Einstein equations via the energy-momentum tensor, and $\mathcal{R}_Q$ is gauge invariant. If we refer to R_τ as spatial curvature, the nomenclature of referring to $\mathcal{R}_Q$ as a curvature perturbation is a bit imprecise, however it has established itself. In any case, examples of various 'flavours' of curvature perturbations are

$$\boxed{\begin{aligned}
\mathcal{R}_\varphi &\equiv -\left(h_{\mathrm{D}} + \frac{\nabla^2\vartheta}{3}\right) - \mathcal{H}\frac{\delta\varphi}{\bar{\varphi}'} , && (4.60) \\[2ex]
\mathcal{R}_v &\equiv -\left(h_{\mathrm{D}} + \frac{\nabla^2\vartheta}{3}\right) + \mathcal{H}(h - v) , && (4.61) \\[2ex]
\mathcal{R}_T &\equiv -\left(h_{\mathrm{D}} + \frac{\nabla^2\vartheta}{3}\right) - \mathcal{H}\frac{\delta T}{\bar{T}'} , && (4.62) \\[2ex]
\mathcal{R}_e &\equiv -\left(h_{\mathrm{D}} + \frac{\nabla^2\vartheta}{3}\right) - \mathcal{H}\frac{\delta e}{\bar{e}'} . && (4.63)
\end{aligned}}$$

As follows from Eqs. (4.49)–(4.54), these are gauge invariant, and for $\mathcal{R}_\varphi$ we also demonstrate this explicitly in Eq. (4.66). We note that if δe can be expressed as a function of the other dynamical variables, like $\delta\varphi$ and δT, then $\mathcal{R}_e$ can be represented as a linear combination of $\mathcal{R}_\varphi$ and $\mathcal{R}_T$. To streamline the notation, we set $\bar{T} \to T$.

A further class of gauge-invariant quantities is obtained by taking differences of Eqs. (4.60)–(4.63). In particular, a *relative energy density perturbation* is defined as

$$\Delta \;\equiv\; \frac{\delta e}{\bar{e}} + \frac{\bar{e}'}{\bar{e}}(h - v) \;=\; \frac{\bar{e}'}{\bar{e}\mathcal{H}}(\mathcal{R}_v - \mathcal{R}_e)\,. \tag{4.64}$$

We also note that in the classic literature, following Ref. [4], the *curvature perturbation related to energy density*, $\mathcal{R}_e$, is often denoted by

$$\zeta \;\equiv\; \mathcal{R}_e \;\overset{(4.28)}{\underset{(4.31)}{=}}\; -\psi - \mathcal{H}\frac{\delta e^{\mathrm{N}}}{\bar{e}'}\,, \tag{4.65}$$

where we also indicated the frequently appearing Newtonian gauge representation.

For us, the task now is to re-organize the Einstein equations so that the dynamical variables are ϕ, ψ, $\mathcal{R}_\varphi$, $\mathcal{R}_v$ and $\mathcal{R}_T$. Given that there are five variables, we need five independent equations; one was already given in Eq. (4.56). We will return to the remaining four in the later chapters (cf. Eqs. (7.81), (7.90), (7.97), (7.102)).

Let us end this section by demonstrating explicitly the gauge independence of $\mathcal{R}_\varphi$ from Eq. (4.60) and Δ from Eq. (4.64). For the former, the gauge transformation properties in Eqs. (4.51)–(4.53) yield

$$\tilde{\mathcal{R}}_\varphi \;\overset{(4.60)}{\underset{(4.51)\text{-}(4.53)}{=}}\; \mathcal{R}_\varphi + \overbrace{\frac{1}{3}\nabla^2\xi - \mathcal{H}\xi^0}^{\text{from }\tilde{h}_{\mathrm{D}}} - \overbrace{\frac{1}{3}\nabla^2\xi}^{\text{from }\tilde{\vartheta}} - \overbrace{\mathcal{H}(-\xi^0)}^{\text{from }\delta\tilde{\varphi}} \;=\; \mathcal{R}_\varphi \;\overset{(4.47)}{=}\; -\frac{\mathcal{H}}{\bar{\varphi}'}\delta\varphi^{\mathrm{f}}\,. \tag{4.66}$$

For the latter, Eqs. (4.36), (4.50) and (4.54) imply

$$\tilde{\Delta} \;\overset{(4.64),\,(4.36)}{\underset{(4.50),\,(4.54)}{=}}\; \underbrace{\frac{\delta e}{\bar{e}} - \frac{\bar{e}'}{\bar{e}}\xi^0}_{\text{from }\delta\tilde{e}} + \frac{\bar{e}'}{\bar{e}}\underbrace{(h - v + \xi^0)}_{\text{from }\tilde{h}-\tilde{v}} \;=\; \Delta \;\overset{(4.45)}{=}\; \frac{\delta e^{\mathrm{c}}}{\bar{e}}\,. \tag{4.67}$$

4.4   Deriving the Einstein Tensor in the Newtonian Gauge

If we choose the conformal Newtonian gauge from Eq. (4.35), implying

$$g^{\mathrm{N}}_{\mu\nu} \;\overset{(3.40)}{\underset{(3.34),\,(4.35)}{=}}\; a^2(\tau) \begin{pmatrix} -1 - 2\phi & -h^{\mathrm{v}}_j \\ -h^{\mathrm{v}}_i & (1 - 2\psi)\,\delta_{ij} + 2\vartheta^{\mathrm{t}}_{ij} \end{pmatrix}, \tag{4.68}$$

$$g_{\mathrm{N}}^{\mu\nu} \;\overset{(3.41)}{=}\; \frac{1}{a^2(\tau)} \begin{pmatrix} -1 + 2\phi & -h^{\mathrm{v}}_j \\ -h^{\mathrm{v}}_i & (1 + 2\psi)\,\delta_{ij} - 2\vartheta^{\mathrm{t}}_{ij} \end{pmatrix} + \mathcal{O}(\delta^2)\,, \tag{4.69}$$

then the Einstein tensor can be derived by hand. Here we show how this goes. To streamline the notation, the superscript $(...)^N$ is suppressed until Eq. (4.85).

Starting with the perturbed Christoffel symbols from Eq. (3.48), *viz.*

$$\delta\Gamma^{\rho}_{\mu\nu} \overset{(3.48)}{=} \frac{1}{2}\bar{g}^{\rho\sigma}(\delta g_{\sigma\mu,\nu} + \delta g_{\sigma\nu,\mu} - \delta g_{\mu\nu,\sigma}) + \frac{1}{2}\delta g^{\rho\sigma}(\bar{g}_{\sigma\mu,\nu} + \bar{g}_{\sigma\nu,\mu} - \bar{g}_{\mu\nu,\sigma}),$$
(4.70)

they become

$$\delta\Gamma^{0}_{00} = \frac{a^{-2}}{2}\Big\{ -\overbrace{(-2a^{2}\phi)'}^{\delta g_{00,0}} + 2\phi \overbrace{[2(-a^{2})' - (-a^{2})']}^{\bar{g}_{00,0}} \Big\} = \cancel{2\mathcal{H}\phi} + \phi' - \cancel{2\mathcal{H}\phi} = \phi',$$
(4.71)

$$\delta\Gamma^{0}_{0i} = \frac{a^{-2}}{2}\Big[-\overbrace{(-2a^{2}\phi)_{,i}}^{\delta g_{00,i}} - h_{j}\overbrace{(a^{2}\delta_{ij})'}^{\bar{g}_{ij,0}} \Big] = \phi_{,i} - \mathcal{H}h_{i},$$
(4.72)

$$\delta\Gamma^{0}_{ij} = \frac{a^{-2}}{2}\Big\{ -[\overbrace{(-a^{2}h_{i})_{,j}}^{\delta g_{0i,j}} + \overbrace{(-a^{2}h_{j})_{,i}}^{\delta g_{0j,i}} - \overbrace{(-2a^{2}\psi\,\delta_{ij} + 2a^{2}\vartheta_{ij})'}^{\delta g_{ij,0}}] + 2\phi\overbrace{[-(a^{2}\delta_{ij})']}^{-\bar{g}_{ij,0}} \Big\}$$
$$= \frac{1}{2}(h_{i,j} + h_{j,i}) - \delta_{ij}[\psi' + 2\mathcal{H}(\phi + \psi)] + \vartheta'_{ij} + 2\mathcal{H}\vartheta_{ij},$$
(4.73)

$$\delta\Gamma^{i}_{00} = \frac{a^{-2}}{2}\Big\{ [\,2\overbrace{(-a^{2}h_{i})'}^{\delta g_{i0,0}} - \overbrace{(-2a^{2}\phi)_{,i}}^{\delta g_{00,i}}\,] - h_{i}\overbrace{(-a^{2})'}^{\bar{g}_{00,0}} \Big\} = \phi_{,i} - \mathcal{H}h_{i} - h'_{i},$$
(4.74)

$$\delta\Gamma^{i}_{0j} = \frac{a^{-2}}{2}\Big\{ [\,\overbrace{(-a^{2}h_{i})_{,j}}^{\delta g_{i0,j}} + \overbrace{(-2a^{2}\psi\,\delta_{ij} + 2a^{2}\vartheta_{ij})'}^{\delta g_{ij,0}} - \overbrace{(-a^{2}h_{j})_{,i}}^{\delta g_{0j,i}}\,] + 2(\psi\,\delta_{ik} - \vartheta_{ik})\overbrace{(a^{2}\delta_{kj})'}^{\bar{g}_{kj,0}} \Big\}$$
$$= \frac{1}{2}(h_{j,i} - h_{i,j}) + \delta_{ij}(-\psi' - \cancel{2\mathcal{H}\psi} + \cancel{2\mathcal{H}\psi}) + \vartheta'_{ij} + \cancel{2\mathcal{H}\vartheta_{ij}} - \cancel{2\mathcal{H}\vartheta_{ij}}$$
$$= \frac{1}{2}(h_{j,i} - h_{i,j}) - \delta_{ij}\,\psi' + \vartheta'_{ij},$$
(4.75)

$$\delta\Gamma^{i}_{jk} = \frac{1}{2}\Big[\overbrace{2(\vartheta_{ij} - \psi\,\delta_{ij})_{,k}}^{a^{-2}\delta g_{ij,k}} + \overbrace{2(\vartheta_{ik} - \psi\,\delta_{ik})_{,j}}^{a^{-2}\delta g_{ik,j}} - \overbrace{2(\vartheta_{jk} - \psi\,\delta_{jk})_{,i}}^{a^{-2}\delta g_{jk,i}} \Big] - \frac{a^{-2}}{2}h_{i}\,[-\overbrace{(a^{2}\delta_{jk})'}^{\bar{g}_{jk,0}}]$$
$$= -\delta_{ij}\,\psi_{,k} - \delta_{ik}\,\psi_{,j} + \delta_{jk}\,\psi_{,i} + \vartheta_{ij,k} + \vartheta_{ik,j} - \vartheta_{jk,i} + \mathcal{H}h_{i}\,\delta_{jk}.$$
(4.76)

The unperturbed Christoffel symbols can be found in Eq. (3.49). The resulting perturbed Ricci tensor, from Eq. (3.50), is

$$\delta R_{00} \overset{(3.50)}{=} \delta\Gamma^{\alpha}_{00,\alpha} - \delta\Gamma^{\alpha}_{0\alpha,0} + \bar{\Gamma}^{\beta}_{00}\delta\Gamma^{\alpha}_{\beta\alpha} + \bar{\Gamma}^{\alpha}_{\beta\alpha}\delta\Gamma^{\beta}_{00} - \bar{\Gamma}^{\beta}_{0\alpha}\delta\Gamma^{\alpha}_{0\beta} - \bar{\Gamma}^{\alpha}_{0\beta}\delta\Gamma^{\beta}_{0\alpha}$$

$$= \overbrace{\phi''}^{\delta\Gamma^0_{00,0}} + \overbrace{\nabla^2\phi}^{\delta\Gamma^i_{00,i}} - \overbrace{\phi''}^{-\delta\Gamma^0_{00,0}} + \overbrace{3\psi''}^{-\delta\Gamma^i_{0i,0}} + \overbrace{\mathcal{H}\delta\Gamma^\alpha_{0\alpha} + 4\mathcal{H}\delta\Gamma^0_{00} - 2\mathcal{H}\delta\Gamma^0_{00} - 2\mathcal{H}\delta\Gamma^i_{0i}}^{\text{from (3.49)}}$$

$$= 3\psi'' + \nabla^2\phi + 3\mathcal{H}(\phi' + \psi') \,, \tag{4.77}$$

$$\delta R_{0i} \overset{(3.50)}{=} \delta\Gamma^\alpha_{0i,\alpha} - \delta\Gamma^\alpha_{0\alpha,i} + \bar\Gamma^\beta_{0i}\delta\Gamma^\alpha_{\beta\alpha} + \bar\Gamma^\alpha_{\beta\alpha}\delta\Gamma^\beta_{0i} - \bar\Gamma^\beta_{0\alpha}\delta\Gamma^\alpha_{i\beta} - \bar\Gamma^\alpha_{i\beta}\delta\Gamma^\beta_{0\alpha}$$

$$= \overbrace{\phi'_{,i} - \mathcal{H}'h_i - \mathcal{H}h'_i}^{\delta\Gamma^0_{0i,0}} + \overbrace{\frac{1}{2}\nabla^2 h_i - \psi'_{,i}}^{\delta\Gamma^j_{0i,j}} \overbrace{-\phi'_{,i}}^{-\delta\Gamma^0_{00,i}} + \overbrace{3\psi'_{,i}}^{-\delta\Gamma^j_{0j,i}}$$

$$+ \overbrace{\mathcal{H}\delta\Gamma^\alpha_{i\alpha} + 4\mathcal{H}\delta\Gamma^0_{0i} - \mathcal{H}\delta\Gamma^0_{i0} - \mathcal{H}\delta\Gamma^j_{ij} - \mathcal{H}\delta_{ij}\delta\Gamma^0_{00} - \mathcal{H}\delta_{ij}\delta\Gamma^0_{0j}}^{\text{from (3.49)}}$$

$$= -\mathcal{H}'h_i - \mathcal{H}h'_i + \frac{1}{2}\nabla^2 h_i + 2\psi'_{,i} + 3\mathcal{H}(\phi_{,i} - \mathcal{H}h_i) - \mathcal{H}(\phi_{,i} - \mathcal{H}h_i - \psi'_{,i})$$

$$= 2(\psi' + \mathcal{H}\phi)_{,i} + \frac{1}{2}\nabla^2 h_i - (\mathcal{H}' + 2\mathcal{H}^2)h_i \,, \tag{4.78}$$

$$\delta R_{ij} \overset{(3.50)}{=} \delta\Gamma^\alpha_{ij,\alpha} - \delta\Gamma^\alpha_{i\alpha,j} + \bar\Gamma^\beta_{ij}\delta\Gamma^\alpha_{\beta\alpha} + \bar\Gamma^\alpha_{\beta\alpha}\delta\Gamma^\beta_{ij} - \bar\Gamma^\beta_{i\alpha}\delta\Gamma^\alpha_{j\beta} - \bar\Gamma^\alpha_{j\beta}\delta\Gamma^\beta_{i\alpha}$$

$$= \overbrace{\frac{1}{2}(h_{i,j} + h_{j,i})' - \delta_{ij}\psi'' + \vartheta''_{ij} - 2\mathcal{H}'[(\phi + \psi)\delta_{ij} - \vartheta_{ij}] - 2\mathcal{H}[(\phi + \psi)\delta_{ij} - \vartheta_{ij}]'}^{\delta\Gamma^0_{ij,0}}$$

$$\overbrace{-2\psi_{,ij} + \delta_{ij}\nabla^2\psi - \nabla^2\vartheta_{ij}}^{\delta\Gamma^k_{ij,k}} \overbrace{-\phi_{,ij} + \mathcal{H}h_{i,j}}^{-\delta\Gamma^0_{i0,j}} + \overbrace{3\psi_{,ij} - \mathcal{H}h_{i,j}}^{-\delta\Gamma^k_{ik,j}}$$

$$+ \overbrace{\mathcal{H}\delta_{ij}\delta\Gamma^\alpha_{0\alpha} + 4\mathcal{H}\delta\Gamma^0_{ij} - \mathcal{H}(2\delta\Gamma^0_{ij} + \delta\Gamma^i_{0j} + \delta\Gamma^j_{0i})}^{\text{from (3.49)}}$$

$$= \frac{1}{2}(h_{i,j} + h_{j,i})' - \delta_{ij}\psi'' + \vartheta''_{ij} - 2\mathcal{H}'[(\phi + \psi)\delta_{ij} - \vartheta_{ij}] - 2\mathcal{H}[(\phi + \psi)\delta_{ij} - \vartheta_{ij}]'$$

$$+ \psi_{,ij} + \delta_{ij}\nabla^2\psi - \nabla^2\vartheta_{ij} - \phi_{,ij} + \mathcal{H}\delta_{ij}(\phi - 3\psi)' + \mathcal{H}(h_{i,j} + h_{j,i})$$

$$- 2\mathcal{H}(\delta_{ij}\psi - \vartheta_{ij})' - 4\mathcal{H}^2[(\phi + \psi)\delta_{ij} - \vartheta_{ij}] + 2\mathcal{H}(\delta_{ij}\psi - \vartheta_{ij})'$$

$$= -\delta_{ij}\left[\psi'' + \mathcal{H}(\phi' + 5\psi') + 2(\mathcal{H}' + 2\mathcal{H}^2)(\phi + \psi) - \nabla^2\psi\right] + (\psi - \phi)_{,ij}$$

$$+ \frac{1}{2}(h'_{i,j} + h'_{j,i}) + \mathcal{H}(h_{i,j} + h_{j,i})$$

$$+ \vartheta''_{ij} + 2\mathcal{H}\vartheta'_{ij} - \nabla^2\vartheta_{ij} + 2(\mathcal{H}' + 2\mathcal{H}^2)\vartheta_{ij} \,. \tag{4.79}$$

The unperturbed version $\bar{R}_{\mu\nu}$ is given in Eq. (3.51). Let us now raise one index of the perturbed Ricci tensor as in Eq. (3.52),

$$\delta R^{\mu}{}_{\nu} \overset{(3.52)}{=} \bar{g}^{\mu\rho}\delta R_{\rho\nu} + \delta g^{\mu\rho}\bar{R}_{\rho\nu}\,. \tag{4.80}$$

We obtain

$$\delta R^{0}{}_{0} = -a^{-2}\Big[\;\overbrace{3\psi'' + \nabla^2\phi + 3\mathcal{H}(\phi' + \psi')}^{\text{from }\bar{g}^{00}\,\delta R_{00}} + \overbrace{6\mathcal{H}'\phi}^{\text{from }\delta g^{00}\,\bar{R}_{00}}\;\Big]\,, \tag{4.81}$$

$$\delta R^{0}{}_{i} = -a^{-2}\Big[\;\overbrace{2\psi'_{,i} + 2\mathcal{H}\phi_{,i} + \frac{1}{2}\nabla^2 h_i - \cancel{(\mathcal{H}' + 2\mathcal{H}^2)h_i}}^{\text{from }\bar{g}^{00}\,\delta R_{0i}} + \overbrace{\cancel{(\mathcal{H}' + 2\mathcal{H}^2)h_i}}^{\text{from }\delta g^{0j}\,\bar{R}_{ji}}\;\Big]$$

$$= -a^{-2}\Big[2(\psi' + \mathcal{H}\phi)_{,i} + \frac{1}{2}\nabla^2 h_i\Big]\,, \tag{4.82}$$

$$\delta R^{i}{}_{0} = a^{-2}\Big[\;\overbrace{2\psi'_{,i} + 2\mathcal{H}\phi_{,i} + \frac{1}{2}\nabla^2 h_i - (\mathcal{H}' + 2\mathcal{H}^2)h_i}^{\text{from }\bar{g}^{ij}\,\delta R_{j0}} + \overbrace{3\mathcal{H}'h_i}^{\text{from }\delta g^{i0}\,\bar{R}_{00}}\;\Big]$$

$$= a^{-2}\Big[2(\psi' + \mathcal{H}\phi)_{,i} + \frac{1}{2}\nabla^2 h_i + 2(\mathcal{H}' - \mathcal{H}^2)h_i\Big]\,, \tag{4.83}$$

$$\delta R^{i}{}_{j} = a^{-2}\Big[\;\overbrace{\delta R_{ij}}^{\text{from }\bar{g}^{ik}\,\delta R_{kj}} + \overbrace{2(\mathcal{H}' + 2\mathcal{H}^2)(\psi\delta_{ij} - \vartheta_{ij})}^{\text{from }\delta g^{ik}\,\bar{R}_{kj}}\;\Big]$$

$$= -a^{-2}\big[\psi'' + \mathcal{H}(\phi' + 5\psi') + 2(\mathcal{H}' + 2\mathcal{H}^2)\phi - \nabla^2\psi\big]\delta_{ij}$$

$$\quad + a^{-2}\Big[(\psi - \phi)_{,ij} + \frac{1}{2}(h'_{i,j} + h'_{j,i}) + \mathcal{H}(h_{i,j} + h_{j,i})$$

$$\quad + \vartheta''_{ij} + 2\mathcal{H}\vartheta'_{ij} - \nabla^2\vartheta_{ij}\Big]\,. \tag{4.84}$$

The perturbed Ricci scalar follows from Eq. (3.53),

$$\frac{1}{2}\delta R^{\mathrm{N}} \overset{(3.53)}{=} \frac{1}{2}\delta R^{\mu}{}_{\mu} = \frac{1}{2}(\delta R^{0}{}_{0} + \delta R^{i}{}_{i})$$

$$\overset{(4.81)}{\underset{(4.84)}{=}} \frac{a^{-2}}{2}\Big[\;\overbrace{-3\psi'' - \nabla^2\phi - 3\mathcal{H}(\phi' + \psi') - 6\mathcal{H}'\phi}^{\text{from }\delta R^{0}{}_{0}}$$

$$\left. \underbrace{-3\psi'' - 3\mathcal{H}(\phi' + 5\psi') - 6(\mathcal{H}' + 2\mathcal{H}^2)\phi + 3\nabla^2\psi + \nabla^2(\psi - \phi)}_{\text{from } \delta R^i{}_i} \right]$$

$$= -a^{-2}\left[3\psi'' + \nabla^2(\phi - 2\psi) + 3\mathcal{H}(\phi' + 3\psi') + 6(\mathcal{H}' + \mathcal{H}^2)\phi \right]. \tag{4.85}$$

We now have all ingredients to evaluate the components of Eq. (3.55),

$$\delta G^{\mathrm{N}}_{\mu\nu} \overset{(3.55)}{=} \delta R^{\mathrm{N}}_{\mu\nu} - \frac{1}{2}\left(\bar{g}_{\mu\nu}\, \delta R^{\mathrm{N}} + \delta g^{\mathrm{N}}_{\mu\nu}\, \bar{R} \right), \tag{4.86}$$

and distinguish the parts involving scalar (s), vector (v) or tensor (t) metric perturbations. From Eqs. (1.28), (3.40), (4.77) and (4.85),

$$\delta G^{\mathrm{N}}_{00} \overset{(4.86)}{=} \overbrace{3\psi'' + \nabla^2\phi + 3\mathcal{H}(\phi' + \psi')}^{\text{from } \delta R^{\mathrm{N}}_{00} \text{ in (4.77)}} + \overbrace{6(\mathcal{H}' + \mathcal{H}^2)}^{\text{from } \bar{R} \text{ in (1.28)}}\ \overbrace{\phi}^{\text{from } \delta g^{\mathrm{N}}_{00} \text{ in (4.68)}}$$
$$- \underbrace{\left[3\psi'' + \nabla^2(\phi - 2\psi) + 3\mathcal{H}(\phi' + 3\psi') + 6(\mathcal{H}' + \mathcal{H}^2)\phi \right]}_{\text{from } \delta R^{\mathrm{N}} \text{ in (4.85)}}$$

$$= \underbrace{2(\nabla^2 - 3\mathcal{H}\partial_\tau)\psi}_{\text{s}}, \tag{4.87}$$

which agrees with the Newtonian gauge limit of Eq. (3.56). Inserting instead δR_{0i} from Eq. (4.78), we find

$$\delta G^{\mathrm{N}}_{0i} \overset{(4.86)}{=} \overbrace{2(\psi' + \mathcal{H}\phi)_{,i} + \frac{1}{2}\nabla^2 h^{\mathrm{N}}_i - (\mathcal{H}' + 2\mathcal{H}^2)h^{\mathrm{N}}_i}^{\text{from } \delta R^{\mathrm{N}}_{0i} \text{ in (4.78)}} + \overbrace{3(\mathcal{H}' + \mathcal{H}^2)}^{\text{from } \bar{R} \text{ in (1.28)}}\ \overbrace{h^{\mathrm{N}}_i}^{\text{from } \delta g^{\mathrm{N}}_{0i} \text{ in (4.68)}}$$

$$= \underbrace{2(\psi' + \mathcal{H}\phi)_{,i}}_{\text{s}} + \underbrace{\frac{1}{2}\nabla^2 h^{\mathrm{N}}_i + (2\mathcal{H}' + \mathcal{H}^2)h^{\mathrm{N}}_i}_{\text{v}}, \tag{4.88}$$

which agrees with the Newtonian gauge limit of Eq. (3.57). Finally, inserting δR_{ij} from Eq. (4.79), we obtain

$$\delta G^{\mathrm{N}}_{ij} \overset{(4.86)}{=} \overbrace{-\delta_{ij}\left[\psi'' + \mathcal{H}(\phi' + 5\psi') + 2(\mathcal{H}' + 2\mathcal{H}^2)(\phi + \psi) - \nabla^2\psi \right] + (\psi - \phi)_{,ij}}^{\text{from } \delta R^{\mathrm{N}}_{ij} \text{ in (4.79)}}$$

$$+ \overbrace{\frac{1}{2}(\partial_\tau + 2\mathcal{H})(h^{\mathrm{N}}_{i,j} + h^{\mathrm{N}}_{j,i}) + \left[\partial_\tau^2 + 2\mathcal{H}\partial_\tau - \nabla^2 + 2(\mathcal{H}' + 2\mathcal{H}^2)\right]\vartheta^{\mathrm{N}}_{ij}}^{\text{from } \delta R^{\mathrm{N}}_{ij} \text{ in (4.79)}}$$

$$+ \underbrace{\delta_{ij}\left[3\psi'' + \nabla^2(\phi - 2\psi) + 3\mathcal{H}(\phi' + 3\psi') + 6(\mathcal{H}' + \mathcal{H}^2)\phi \right]}_{\text{from } \delta R^{\mathrm{N}} \text{ in (4.85)}}$$

$$+ \underbrace{6(\mathcal{H}' + \mathcal{H}^2)}_{\text{from } \bar{R} \text{ in (1.28)}} \underbrace{(\psi\,\delta_{ij} - \vartheta^{\mathrm{N}}_{ij})}_{\text{from } \delta g^{\mathrm{N}}_{ij} \text{ in (4.68)}}$$

$$
\begin{aligned}
= {}& \delta_{ij} \overbrace{\left[2\psi'' + 2\mathcal{H}(\phi' + 2\psi') + 2(2\mathcal{H}' + \mathcal{H}^2)(\phi + \psi) + \nabla^2(\phi - \psi) \right]}^{s} + (\psi - \phi)_{,ij} \\
& + \underbrace{\frac{1}{2}(\partial_\tau + 2\mathcal{H})(h^{\mathrm{N}}_{i,j} + h^{\mathrm{N}}_{j,i})}_{v} + \underbrace{\left[\partial_\tau^2 + 2\mathcal{H}\partial_\tau - \nabla^2 - 2(2\mathcal{H}' + \mathcal{H}^2) \right] \vartheta^{\mathrm{N}}_{ij}}_{t} .
\end{aligned}
\tag{4.89}
$$

This agrees with the Newtonian gauge limit of Eq. (3.58).

4.5   Einstein Equations for a Fluid in the Newtonian Gauge

In this section we complement $\delta G^{\mathrm{N}}_{\mu\nu}$ from Sect. 4.4 by equating it with $8\pi G \delta T^{\mathrm{N}}_{\mu\nu}$, and thereby setting up the Einstein equations. To simplify the task, we consider a fluid, from Eqs. (3.74) and (3.75), omitting the scalar field contribution from Eqs. (3.83)–(3.85). In addition, we restrict ourselves to scalar perturbations.

If we equate $\delta G^{\mathrm{N}}_{00}$ from Eq. (4.87) with $8\pi G \delta T^{\mathrm{N}}_{00} = 8\pi G a^2 (\delta e^{\mathrm{N}} + 2\bar{e}\phi)$ from Eq. (3.74), and make use of the background identity $8\pi G a^2 \bar{e} = 3\mathcal{H}^2$ from Eq. (3.88), we get

$$
\nabla^2 \psi - 3\mathcal{H}(\psi' + \mathcal{H}\phi) = 4\pi G a^2 \delta e^{\mathrm{N}} ,
\tag{4.90}
$$

which is a gauge-fixed version of Eq. (3.92). The scalar part of $\delta G^{\mathrm{N}}_{0i}$ from Eq. (4.88) and the gauge-fixed value $\delta T^{\mathrm{N}}_{0i} = a^2(\bar{e} + \bar{p})v^{\mathrm{N}}_{,i}$ from Eq. (3.74), where the scalar part of the velocity was obtained from Eq. (3.76), can be integrated into

$$
\boxed{ \psi' + \mathcal{H}\phi = 4\pi G a^2 (\bar{e} + \bar{p})v^{\mathrm{N}} , }
\tag{4.91}
$$

which is a gauge-fixed version of Eq. (3.93). If we insert Eq. (4.91) into Eq. (4.90), and leave only the Laplacian on the left-hand side, we find the constraint equation

$$
\nabla^2 \psi = 4\pi G a^2 \underbrace{\left[\delta e^{\mathrm{N}} + 3\mathcal{H}(\bar{e} + \bar{p})v^{\mathrm{N}} \right]}_{\text{via (1.42): } \bar{e}\,\Delta \text{ from (4.64)}} \qquad \Leftrightarrow \qquad \boxed{ \nabla^2 \psi = 4\pi G a^2 \bar{e}\,\Delta . }
\tag{4.92}
$$

Equation (4.92) has the appearance of Newtonian gravity, with ψ playing the role of the gravitational potential.

The remaining Einstein equations come from the ij-components. The traceless scalar parts of Eqs. (4.89) and (3.75) (via Eq. (3.77)) yield

$$
\left(\partial_i \partial_j - \frac{1}{3}\delta_{ij} \nabla^2 \right) (\psi - \phi) = 8\pi G a^2 \left(\partial_i \partial_j - \frac{1}{3}\delta_{ij} \nabla^2 \right) \Pi .
\tag{4.93}
$$

Integrating with vanishing boundary conditions, we obtain a second constraint equation,

$$
\boxed{ \psi - \phi = 8\pi G a^2 \Pi . }
\tag{4.94}
$$

This corresponds to Eq. (3.94). Finally, after making use of the background identity $-8\pi G a^2 \bar{p} = 2\mathcal{H}' + \mathcal{H}^2$ from Eq. (3.88), the spatial trace parts of Eqs. (4.89) and (3.74), $[\delta T^{\scriptscriptstyle\mathrm{N}}_{ij}]_{\mathrm{pf}} \supset a^2 (\delta p^{\scriptscriptstyle\mathrm{N}} - 2\bar{p}\psi)\delta_{ij}$, give the evolution equation

$$\psi'' + \mathcal{H}(\phi' + 2\psi') + (2\mathcal{H}' + \mathcal{H}^2)\phi + \frac{1}{3}\nabla^2(\phi - \psi) = 4\pi G a^2 \delta p^{\scriptscriptstyle\mathrm{N}}\,. \tag{4.95}$$

This contains equivalent time-evolution information as Eq. (3.95).

For completeness, we also write down the Newtonian gauge versions of the energy-momentum conservation laws from Eqs. (3.119) and (3.121), obtaining

$$\nu = 0: \qquad \delta e^{\scriptscriptstyle\mathrm{N}\prime} + 3\mathcal{H}(\delta e^{\scriptscriptstyle\mathrm{N}} + \delta p^{\scriptscriptstyle\mathrm{N}}) - (\bar{e} + \bar{p})(3\psi' + \nabla^2 v^{\scriptscriptstyle\mathrm{N}}) = 0\,, \tag{4.96}$$

$$\nu = i: \qquad -\delta p^{\scriptscriptstyle\mathrm{N}} + (\bar{e} + \bar{p})(4\mathcal{H}v^{\scriptscriptstyle\mathrm{N}} - \phi) + \left[(\bar{e} + \bar{p})v^{\scriptscriptstyle\mathrm{N}}\right]' - \frac{2}{3}\nabla^2\Pi = 0\,. \tag{4.97}$$

References

1. Malik, K.A., Matravers, D.R.: A concise introduction to perturbation theory in cosmology. Class. Quant. Grav. **25**, 193001 (2008). https://doi.org/10.1088/0264-9381/25/19/193001, 0804.3276
2. Bardeen, J.M.: Gauge-invariant cosmological perturbations. Phys. Rev. D **22**, 1882 (1980). https://doi.org/10.1103/PhysRevD.22.1882
3. Ma, C.-P., Bertschinger, E.: Cosmological Perturbation Theory in the Synchronous and Conformal Newtonian Gauges. Astrophys. J. **455**, 7 (1995). https://doi.org/10.1086/176550, astro-ph/9506072
4. Bardeen, J.M., Steinhardt, P.J., Turner, M.S.: Spontaneous creation of almost scale-free density perturbations in an inflationary universe. Phys. Rev. D **28**, 679 (1983). https://doi.org/10.1103/PhysRevD.28.679

Part II

The Inflationary Paradigm — Generation of Cosmological Perturbations

Idealized Initial Conditions and Solution for Curvature Perturbations

Abstract

Combining Einstein equations and a scalar field evolution equation, obtained in Chap. 3, we derive a Mukhanov-Sasaki equation for a gauge-invariant curvature perturbation, defined in Chap. 4. Proceeding to its solution, we review the case in which the inflationary period corresponds to an almost de Sitter space-time. We show, analytically, how the amplitude and the phase of the perturbations can be fixed in the distant-past vacuum, and how the perturbations evolve as a function of conformal time, until they cross outside of the Hubble horizon. We demonstrate that this produces an almost scale-invariant spectrum of curvature perturbations, which can subsequently be compared with observational data, as they were reviewed in Chap. 2.

Keywords

Mukhanov-Sasaki equation · Bunch-Davies vacuum · Ultra-slow-roll regime · Slow-roll regime · Quasi de Sitter space-time · Exiting or crossing outside of the Hubble horizon · Scale-invariant spectrum of curvature perturbations

5.1   Deriving the Mukhanov-Sasaki Equation

In Chap. 3, we have derived a general set of evolution equations for first-order cosmological perturbations. The goal of the present chapter is to solve these in a special limit. Until Chap. 7, we assume the absence of a fluid, so that the energy-momentum tensor is dominated by a scalar field, cf. Eqs. (3.83)–(3.85). The first step is to rephrase the equations in a gauge-invariant form, as outlined in Sect. 4.3. This leads to a single equation, cf. Eq. (5.28), for a gauge-invariant variable, $\widehat{Q}_\varphi$, which according to Eqs. (5.16) and (5.17) is proportional to the curvature perturbation, $\mathcal{R}_\varphi$, that we have introduced in Eq. (4.60).

Our starting point is the scalar field evolution equation in vacuum, originally given in Eq. (1.48). In terms of Eq. (1.62), we drop the terms induced by interactions with

M. Laine and S. Procacci, *From Inflation to Hot Big Bang*, Lecture Notes in Physics 1047, https://doi.org/10.1007/978-3-032-09893-1_5

a fluid, namely the friction Υ and the noise ϱ (the computation is repeated in their presence in Sect. 7.5), considering then just

$$\varphi^{;\mu}{}_{;\mu} - V_{,\varphi} \overset{(1.62)}{\underset{\Upsilon,\,\varrho \to 0}{=}} 0. \tag{5.1}$$

Writing the scalar field as $\varphi = \bar{\varphi} + \delta\varphi$, where $\bar{\varphi}$ only depends on time, the background (i.e. zeroth order) field equation, governing the behaviour of $\bar{\varphi}$, was derived in Eq. (1.53),

$$\bar{\varphi}'' + 2\mathcal{H}\bar{\varphi}' + a^2 V_{,\varphi} \overset{(1.53)}{=} 0, \tag{5.2}$$

where $(\dots)' \equiv \partial_\tau(\dots)$ and $\mathcal{H} \equiv a'/a$. Higher derivatives of $\bar{\varphi}$ can be obtained by taking time derivatives of Eq. (5.2). In particular, taking one time derivative, and then re-substituting $\bar{\varphi}''$ via Eq. (5.2), we find a relation that is needed later in this section,

$$(5.2)' \Rightarrow \quad \bar{\varphi}''' + 2(\mathcal{H}'\bar{\varphi}' + \mathcal{H}\bar{\varphi}'') + a^2(2\mathcal{H}V_{,\varphi} + V_{,\varphi\varphi}\bar{\varphi}') = 0$$

$$\overset{(5.2)}{\Rightarrow} \quad \bar{\varphi}''' + (2\mathcal{H}' - 4\mathcal{H}^2 + a^2 V_{,\varphi\varphi})\bar{\varphi}' = 0. \tag{5.3}$$

Going to first order in perturbations, the equation satisfied by $\delta\varphi$ is derived in Sect. 3.5. Dropping again Υ and ϱ, and writing $\delta V_{,\varphi} = V_{,\varphi\varphi}\delta\varphi$, Eq. (3.114) reduces to

$$\delta\varphi'' + 2\mathcal{H}\delta\varphi' - \nabla^2\delta\varphi - (h_0' + 3h_{\mathrm{D}}' + \nabla^2 h)\bar{\varphi}' + a^2 V_{,\varphi\varphi}\delta\varphi + 2h_0 a^2 V_{,\varphi} \overset{(3.114)}{\underset{\Upsilon,\,\varrho \to 0}{=}} 0. \tag{5.4}$$

It will be convenient to rewrite this for the rescaled variable $\widehat{\delta\varphi} \equiv a\delta\varphi$. Noting that

$$\delta\varphi' = \frac{1}{a}\left(\widehat{\delta\varphi}' - \mathcal{H}\widehat{\delta\varphi}\right), \tag{5.5}$$

$$\delta\varphi'' + 2\mathcal{H}\delta\varphi' = \frac{1}{a}\left(\widehat{\delta\varphi}'' - \frac{a''}{a}\widehat{\delta\varphi}\right), \tag{5.6}$$

and moving the terms involving metric perturbations to the right-hand side, we get

$$\boxed{\widehat{\delta\varphi}'' - \nabla^2\widehat{\delta\varphi} + \left(a^2 V_{,\varphi\varphi} - \frac{a''}{a}\right)\widehat{\delta\varphi} \overset{(5.4)}{\underset{(5.6)}{=}} (h_0' + 3h_{\mathrm{D}}' + \nabla^2 h)a\bar{\varphi}' - 2h_0 a^3 V_{,\varphi}.}$$

$$\tag{5.7}$$

From Eq. (5.7), we see that the metric perturbations $h_0' + 3h_{\mathrm{D}}' + \nabla^2 h$ and h_0 need to be eliminated, if we want to obtain a closed equation for $\widehat{\delta\varphi}$. For this we need to make use of perturbed Einstein equations.

At the background level, the Einstein equations for our system read, from Eqs. (1.56) and (1.60),

$$\mathcal{H}^2 \overset{(1.56)}{\underset{\kappa=0}{=}} \frac{4\pi G}{3}\left[(\bar{\varphi}')^2 + 2a^2\, V\right], \quad \mathcal{H}' \overset{(1.60)}{=} \frac{8\pi G}{3}\left[-(\bar{\varphi}')^2 + a^2\, V\right]. \quad (5.8)$$

Here we have set $\kappa = 0$, as also done when deriving the perturbed Einstein equations (cf. the discussion around Eq. (1.80), and Sect. 1.5). Taking a linear combination and then a time derivative, yields the further useful relations

$$(5.8) \Rightarrow \mathcal{H}^2 - \mathcal{H}' = 4\pi G(\bar{\varphi}')^2 \quad \Rightarrow \quad 2\mathcal{H}\mathcal{H}' - \mathcal{H}'' = 8\pi G\bar{\varphi}'\bar{\varphi}''. \quad (5.9)$$

At the first order in perturbations, the variable h_0 is conveniently eliminated via the $0i$ components of the Einstein equations, cf. Eq. (3.93). Dropping the medium part, this leads to

$$\mathcal{H}h_0 + \left(h_{\text{D}} + \frac{\nabla^2\vartheta}{3}\right)' \overset{(3.93)}{\underset{T\to 0}{=}} \frac{4\pi G\bar{\varphi}'}{a}\,\widehat{\delta\varphi}. \quad (5.10)$$

For eliminating $h_0' + 3h_{\text{D}}' + \nabla^2 h$, we make use of the linear combination of the Einstein equations given in Eq. (3.96). Recalling the $e \supset V$ and $p \supset -V$ (cf. Eq. (3.86)), this yields

$$2\left(\mathcal{H}' + 2\mathcal{H}^2\right)h_0 + \mathcal{H}\left(h_0' + 3h_{\text{D}}' + \nabla^2 h\right)$$

$$+ \left(\partial_\tau^2 + 2\mathcal{H}\partial_\tau - \nabla^2\right)\left(h_{\text{D}} + \frac{\nabla^2\vartheta}{3}\right) \overset{(3.93)}{\underset{T\to 0}{=}} -8\pi GaV_{,\varphi}\,\widehat{\delta\varphi}. \quad (5.11)$$

Therefore, we can solve for h_0 from Eq. (5.10), and for $h_0' + 3h_{\text{D}}' + \nabla^2 h$ from Eq. (5.11). It is helpful to introduce the shorthand notation

$$\boxed{X \equiv h - \vartheta', \quad Y \equiv h_{\text{D}} + \frac{\nabla^2\vartheta}{3},} \quad (5.12)$$

where X has already been employed to derive Eqs. (4.56) and (4.57). We remark in passing that X and Y transform only with ξ^0 under gauge transformations (cf. Eqs. (4.50)–(4.52)), $\tilde{X} = X + \xi^0$, $\tilde{Y} = Y + \mathcal{H}\xi^0$. In any case, the results for h_0 and $h_0' + 3h_{\text{D}}' + \nabla^2 h$ become

$$h_0 \overset{(5.10)}{=} -\frac{Y'}{\mathcal{H}} + \frac{4\pi G\bar{\varphi}'}{a\mathcal{H}}\,\widehat{\delta\varphi}, \quad (5.13)$$

$$h_0' + 3h_{\text{D}}' + \nabla^2 h \overset{(5.11)}{=} -2\left(2\mathcal{H} + \frac{\mathcal{H}'}{\mathcal{H}}\right)h_0 - \frac{\left(\partial_\tau^2 + 2\mathcal{H}\partial_\tau - \nabla^2\right)Y}{\mathcal{H}} - \frac{8\pi GaV_{,\varphi}}{\mathcal{H}}\,\widehat{\delta\varphi}.$$

$$(5.14)$$

We now express the inflaton perturbation as

$$\delta\widehat{\varphi} \;=\; \widehat{\mathcal{Q}}_{\varphi} - \frac{a\bar{\varphi}'}{\mathcal{H}}\, Y \;,\tag{5.15}$$

where we have defined (cf. Eqs. (4.47) and (4.48))

$$\widehat{\mathcal{Q}}_{\varphi} \equiv a\mathcal{Q}_{\varphi}\;,\tag{5.16}$$

$$\mathcal{Q}_{\varphi} \stackrel{(4.48)}{\equiv} \delta\varphi + \frac{\bar{\varphi}'}{\mathcal{H}}\underbrace{\left(h_{\mathrm{D}} + \frac{\nabla^2\vartheta}{3}\right)}_{(5.12):\,Y} \stackrel{(4.47)}{=} -\frac{\bar{\varphi}'}{\mathcal{H}}\mathcal{R}_{\varphi}\;.\tag{5.17}$$

Here $\mathcal{R}_{\varphi}$ is the gauge-invariant curvature perturbation, defined in Eq. (4.60).

When we insert Eqs. (5.13)–(5.15) into Eq. (5.7), there are many appearances of Y. On the left-hand side (L),

$$(5.7)_{\mathrm{L}} \supset -\left(\frac{a\bar{\varphi}'}{\mathcal{H}}\right)'' Y - 2\left(\frac{a\bar{\varphi}'}{\mathcal{H}}\right)' Y' - \frac{a\bar{\varphi}'}{\mathcal{H}}\left(\partial_\tau^2 - \nabla^2 + a^2\, V_{,\varphi\varphi} - \frac{a''}{a}\right) Y\;.\tag{5.18}$$

On the right-hand side (R),

$$(5.7)_{\mathrm{R}} \stackrel{(5.13)-(5.15)}{\supset} 2\overbrace{\left[\frac{a\bar{\varphi}'}{\mathcal{H}}\left(2\mathcal{H} + \frac{\mathcal{H}'}{\mathcal{H}}\right) + \frac{a^3 V_{,\varphi}}{\mathcal{H}}\right]}^{\text{coefficient of }(-h_0)H}\overbrace{\left[Y' + \frac{4\pi G(\bar{\varphi}')^2}{\mathcal{H}}Y\right]}^{\text{from }(-h_0)H\text{ via }(5.13)}$$
$$-\frac{a\bar{\varphi}'}{\mathcal{H}}(\partial_\tau^2 + 2\mathcal{H}\partial_\tau - \nabla^2)Y + \left(\frac{a\bar{\varphi}'}{\mathcal{H}}\right)^2 8\pi G a V_{,\varphi}\, Y\;.\tag{5.19}$$

Subtracting the terms and inserting $a''/a = \mathcal{H}' + \mathcal{H}^2$, $(\partial_\tau^2 - \nabla^2)Y$ cancels, and we find

$$(5.7)_{\mathrm{R}} - (5.7)_{\mathrm{L}} \stackrel{\substack{(5.18)\\(5.19)}}{\supset} \left[2\left(\frac{a\bar{\varphi}'}{\mathcal{H}}\right)' + a\bar{\varphi}'\left(4 - 2 + \frac{2\mathcal{H}'}{\mathcal{H}^2}\right) + \frac{2a^3 V_{,\varphi}}{\mathcal{H}}\right] Y'$$
$$+ \left\{\left(\frac{a\bar{\varphi}'}{\mathcal{H}}\right)'' + \frac{a\bar{\varphi}'}{\mathcal{H}}\left[a^2\, V_{,\varphi\varphi} - \mathcal{H}' - \mathcal{H}^2\right.\right.$$
$$\left.\left. + \left(2\mathcal{H} + \frac{\mathcal{H}'}{\mathcal{H}}\right)\frac{8\pi G(\bar{\varphi}')^2}{\mathcal{H}} + \frac{16\pi G a^2 V_{,\varphi}\bar{\varphi}'}{\mathcal{H}}\right]\right\} Y\;.\tag{5.20}$$

We now use the background identity in Eq. (5.2) to write

$$\left(\frac{a\bar{\varphi}'}{\mathcal{H}}\right)' \;=\; \frac{a\bar{\varphi}''}{\mathcal{H}} + a\bar{\varphi}'\left(1 - \frac{\mathcal{H}'}{\mathcal{H}^2}\right) \stackrel{(5.2)}{=} a\bar{\varphi}'\left(-1 - \frac{\mathcal{H}'}{\mathcal{H}^2}\right) - \frac{a^3 V_{,\varphi}}{\mathcal{H}}\;,\tag{5.21}$$

and see that the coefficient of Y' cancels in Eq. (5.20). So we are on the right track!

To simplify the coefficient of Y, we take the second derivative of Eq. (5.21), obtaining

$$\left(\frac{a\bar{\varphi}'}{\mathcal{H}}\right)'' \overset{(5.21)}{=} a\bar{\varphi}''\left(-1 - \frac{\mathcal{H}'}{\mathcal{H}^2}\right) + a\bar{\varphi}'\left[-\mathcal{H} - \frac{\mathcal{H}'}{\mathcal{H}} - \frac{\mathcal{H}''}{\mathcal{H}^2} + \frac{2(\mathcal{H}')^2}{\mathcal{H}^3} - \frac{a^2 V_{,\varphi\varphi}}{\mathcal{H}}\right]$$

$$- a^3 V_{,\varphi}\left(3 - \frac{\mathcal{H}'}{\mathcal{H}^2}\right) \tag{5.22}$$

$$\overset{(5.2)}{\underset{(5.9)}{=}} \frac{a\bar{\varphi}'}{\mathcal{H}}\left[\overbrace{\left(\mathcal{H} + \frac{\mathcal{H}'}{\mathcal{H}}\right)\left(2\mathcal{H} + \frac{a^2 V_{,\varphi}}{\bar{\varphi}'}\right)}^{\text{from }\bar{\varphi}''} - \mathcal{H}^2 - \mathcal{H}' + \overbrace{\frac{8\pi G\bar{\varphi}'\bar{\varphi}'' - 2\mathcal{H}\mathcal{H}'}{\mathcal{H}}}^{\text{from }-\mathcal{H}''}\right.$$

$$\left. + \frac{2(\mathcal{H}')^2}{\mathcal{H}^2} - a^2 V_{,\varphi\varphi} - \frac{a^2 V_{,\varphi}}{\bar{\varphi}'}\left(3\mathcal{H} - \frac{\mathcal{H}'}{\mathcal{H}}\right)\right] \tag{5.23}$$

$$\overset{(5.2)}{\underset{(5.9)}{=}} \frac{a\bar{\varphi}'}{\mathcal{H}}\left[\mathcal{H}^2 - \mathcal{H}' + \frac{2(\mathcal{H}')^2}{\mathcal{H}^2} + \overbrace{\left(\mathcal{H} + \frac{\mathcal{H}'}{\mathcal{H}} - 3\mathcal{H} + \frac{\mathcal{H}'}{\mathcal{H}}\right)\frac{a^2 V_{,\varphi}}{\bar{\varphi}'}}^{\text{becomes }-8\pi G(\bar{\varphi}')^2/\mathcal{H}}\right.$$

$$\left. - a^2 V_{,\varphi\varphi} - \frac{8\pi G(\bar{\varphi}')^2}{\mathcal{H}}\overbrace{\left(2\mathcal{H} + \frac{a^2 V_{,\varphi}}{\bar{\varphi}'}\right)}^{\text{from }\bar{\varphi}''}\right] \tag{5.24}$$

$$\overset{(5.9)}{=} \frac{a\bar{\varphi}'}{\mathcal{H}}\left[\mathcal{H}^2 + \mathcal{H}' + \overbrace{\frac{2\mathcal{H}'(\mathcal{H}' - \mathcal{H}^2)}{\mathcal{H}^2}}^{\text{becomes }-\mathcal{H}'8\pi G(\bar{\varphi}')^2/\mathcal{H}^2} - 16\pi G(\bar{\varphi}')^2 - \frac{16\pi G a^2 V_{,\varphi}\bar{\varphi}'}{\mathcal{H}} - a^2 V_{,\varphi\varphi}\right].$$

$$\tag{5.25}$$

From here, we see that the coefficient of Y also cancels in Eq. (5.20).

The vanishing of Eq. (5.20) constitutes a check of gauge invariance in the sense of Sect. 4.3. To summarize, once we replace $\delta\widehat{\varphi}$ by the gauge-invariant $\widehat{\mathcal{Q}}_\varphi$, and make use of background identities, all appearances of $Y = h_{\mathrm{D}} + \frac{\nabla^2\vartheta}{3}$ cancel. Of course, had we chosen the spatially flat case, cf. Eq. (4.46), then appearances of $h_{\mathrm{D}} + \frac{\nabla^2\vartheta}{3}$ would have cancelled all along. Thus the end result would have been the same, and the computation would have been simpler, but we would have missed a powerful crosscheck. Also, gauge-fixed computations run the danger that we might inadvertently consider gauge-variant quantities; if this happens, and we combine results from different gauges, the results are no longer correct.

Let us finally work out the physical (gauge-invariant) evolution equation. We again insert Eqs. (5.13)–(5.15) into (5.7), but now we can omit all appearances of Y, since they have been verified to cancel. Then

$$(5.7)_{\mathrm{L}} \overset{(5.15)}{\underset{Y \to 0}{=}} \widehat{Q}_\varphi'' - \nabla^2 \widehat{Q}_\varphi + \left(a^2 V_{,\varphi\varphi} - \frac{a''}{a} \right) \widehat{Q}_\varphi \,, \tag{5.26}$$

$$(5.7)_{\mathrm{R}} \overset{(5.13)-(5.15)}{\underset{Y \to 0}{=}} \overbrace{-2\left(2\mathcal{H} + \frac{\mathcal{H}'}{\mathcal{H}} + \frac{a^2 V_{,\varphi}}{\bar{\varphi}'} \right)}^{\text{coefficient of } h_0\, a\bar{\varphi}'} \overbrace{\frac{4\pi G (\bar{\varphi}')^2}{\mathcal{H}} \widehat{Q}_\varphi}^{\text{from } h_0\, a\bar{\varphi}' \text{ via (5.13)}} - \frac{8\pi G a^2 V_{,\varphi}\bar{\varphi}'}{\mathcal{H}} \widehat{Q}_\varphi \,. \tag{5.27}$$

Moving all the terms on the left-hand side, this can be expressed as

$$(5.7)_{\mathrm{L}} - (5.7)_{\mathrm{R}} = \boxed{\left[\partial_\tau^2 - \nabla^2 + \widehat{m}^2(\tau) \right] \widehat{Q}_\varphi = 0 \,.} \tag{5.28}$$

This is known as the *Mukhanov-Sasaki equation* [1,2]. Furthermore, the mass parameter can be simplified by making use of the same information that was used for showing gauge invariance in Eq. (5.25) (inserting again $a''/a = \mathcal{H}' + \mathcal{H}^2$),

$$\widehat{m}^2(\tau) \overset{(5.26)}{\underset{(5.27)}{=}} a^2 V_{,\varphi\varphi} - \mathcal{H}' - \mathcal{H}^2 + 2\left(2\mathcal{H} + \frac{\mathcal{H}'}{\mathcal{H}} + \frac{a^2 V_{,\varphi}}{\bar{\varphi}'} \right) \frac{4\pi G (\bar{\varphi}')^2}{\mathcal{H}} + \frac{8\pi G a^2 V_{,\varphi}\bar{\varphi}'}{\mathcal{H}}$$

$$= a^2 V_{,\varphi\varphi} - \mathcal{H}' - \mathcal{H}^2 + \frac{\mathcal{H}' 8\pi G (\bar{\varphi}')^2}{\mathcal{H}^2} + 16\pi G (\bar{\varphi}')^2 + \frac{16\pi G a^2 V_{,\varphi}\bar{\varphi}'}{\mathcal{H}}$$

$$\overset{(5.25)}{\Rightarrow} \boxed{\widehat{m}^2(\tau) = -\frac{\mathcal{H}}{a\bar{\varphi}'}\left(\frac{a\bar{\varphi}'}{\mathcal{H}} \right)''.} \tag{5.29}$$

We conclude this section by remarking that even though the inflaton potential $V(\varphi)$ does not appear explicitly in Eq. (5.29), the effective mass squared, $\widehat{m}^2$, still depends on it via the background solution. For instance, suppose that we go to Minkowskian space-time by setting $a = a_0 + a_1\tau$ and considering the limit $a_1 \to 0$. Then $\mathcal{H} \to a_1/a_0$, and $\widehat{m}^2 \to -\bar{\varphi}'''/\bar{\varphi}'$ from Eq. (5.29). The third derivative can be obtained from Eq. (5.3), where $\lim_{a_1 \to 0}(\mathcal{H}' - 2\mathcal{H}^2) = 0$. Therefore $\lim_{a_1 \to 0} \widehat{m}^2 = a^2 V_{,\varphi\varphi}$, reproducing the usual Minkowskian definition of mass squared as the curvature of a potential.

5.2　Initial Conditions from the Bunch-Davies Vacuum

Equation (5.28) is a second-order partial differential equation. We can Fourier transform it in spatial directions, and then we are faced with a second-order ordinary differential equation in time. In order to specify a solution, two initial conditions are needed, the value of $\widehat{Q}_\varphi$ and its time derivative, $\widehat{Q}_\varphi'$. The idea of the inflationary paradigm is that when a given mode is deep inside the Hubble horizon, its perturbations are localized quantum fluctuations, not affected by the expansion of the universe. In terms of Eq. (5.30), this means that $\lim_{\tau \to -\infty} \widehat{m}^2(\tau) = 0$ (cf. Eq. (5.44)), so that we are faced with a time-independent Minkowskian-looking equation. Thus we require that $\lim_{\tau \to -\infty} \widehat{\mathcal{Q}}_\varphi$ corresponds to a canonically normalized *quantum field*. The corresponding *ground state* at $\tau \to -\infty$ is referred to as a Bunch-Davies vacuum [3]. Effectively, what we do is to replace the classical $\widehat{\mathcal{Q}}_\varphi$ by a quantum-mechanical operator, and re-interpret the time-dependent problem as a determination of the corresponding *mode function*.

To proceed, let us simplify the notation, dropping the subscript from $\widehat{\mathcal{Q}}_\varphi$, so that the equation reads

$$\widehat{\mathcal{Q}}'' + \underbrace{\left[-\nabla^2 + \widehat{m}^2(\tau) \right]}_{\equiv \, \hat{\epsilon}_{\mathbf{x}}^2(\tau)} \widehat{\mathcal{Q}} \overset{(5.28)}{=} 0 \; . \tag{5.30}$$

In general, a solution of Eq. (5.30) can be found with a *mode expansion*,

$$\widehat{\mathcal{Q}}(\tau, \mathbf{x}) = \int \frac{\mathrm{d}^3\mathbf{k}}{\sqrt{(2\pi)^3}} \left[w_{\mathbf{k}} \, \widehat{\mathcal{Q}}_k(\tau) \, e^{i\mathbf{k}\cdot\mathbf{x}} + w_{\mathbf{k}}^\dagger \, \widehat{\mathcal{Q}}_k^*(\tau) \, e^{-i\mathbf{k}\cdot\mathbf{x}} \right] , \tag{5.31}$$

where $k \equiv |\mathbf{k}|$ and the mode function, $\widehat{\mathcal{Q}}_k$, satisfies

$$\widehat{\mathcal{Q}}_k'' + \underbrace{\left[k^2 + \widehat{m}^2(\tau) \right]}_{\equiv \, \hat{\epsilon}_k^2(\tau)} \widehat{\mathcal{Q}}_k \overset{(5.30)}{\underset{(5.31)}{=}} 0 \; . \tag{5.32}$$

The coefficients $w_{\mathbf{k}}$ and $w_{\mathbf{k}}^\dagger$ are quantum-mechanical operators, normalized as

$$[\, w_{\mathbf{k}} \, , \, w_{\mathbf{q}} \,] = [\, w_{\mathbf{k}}^\dagger \, , \, w_{\mathbf{q}}^\dagger \,] = 0 \, , \quad [\, w_{\mathbf{k}} \, , \, w_{\mathbf{q}}^\dagger \,] = \delta^{(3)}(\mathbf{k} - \mathbf{q}) \; . \tag{5.33}$$

To specify the full solution, the task is to fix the normalization and first derivative of the mode function at some initial time.

Let us define a *Wronskian* between two mode functions as

$$\mathcal{W}[\widehat{\mathcal{Q}}_k^{(1)}, \widehat{\mathcal{Q}}_k^{(2)}] \equiv \widehat{\mathcal{Q}}_k^{(1)} (\widehat{\mathcal{Q}}_k^{(2)})' - (\widehat{\mathcal{Q}}_k^{(1)})' \, \widehat{\mathcal{Q}}_k^{(2)} \; . \tag{5.34}$$

It follows from Eq. (5.32) that the Wronskian is independent of time,

$$\mathcal{W}' \overset{(5.34)}{=} \widehat{\mathcal{Q}}_k^{(1)} (\widehat{\mathcal{Q}}_k^{(2)})'' - (\widehat{\mathcal{Q}}_k^{(1)})'' \, \widehat{\mathcal{Q}}_k^{(2)} \overset{(5.32)}{=} -\hat{\epsilon}_k^2(\tau) \left[\widehat{\mathcal{Q}}_k^{(1)} \, \widehat{\mathcal{Q}}_k^{(2)} - \widehat{\mathcal{Q}}_k^{(1)} \, \widehat{\mathcal{Q}}_k^{(2)} \right] = 0 \; . \tag{5.35}$$

Its correct value can be deduced by inspecting the *canonical commutation relation* following from Eq. (5.31),

$$\left[\widehat{\mathcal{Q}}(\tau, \mathbf{x}) \, , \, \partial_\tau \widehat{\mathcal{Q}}(\tau, \mathbf{y}) \right]$$

$$\stackrel{(5.31)}{=} \int \frac{\mathrm{d}^3 \mathbf{k} \, \mathrm{d}^3 \mathbf{q}}{(2\pi)^3} \left\{ \left[w_{\mathbf{k}} \, , \, w_{\mathbf{q}}^\dagger \right] \widehat{\mathcal{Q}}_k(\tau) \, \widehat{\mathcal{Q}}_q^{*\prime}(\tau) \, e^{i\mathbf{k}\cdot\mathbf{x} - i\mathbf{q}\cdot\mathbf{y}} \right.$$

$$\left. + \left[w_{\mathbf{k}}^\dagger \, , \, w_{\mathbf{q}} \right] \widehat{\mathcal{Q}}_k^*(\tau) \, \widehat{\mathcal{Q}}_q^{\prime}(\tau) \, e^{i\mathbf{q}\cdot\mathbf{y} - i\mathbf{k}\cdot\mathbf{x}} \right\}$$

$$\stackrel{(5.33)}{=} \int \frac{\mathrm{d}^3 \mathbf{k}}{(2\pi)^3} \left\{ \widehat{\mathcal{Q}}_k(\tau) \, \widehat{\mathcal{Q}}_k^{*\prime}(\tau) \, e^{i\mathbf{k}\cdot(\mathbf{x}-\mathbf{y})} - \widehat{\mathcal{Q}}_k^{\prime}(\tau) \, \widehat{\mathcal{Q}}_k^*(\tau) \, e^{i\mathbf{k}\cdot(\mathbf{y}-\mathbf{x})} \right\}$$

$$\stackrel[(5.32)]{\mathbf{k}\to-\mathbf{k}}{=} \int \frac{\mathrm{d}^3 \mathbf{k}}{(2\pi)^3} \, \mathcal{W}[\widehat{\mathcal{Q}}_k(\tau), \widehat{\mathcal{Q}}_k^*(\tau)] \, e^{i\mathbf{k}\cdot(\mathbf{x}-\mathbf{y})} \ . \tag{5.36}$$

For the last equality, we observe from Eq. (5.32) that the mode functions are even in $\mathbf{k} \to -\mathbf{k}$. It then follows that we obtain the canonical value, $i\delta^{(3)}(\mathbf{x} - \mathbf{y})$, if we set

$$\mathcal{W}[\widehat{\mathcal{Q}}_k(\tau), \widehat{\mathcal{Q}}_k^*(\tau)] = i \quad \forall k \ . \tag{5.37}$$

This fixes one of the integration constants.

The other initial condition is that the mode function $\widehat{\mathcal{Q}}_k$, associated with $w_{\mathbf{k}}$, should correspond to the '*forward-propagating*' or '*positive-energy*' mode at early times. The logic is that $w_{\mathbf{k}}$ corresponds to an annihilation operator, and a vacuum state is defined by $w_{\mathbf{k}}|0\rangle = 0$ for every $\mathbf{k}$. Since $\widehat{\mathcal{Q}}_k$ multiplies $w_{\mathbf{k}}$ in Eq. (5.31), putting the positive-energy modes in $\widehat{\mathcal{Q}}_k$ guarantees that there are no forward-propagating particles in the initial state $|0\rangle$. In concrete terms, recalling the Schrödinger equation, this is implemented as a 'root' of Eq. (5.32),

$$\lim_{\tau\to-\infty} i\partial_\tau \widehat{\mathcal{Q}}_k = + \lim_{\tau\to-\infty} \underbrace{\hat{\epsilon}_k(\tau)}_{\text{from (5.32)}} \widehat{\mathcal{Q}}_k \ . \tag{5.38}$$

Thereby the value and the first derivative of the mode function are related to each other. With this information, the solution is fully specified. We note that $\widehat{\mathcal{Q}}_k \in \mathbb{C}$.

5.3　Solution during Slow-Roll Expansion

In Sect. 5.2 we have determined the initial conditions satisfied by the mode functions, $\widehat{\mathcal{Q}}_k(\tau)$. We can then solve for their time dependence, from Eq. (5.32). In this section we do this by assuming approximately de Sitter expansion, $H \approx$ constant, which in physical time is expressed as

$$\left| \frac{\Delta H}{H} \right| \approx \left| \frac{\Delta t \, \dot{H}}{H} \right| \stackrel{\Delta t \equiv H^{-1}}{=} \left| \frac{\dot{H}}{H^2} \right| \ll 1 \ . \tag{5.39}$$

We remark that among the earliest examples of a quantum-mechanical computation in such a context, though not explicitly for inflation, can be found in Ref. [4], whereas some of the first inflationary computations were presented in Refs. [5–8].

For an exactly constant H, the background solution was determined around Eq. (1.96),

$$a \overset{(1.96)}{=} -\frac{1}{H\tau} , \quad \tau \in (-\infty, \tau_e) . \tag{5.40}$$

Here we adopt this functional form also for the case that H may change adiabatically with time. We refer to this scenario as a *quasi de Sitter space-time*. Let us first elaborate on the conceptual foundation of this approximation in our context.

In general, the background solution is determined by Eq. (5.2),

$$\bar{\varphi}'' + 2\mathcal{H}\bar{\varphi}' + a^2 V_{,\varphi}(\bar{\varphi}) \overset{(5.2)}{=} 0 \quad \overset{\tau \leftrightarrow t}{\underset{(1.53)}{\Longleftrightarrow}} \quad \ddot{\bar{\varphi}} + 3H\dot{\bar{\varphi}} + V_{,\varphi}(\bar{\varphi}) = 0 . \tag{5.41}$$

Given that H is approximately constant but $\mathcal{H}$ is not, it is easier to inspect the version on the right. The Hubble rate H is itself a function of $\bar{\varphi}$ and $\dot{\bar{\varphi}}$, as dictated by Eq. (1.57), so the equation is non-linear.

Solutions of non-linear differential equations are often classified according to their asymptotic behaviour at late times, known as *fixed points*. It may be observed that Eq. (5.41) does possess a fixed point corresponding to exact de Sitter expansion. Namely, a constant H could be obtained if we set $V_{,\varphi} = 0$, i.e. the potential is just a cosmological constant. The equation $\ddot{\bar{\varphi}} + 3H\dot{\bar{\varphi}} = 0$ then has two solutions, $\dot{\bar{\varphi}}(t) = 0$ and $\dot{\bar{\varphi}}(t) = \dot{\bar{\varphi}}(t_i)e^{-3H(t-t_i)}$. The second solution approaches the first one exponentially fast. Afterwards, we find a static background field, $\dot{\bar{\varphi}} = 0$, implying eternal de Sitter expansion. However, strictly speaking exact de Sitter expansion is *not* a cosmologically relevant trajectory, because observations point to a power-like rather than exponential expansion, during the period when CMB was formed.

We note in passing that if a solution satisfying $\ddot{\bar{\varphi}} + 3H\dot{\bar{\varphi}} \approx 0$ is present for a finite period of time, this dynamics is known as *ultra-slow-roll expansion*. Cosmological models having this property have been built (cf., e.g., Refs. [9, 10]), however mostly not for the inflationary dynamics affecting CMB modes, but rather for some later period, influencing shorter wavelengths.

Another simple solution can be found if we look at a situation in which $\dot{\bar{\varphi}}$ and $V_{,\varphi}$ do not vanish but are small. Adopting the language from the harmonic oscillator, we envisage finding ourselves in an *overdamped regime*, with $|\ddot{\bar{\varphi}}| \ll |3H\dot{\bar{\varphi}}|$. Then Eq. (5.41) reduces to a first-order differential equation,

$$3H\dot{\bar{\varphi}} + V_{,\varphi}(\bar{\varphi}) \approx 0 . \tag{5.42}$$

From here it follows that

$$\frac{\bar{\varphi}'}{\mathcal{H}} \overset{(1.3)}{=} \frac{\dot{\bar{\varphi}}}{H} \overset{(5.42)}{\approx} -\frac{V_{,\varphi}}{3H^2} . \tag{5.43}$$

If the potential and the initial value of $\bar{\varphi}$ satisfy certain conditions, known as *slow-roll* constraints (cf. Sect. 6.2), then the condition $|\ddot{\bar{\varphi}}| \ll |3H\dot{\bar{\varphi}}|$ indeed remains valid for a relatively long period of time.

Furthermore, it turns out that the ratio in Eq. (5.43) only evolves slowly, and in a first approximation can be treated as a constant (cf. Sect. 6.2). If $\bar{\varphi}'/\mathcal{H}$ is approximately constant, and a evolves according to Eq. (5.40), then it follows that the effective mass parameter from Eq. (5.29) becomes

$$\widehat{m}^2(\tau) \overset{(5.29)}{=} -\frac{\mathcal{H}}{a\bar{\varphi}'}\left(\frac{a\bar{\varphi}'}{\mathcal{H}}\right)'' \overset{\bar{\varphi}'/\mathcal{H}\approx\text{constant}}{\approx} -\frac{a''}{a} \overset{(5.40)}{=} -\frac{2}{\tau^2} . \tag{5.44}$$

Even though this is an approximation for curvature perturbations, the same effective mass squared, $-a''/a$, is obtained in an exact form for some other excitations in de Sitter space-time, such as gravitons, or massless minimally coupled 'spectator' scalar fields, which have no background value. With a slight abuse of notation, we therefore write the equation for mode functions, Eq. (5.32), as an equality,

$$\left(\partial_\tau^2 + \underbrace{k^2 - \frac{2}{\tau^2}}_{\approx\,\hat{\epsilon}_k^2(\tau)}\right)\widehat{Q}_k = 0 . \tag{5.45}$$

An explicit solution can be found,

$$\widehat{Q}_k = \frac{e^{-ik\tau}}{\sqrt{2k}}\left(1 - \frac{i}{k\tau}\right) , \tag{5.46}$$

with another solution given by the complex conjugate, $\widehat{Q}_k^*$.

Let us check that the integration constants have been fixed correctly in Eq. (5.46). Taking time derivatives, we find

$$\widehat{Q}_k' = \frac{e^{-ik\tau}}{\sqrt{2k}}\left(-ik - \frac{1}{\tau} + \frac{i}{k\tau^2}\right) , \quad \widehat{Q}_k^{*\prime} = \frac{e^{ik\tau}}{\sqrt{2k}}\left(ik - \frac{1}{\tau} - \frac{i}{k\tau^2}\right) . \tag{5.47}$$

The Wronskian from Eq. (5.34) then becomes

$$W[\widehat{Q}_k, \widehat{Q}_k^*] \overset{(5.34)}{=} \frac{1}{2k}\left[\overbrace{\left(1 - \frac{i}{k\tau}\right)}^{\text{from (5.46)}}\overbrace{\left(ik - \frac{1}{\tau} - \frac{i}{k\tau^2}\right)}^{\text{from (5.47)}} - \left(1 + \frac{i}{k\tau}\right)\left(-ik - \frac{1}{\tau} + \frac{i}{k\tau^2}\right)\right]$$

$$= \frac{i}{k}\,\text{Im}\left[\left(1 - \frac{i}{k\tau}\right)\left(ik - \frac{1}{\tau} - \frac{i}{k\tau^2}\right)\right]$$

$$= \frac{i}{k}\,\text{Im}\left[ik - \frac{1}{\tau} - \frac{i}{k\tau^2} + \frac{1}{\tau} + \frac{i}{k\tau^2} - \frac{1}{k^2\tau^3}\right] = i , \tag{5.48}$$

indicating that the normalization condition from Eq. (5.37) is respected.

Furthermore, if $k|\tau| \gg 1$, Eq. (5.45) implies

$$-i\,\hat{\epsilon}_k(\tau) \overset{(5.45)}{=} -i\sqrt{k^2 - \frac{2}{\tau^2}} \overset{k \gg 1/|\tau|}{\approx} -ik\left(1 - \frac{1}{k^2\tau^2} + \dots\right) = -ik + \frac{i}{k\tau^2} + \dots .$$

$$(5.49)$$

This leads to

$$-i\,\hat{\epsilon}_k(\tau)\,\widehat{\mathcal{Q}}_k = \frac{e^{-ik\tau}}{\sqrt{2k}} \overbrace{\left(-ik + \frac{i}{k\tau^2} + \dots\right)}^{\text{from (5.49)}} \overbrace{\left(1 - \frac{i}{k\tau}\right)}^{\text{from (5.46)}}$$

$$= \frac{e^{-ik\tau}}{\sqrt{2k}} \underbrace{\left(-ik - \frac{1}{\tau} + \frac{i}{k\tau^2} + \frac{1}{k^2\tau^3} + \dots\right)}_{\widehat{\mathcal{Q}}'_k \text{ from (5.47)}} . \qquad (5.50)$$

Therefore, Eq. (5.38) is satisfied in the domain $k|\tau| \gg 1$, and we conclude that Eq. (5.46) represents a correctly normalized forward-propagating mode function.

The next task is to compute the variance of the perturbations, defined like in Eq. (2.51). The expectation value is taken in the Bunch-Davies vacuum, making use of the creation and annihilation operators from Eq. (5.33). We obtain

$$\left\langle \widehat{\mathcal{Q}}_\varphi^2(\tau, \mathbf{x}) \right\rangle \overset{(2.55)}{=} \left\langle 0 \middle| \widehat{\mathcal{Q}}_\varphi(\tau, \mathbf{x}) \widehat{\mathcal{Q}}_\varphi(\tau, \mathbf{x}) \middle| 0 \right\rangle \qquad (5.51)$$

$$\overset{(5.31)}{=} \int \frac{d^3\mathbf{k}\,d^3\mathbf{q}}{(2\pi)^3} \langle 0| \, w_{\mathbf{k}}\,\widehat{\mathcal{Q}}_k(\tau)\,e^{i\mathbf{k}\cdot\mathbf{x}}\, w_{\mathbf{q}}^\dagger\,\widehat{\mathcal{Q}}_q^*(\tau)\,e^{-i\mathbf{q}\cdot\mathbf{x}} \, |0\rangle$$

$$\overset{(5.33)}{=} \int_{\mathbf{k}} |\widehat{\mathcal{Q}}_k(\tau)|^2 \overset{(5.46)}{=} \int_{-\infty}^{+\infty} d\ln k \, \underbrace{\frac{k^3}{2\pi^2}\frac{1}{2k}\left(1 + \frac{1}{k^2\tau^2}\right)}_{\text{from (2.51): } \mathcal{P}_{\widehat{\mathcal{Q}}_\varphi}(\tau,k)} . \qquad (5.52)$$

Subsequently, the *power spectrum for curvature perturbations* from Eq. (5.17), noting that $\mathcal{R}_\varphi = -H\widehat{\mathcal{Q}}_\varphi/(a\dot{\bar{\varphi}})$, is obtained as

$$\mathcal{P}_{\mathcal{R}_\varphi}(\tau, k) \overset{(5.52)}{\underset{(5.16),(5.17)}{=}} \left(\frac{H}{2\pi\dot{\bar{\varphi}}}\right)^2\left(\frac{k^2}{a^2} + \frac{1}{a^2\tau^2}\right) \overset{(5.40)}{=} \left(\frac{H}{2\pi\dot{\bar{\varphi}}}\right)^2\left(\frac{k^2}{a^2} + H^2\right) . \qquad (5.53)$$

Now, as illustrated in Fig. 1.1 in Sect. 1.4, the physical momentum k/a equals the Hubble rate H at a certain moment, which is denoted by τ_*, and decreases very fast below it afterwards, $k/a \ll H$. In terms of Eq. (5.45), this means that the time-dependent term $-2/\tau^2$ becomes more important than k^2, and vacuum-like oscillations cease. As will be illustrated numerically in Fig. 6.1 in Sect. 6.5, and shown in Chap. 8, the curvature power spectrum *freezes out* at this moment, becoming a constant. So, if we denote $H_* \equiv H(\tau_*)$, $a_* \equiv a(\tau_*)$, and $\dot{\bar{\varphi}}_* \equiv \dot{\bar{\varphi}}(\tau_*)$, then the prediction for the curvature power spectrum can be expressed as

$$\mathcal{P}_{\mathcal{R}_{\varphi}}(k) \overset{(5.53)}{\approx} \left(\frac{H_*^2}{2\pi\dot{\varphi}_*}\right)^2\Bigg|_{H_*=k/a_*} . \tag{5.54}$$

The prediction for the curvature power spectrum can now be compared with the experimentally measured power spectrum. In particular, at the pivot scale, k_*, we can extract the *scalar amplitude* as (cf. Eq. (2.7))

$$A_{\mathrm{s}} \overset{(2.7)}{=} \mathcal{P}_{\mathcal{R}_{\varphi}}(k_*) . \tag{5.55}$$

This can be compared with Eq. (2.8). We remark in passing that the asterisk in k_* (signifying the pivot scale) has no relation to that in τ_* (signifying the moment of horizon crossing for any k). For the *spectral tilt*, we write

$$n_{\mathrm{s}} \overset{(2.7)}{=} 1 + \frac{\mathrm{d}\ln\mathcal{P}_{\mathcal{R}_{\varphi}}(k)}{\mathrm{d}\ln k}\Bigg|_{k=k_*} = 1 + \frac{\mathrm{d}\ln\mathcal{P}_{\mathcal{R}_{\varphi}}(k)}{\mathrm{d}t}\frac{\mathrm{d}t}{\mathrm{d}\ln k}\Bigg|_{k=k_*} . \tag{5.56}$$

For the relation between the crossing time and the momentum mode k, the definition $k(t_*) = a_* H_*$ in Eq. (5.54) yields

$$\frac{\mathrm{d}\ln k}{\mathrm{d}t} \overset{(5.54)}{=} \frac{\dot{a}_* H_* + a_* \dot{H}_*}{k} \overset{k=a_* H_*}{=} \frac{H_*^2 + \dot{H}_*}{H_*} . \tag{5.57}$$

This then results in

$$\boxed{n_{\mathrm{s}} \underset{(5.56),\,(5.57)}{\overset{(5.54)}{\approx}} 1 + \frac{2H_*}{H_*^2 + \dot{H}_*}\left(\frac{2\dot{H}_*}{H_*} - \frac{\ddot{\varphi}_*}{\dot{\varphi}_*}\right),} \tag{5.58}$$

which can be compared with Eq. (2.9). Given that we have assumed $|\dot{H}_*| \ll H_*^2$ (cf. Eq. (5.39)) and $|\ddot{\varphi}_*| \ll |H\dot{\varphi}_*|$ (cf. Eq. (5.42)), we find an almost flat spectrum, $n_{\mathrm{s}} \approx 1$, corresponding to a *Harrison–Zeldovich spectrum* as anticipated below Eq. (2.7). We end this chapter with three remarks:

- Tensor perturbations generated by vacuum fluctuations undergo a similar dynamics as the curvature perturbations that have been discussed above. We return to this topic in Sect. 10.2, obtaining then a prediction also for the tensor-to-scalar ratio r from Eq. (2.17). Thereby we can contrast the predictions originating from a given inflationary potential V with all stringent observational bounds.

- Vector perturbations obey a diffusive rather than a (second-order) wave equation, cf. the discussion around Eqs. (3.100) and (3.101). Therefore their dynamics differs qualitatively from that of the curvature and tensor perturbations.

- If the inflationary potential V is not quadratic, we may treat interactions perturbatively, and proceed to compute also higher-point correlation functions of the type in Eq. (5.51) [11,12] (see also Ref. [13] for the simple case of a 'spectator' scalar field). This has evolved into a large field of research in the meanwhile. Such corrections represent a primordial source for the *non-Gaussianities* and *bispectrum* that were discussed around Eq. (2.62).

References

1. Sasaki, M.: Large Scale Quantum Fluctuations in the Inflationary Universe. Prog. Theor. Phys. **76**, 1036 (1986). https://doi.org/10.1143/PTP.76.1036
2. Mukhanov, V.F.: Quantum Theory of Gauge Invariant Cosmological Perturbations. Sov. Phys. JETP **67**, 1297 (1988) [Zh. Eksp. Teor. Fiz. 94N7 (1988) 1]
3. Bunch, T.S., Davies, P.C.W.: Quantum field theory in de Sitter space: renormalization by point-splitting. Proc. R. Soc. Lond. A **360**, 117 (1978). https://doi.org/10.1098/rspa.1978.0060
4. Mukhanov, V.F., Chibisov, G.V.: Quantum fluctuations and a nonsingular universe. JETP Lett. **33**, 532 (1981) [Pisma Zh. Eksp. Teor. Fiz. 33 (1981) 549]
5. Hawking, S.W.: The development of irregularities in a single bubble inflationary universe. Phys. Lett. B **115**, 295 (1982). https://doi.org/10.1016/0370-2693(82)90373-2
6. Starobinsky, A.A.: Dynamics of phase transition in the new inflationary universe scenario and generation of perturbations. Phys. Lett. B **117**, 175 (1982). https://doi.org/10.1016/0370-2693(82)90541-X
7. Guth, A.H., Pi, S.Y.: Fluctuations in the New Inflationary Universe. Phys. Rev. Lett. **49**, 1110 (1982). https://doi.org/10.1103/PhysRevLett.49.1110
8. Bardeen, J.M., Steinhardt, P.J., Turner, M.S.: Spontaneous creation of almost scale-free density perturbations in an inflationary universe. Phys. Rev. D **28**, 679 (1983). https://doi.org/10.1103/PhysRevD.28.679
9. Inoue, S., Yokoyama, J.: Curvature perturbation at the local extremum of the inflaton's potential. Phys. Lett. B **524**, 15 (2002). https://doi.org/10.1016/S0370-2693(01)01369-7, hep-ph/0104083
10. Kinney, W.H.: Horizon crossing and inflation with large η. Phys. Rev. D **72**, 023515 (2005). https://doi.org/10.1103/PhysRevD.72.023515, gr-qc/0503017
11. Maldacena, J.M.: Non-Gaussian features of primordial fluctuations in single field inflationary models. JHEP **05**, 013 (2003). https://doi.org/10.1088/1126-6708/2003/05/013, astro-ph/0210603
12. Weinberg, S.: Quantum contributions to cosmological correlations. Phys. Rev. D **72**, 043514 (2005). https://doi.org/10.1103/PhysRevD.72.043514, hep-th/0506236
13. Bernardeau, F., Brunier, T., Uzan, J.P.: High order correlation functions for self-interacting scalar field in de Sitter space. Phys. Rev. D **69**, 063520 (2004). https://doi.org/10.1103/PhysRevD.69.063520, astro-ph/0311422

Scalar Power Spectrum in a General Cold Inflation Scenario

6

Abstract

Moving on from the idealized quasi de Sitter space-time to a more general situation in which the Hubble rate is a function of time, we show how the equation for the curvature perturbation can be set up in a manageable form. The concept of an overdamped 'slow-roll' regime is elaborated upon, and we show how observable quantities can be derived as a power series in small parameters. We demonstrate how the basic equations can be rephrased through the so-called stochastic formalism, introducing the notion of quantum noise. We also explain how the stochastic formalism can be simplified under a number of further assumptions, leading to a framework that has been used in the literature for simulating non-linear aspects of inflationary dynamics.

Keywords

Cold inflation · Mukhanov-Sasaki equation in physical time · Slow-roll parameters · Stochastic formalism · Quantum noise · Numerical solution · Non-linear simulations

6.1 Setup in Physical Time

Let us start by recapitulating the equations that we are trying to solve. The unknown variable of the Mukhanov-Sasaki equation is the dimensionless $\widehat{Q}_\varphi$ (cf. Eq. (5.28)), obtained by rescaling the curvature perturbation $\mathcal{R}_\varphi$ (cf. Eq. (4.60)),

$$\mathcal{R}_\varphi \overset{(5.16)}{\underset{(5.17)}{=}} -\frac{\mathcal{H}}{a\bar{\varphi}'}\widehat{Q}_\varphi \,. \tag{6.1}$$

© The Author(s) 2026

M. Laine and S. Procacci, *From Inflation to Hot Big Bang*, Lecture Notes in Physics 1047, https://doi.org/10.1007/978-3-032-09893-1_6

Representing $\widehat{\mathcal{Q}}_\varphi$ via the corresponding mode function $\widehat{\mathcal{Q}}_k$ (cf. Eq. (5.31)), the power spectrum of $\widehat{\mathcal{Q}}_\varphi$ is proportional to $|\widehat{\mathcal{Q}}_k|^2$ (cf. Eq. (5.52)). The mode function satisfies

$$\widehat{\mathcal{Q}}_k'' + \left[k^2 + \widehat{m}^2(\tau) \right]\widehat{\mathcal{Q}}_k \overset{(5.32)}{=} 0 \,, \quad \widehat{m}^2(\tau) \overset{(5.29)}{\equiv} -\frac{\mathcal{H}}{a\bar{\varphi}'}\left(\frac{a\bar{\varphi}'}{\mathcal{H}}\right)'' . \tag{6.2}$$

One benefit of the rescaling in Eq. (6.1) is that no first-order time derivatives appear in Eq. (6.2), and therefore one of the initial conditions can be expressed as a normalization condition that is satisfied at all times,

$$\mathcal{W}[\widehat{\mathcal{Q}}_k(\tau), \widehat{\mathcal{Q}}_k^*(\tau)] \overset{(5.37)}{=} i \quad \forall k, \tau \,, \tag{6.3}$$

where $\mathcal{W}$ denotes the Wronskian (cf. Eq. (5.34)). The other initial condition, originating from the interpretation of $\widehat{\mathcal{Q}}_\varphi$ as a quantum-mechanical vacuum fluctuation, can be viewed as a 'root' of Eq. (6.2),

$$\widehat{\mathcal{Q}}_k' \underset{\tau \to -\infty}{\overset{(5.38)}{=}} \underbrace{-i \sqrt{k^2 + \widehat{m}^2(\tau)}\; \widehat{\mathcal{Q}}_k}_{\equiv \hat{\epsilon}_k(\tau)} \,, \tag{6.4}$$

where we furthermore expect $\lim_{\tau \to -\infty} \widehat{m}^2(\tau) = 0$ (cf. Eq. (5.44)). We note that according to Eqs. (6.3) and (6.4), $\widehat{\mathcal{Q}}_k \in \mathbb{C}$.

Now, even if the equations above are not too complicated, corresponding just to a harmonic oscillator with a time-dependent mass, their solution requires in general numerical methods. The reason is that the parameter $\widehat{m}^2(\tau)$ is obtained from a solution of *non-linear* differential equations, and therefore cannot be represented in closed form. A practical procedure is therefore to solve for $\widehat{m}^2(\tau)$ and $\widehat{\mathcal{Q}}_k(\tau)$ simultaneously, starting from given initial values. However, in order to do this efficiently, it turns out to be helpful to change variables, as we describe in the following.

In a first step, we go back to the curvature perturbation $\mathcal{R}_\varphi$ from Eq. (6.1), but now with the mode functions, getting

$$\widehat{\mathcal{Q}}_k = -\frac{a\bar{\varphi}'}{\mathcal{H}}\mathcal{R}_k \,, \tag{6.5}$$

$$\widehat{\mathcal{Q}}_k' = -\left(\frac{a\bar{\varphi}'}{\mathcal{H}}\right)'\mathcal{R}_k - \frac{a\bar{\varphi}'}{\mathcal{H}}\mathcal{R}_k' \,, \tag{6.6}$$

$$\widehat{\mathcal{Q}}_k'' = -\left(\frac{a\bar{\varphi}'}{\mathcal{H}}\right)''\mathcal{R}_k - 2\left(\frac{a\bar{\varphi}'}{\mathcal{H}}\right)'\mathcal{R}_k' - \frac{a\bar{\varphi}'}{\mathcal{H}}\mathcal{R}_k'' \,. \tag{6.7}$$

Substituting these in Eq. (6.2), and multiplying the whole with $-\mathcal{H}/(a\bar{\varphi}')$, yields

$$\mathcal{R}_k'' + 2\frac{\mathcal{H}}{a\bar{\varphi}'}\left(\frac{a\bar{\varphi}'}{\mathcal{H}}\right)'\mathcal{R}_k' + \left[\frac{\mathcal{H}}{a\bar{\varphi}'}\left(\frac{a\bar{\varphi}'}{\mathcal{H}}\right)'' + k^2 - \frac{\mathcal{H}}{a\bar{\varphi}'}\left(\frac{a\bar{\varphi}'}{\mathcal{H}}\right)''\right]\mathcal{R}_k = 0 \,. \tag{6.8}$$

In a second step, like for the numerical solution of the background equations in Sect. 1.7, it is beneficial to transform to physical time. From Eq. (1.3), we recall that $d\tau = a^{-1}(t)dt$, and physical time derivatives are denoted by

$$(...)^{\cdot} \;\equiv\; \partial_t(...) \;=\; \frac{1}{a}(...)' \;\Leftrightarrow\; (...)' \stackrel{(1.7)}{=} a\,(...)^{\cdot}\,. \tag{6.9}$$

For the derivatives of $\mathcal{R}_k$ we thus obtain $\mathcal{R}_k' = a\dot{\mathcal{R}}_k$, and

$$\mathcal{R}_k'' \stackrel{(6.9)}{=} a\left(a\dot{\mathcal{R}}_k\right)^{\cdot} \;=\; a^2\left(\ddot{\mathcal{R}}_k + \underbrace{\frac{\dot{a}}{a}}_{H}\dot{\mathcal{R}}_k\right). \tag{6.10}$$

Substituting these in Eq. (6.8), and multiplying the whole equation with $1/a^2$, yields

$$\ddot{\mathcal{R}}_k + \underbrace{\left[\frac{\dot{a}}{a} + 2\frac{H}{a\dot{\bar{\varphi}}}\left(\frac{a\dot{\bar{\varphi}}}{H}\right)^{\cdot}\right]}_{\equiv\,2\mathcal{F}+3H}\dot{\mathcal{R}}_k + \frac{k^2}{a^2}\mathcal{R}_k \;=\; 0\,, \tag{6.11}$$

where we have defined

$$\mathcal{F}(t) \;\equiv\; \frac{H}{\dot{\bar{\varphi}}}\left(\frac{\dot{\bar{\varphi}}}{H}\right)^{\cdot} \;=\; \frac{\ddot{\bar{\varphi}}}{\dot{\bar{\varphi}}} - \frac{\dot{H}}{H}\,. \tag{6.12}$$

Here $\ddot{\bar{\varphi}}$ is given by Eq. (1.53), H by Eq. (1.57), and $\dot{H}$ by Eq. (1.61). For a physical interpretation, it is useful to note that Eq. (6.12) represents the relative rate of change of the variable $\dot{\bar{\varphi}}/H$ from Eq. (5.43), which is approximately constant in the slow-roll regime. Equation (6.11) is the *Mukhanov-Sasaki equation in physical time*.

Let us remark that both Eq. (6.2) and Eq. (6.11) have important special properties which underline the value of using $\widehat{\mathcal{Q}}_k$ as a variable in conformal time and $\mathcal{R}_k$ in physical time. Specifically, Eq. (6.11) includes a first-order time derivative, implying that in physical time, the Wronskian of $\mathcal{R}_k$ is not constant (cf. Eq. (5.34)). On the other hand, the mass term is absent from Eq. (6.11), which turns out to have a very helpful implication for the late-time evolution, as discussed in Sects. 6.2 and 8.2.

We still need to transcribe the initial conditions from Eqs. (6.3) and (6.4) into the new variables. Given that Eq. (6.3) concerns overall normalization, let us recall from Eq. (5.52) that the power spectrum corresponding to $\mathcal{R}_\varphi$ is obtained from the corresponding mode function, $\mathcal{R}_k$, via

$$\mathcal{P}_{\mathcal{R}_\varphi} \;=\; \frac{k^3}{2\pi^2}|\mathcal{R}_k|^2 \;\equiv\; \left|\left[\mathcal{R}_k\right]_{\text{rescaled}}\right|^2. \tag{6.13}$$

In the second step, we noted that it is practical to 'rescale' the mode function by $k^{3/2}/(\sqrt{2}\pi)$, so that its absolute value squared yields directly the power spectrum.

Let t_i denote a moment at which initial conditions are set, and $f_i \equiv f(t_i)$ the values of various functions at that time. Given that the variable k/a in Eq. (6.11) is exponentially large at early times (cf. Fig. 1.1 (right) in Sect. 1.4), it is not possible to take t_i arbitrarily small in practice. Rather, it is sufficient to take t_i from a domain in which $k/a_i \gg H_i$, so that k/a_i represents the dominant scale in the evolution, and Eq. (6.4) becomes

$$\partial_t \widehat{\mathcal{Q}}_k(t_i) \overset{(6.4)}{\underset{(6.9)}{\approx}} -i \frac{k}{a_i} \widehat{\mathcal{Q}}_k(t_i) \, . \tag{6.14}$$

We rewrite Eq. (6.14) in terms of $\mathcal{R}_k$ by making use of Eq. (6.5). Furthermore, we note that we can approximate $a\dot{\varphi}/H$ as constant, because $\dot{a}_i/a_i = H_i \ll k/a_i$, and similarly for the relative time dependence of $\dot{\varphi}/H$, given by $\mathcal{F}$ from Eq. (6.12). Then Eq. (6.14) implies

$$\boxed{\left[\partial_t \mathcal{R}_k\right]_{\text{rescaled}}(t_i) \overset{k/a_i \gg H_i}{\approx} -i \frac{k}{a_i} \left[\mathcal{R}_k\right]_{\text{rescaled}}(t_i) \, .} \tag{6.15}$$

Under the same approximation, the Wronskian from Eq. (6.3) can be written as

$$i = \mathcal{W}[\widehat{\mathcal{Q}}_k, \widehat{\mathcal{Q}}_k^*] \approx \left(\frac{a_i \dot{\varphi}_i}{H_i}\right)^2 \times 2i a_i \, \text{Im}\big(\mathcal{R}_k \dot{\mathcal{R}}_k^*\big)$$

$$\overset{(6.15)}{\underset{(6.13)}{=}} \left(\frac{a_i \dot{\varphi}_i}{H_i}\right)^2 \times 2i k \frac{2\pi^2}{k^3} \left|\left[\mathcal{R}_k\right]_{\text{rescaled}}(t_i)\right|^2 \, . \tag{6.16}$$

Choosing the overall phase so that the $R_k(t_i) > 0$ (assuming $\dot{\varphi}_i < 0$), we obtain

$$\boxed{\left[\mathcal{R}_k\right]_{\text{rescaled}}(t_i) \overset{k/a_i \gg H_i}{\approx} -\frac{H_i}{\dot{\varphi}_i} \frac{1}{2\pi} \frac{k}{a_i} \, .} \tag{6.17}$$

An algorithm for solving Eq. (6.11) numerically, with initial values fixed according to Eqs. (6.15) and (6.17), is given in Sect. 6.5.

6.2   Simplifications in the Slow-Roll Regime

Apart from numerical solutions, Eq. (6.11) can also be used as a starting point for analytic approximations. We already discussed them in Sect. 5.3, where a background solution was obtained from Eq. (5.42), and subsequently the power spectrum was determined, leading to Eq. (5.54). Here we want to identify small parameters, which will be called *slow-roll parameters*, and formalize the computation as a Taylor expansion in them [1].

In the present section, we remain at the zeroth or first order in the slow-roll expansion, showing how predictions such as Eq. (5.54) can be justified. However, we remark in passing that it is possible to go up to high orders in the slow-roll expansion, even though formulating this systematically requires modified definitions of the expansion parameters [2].

While the curvature power spectrum originates by solving for scalar perturbations, the slow-roll approximation operates mostly on the side of the (non-linear) background equations. The conventional logic for determining the slow-roll predictions is to first assume ourselves to be in a specific regime, and then work out consistency criteria for when this assumption is self-consistent. So, we start by *assuming* that $\ddot{\varphi} \approx 0$ in Eq. (5.41), implying

$$\dot{\varphi} \stackrel{(5.42)}{\approx} -\frac{V_{,\varphi}}{3H}.$$
(6.18)

The slow-roll parameters are defined as

$$\epsilon_V \equiv \frac{1}{16\pi G}\left(\frac{V_{,\varphi}}{V}\right)^2, \quad \eta_V \equiv \frac{1}{8\pi G}\frac{V_{,\varphi\varphi}}{V},$$
(6.19)

where the subscript $(\ldots)_V$ refers to the potential V (there is another convention in which slow-roll parameters are determined in terms of the Hubble rate). The time moment at which the parameters are evaluated is left implicit for now, and they are treated as being approximately constant around that moment.

We note that, since $G = 1/m_{\mathrm{pl}}^2$ (cf. Eq. (1)), ϵ_V and η_V in Eq. (6.19) can be small only if variations in the field φ are at the Planck scale, $m_{\mathrm{pl}}\partial_\varphi < 1$. Naively this requires *trans-Planckian field values*, $\varphi > m_{\mathrm{pl}}$, which is considered a foundational challenge for inflationary model building.

For the Hubble rate, it now follows from Eq. (1.57) that

$$H^2 \underset{\kappa=0}{\overset{(1.57)}{=}} \frac{8\pi G V}{3}\left(1+\frac{\dot{\varphi}^2}{2V}\right) \stackrel{(6.18)}{\approx} \frac{8\pi G V}{3}\left(1+\frac{V_{,\varphi}^2}{18VH^2}\right)$$

$$\stackrel{(6.19)}{=} \frac{8\pi G V}{3}\left(1+\frac{8\pi G V}{3H^2}\frac{\epsilon_V}{3}\right) \stackrel{\epsilon_V \ll 1}{\approx} \frac{8\pi G V}{3}\left(1+\frac{\epsilon_V}{3}\right).$$
(6.20)

The last equality shows an approximate solution of the quadratic equation for H^2.

To verify the consistency of our approximation, we need to determine the time derivatives of the key quantities. From Eqs. (1.61) and (1.57), we obtain

$$\dot{H} \overset{(1.57)}{\underset{(1.61)}{=}} -4\pi G\dot{\varphi}^2 \stackrel{(6.18)}{\approx} -\frac{4\pi G V_{,\varphi}^2}{9H^2} \stackrel{(6.19)}{=} -\left(\frac{8\pi G V}{3H}\right)^2\epsilon_V \stackrel{(6.20)}{\approx} -H^2\epsilon_V.$$
(6.21)

Taking a time derivative of Eq. (6.18) yields

$$\ddot{\varphi} \stackrel{(6.18)}{\approx} -\frac{V_{,\varphi\varphi}\dot{\varphi}}{3H}+\frac{V_{,\varphi}\dot{H}}{3H^2} \overset{(6.21)}{\underset{(6.19)}{\approx}} -\frac{8\pi G V\dot{\varphi}}{3H}\eta_V - \frac{V_{,\varphi}}{3}\epsilon_V \overset{(6.18)}{\underset{(6.20)}{\approx}} (\epsilon_V - \eta_V)H\dot{\varphi}.$$
(6.22)

Now, comparing $\ddot{\ddot{\varphi}}$ with $3H\dot{\varphi}$, Eq. (6.22) shows that $\ddot{\ddot{\varphi}}$ is indeed subdominant in the regime $\epsilon_V, |\eta_V| \ll 1$. Therefore, the assumption in Eq. (6.18) is self-consistent. For $\mathcal{F}$ from Eq. (6.12), Eqs. (6.21) and (6.22) imply that $\mathcal{F} \approx (2\epsilon_V - \eta_V)H \ll H$. So, Eq. (6.11) becomes

$$\ddot{\mathcal{R}}_k + \left[3 + 2(2\epsilon_V - \eta_V) \right] H \dot{\mathcal{R}}_k + \frac{k^2}{a^2} \mathcal{R}_k \overset{(6.11)}{\underset{(6.21),\,(6.22)}{\approx}} 0 \,. \tag{6.23}$$

Furthermore, Eq. (6.21) shows that the change of H in a *Hubble time*, $\Delta t \equiv H^{-1}$, is $\Delta H = H^{-1}\dot{H} \approx -H\epsilon_V \ll H$, so that H is approximately constant. In other words, we are in quasi de Sitter space-time, in accordance with Eq. (5.39).

We can then ask what is the curvature power spectrum in the regime governed by Eqs. (6.23), (6.15) and (6.17). It is non-trivial to do this up to first order in the slow-roll parameters, because then the first-order time evolution of H needs to be accounted for. Let us instead do this up to *zeroth order* in the slow-roll parameters.

The key insight is that according to Fig. 1.1 (right) in Sect. 1.4, k/a decreases very rapidly below the Hubble rate at a certain moment, which is denoted by t_* and called *horizon exit* (by convention, $k/a_* = H_*$ at t_*). If the horizon exit happens in the slow-roll regime, and we evaluate H at zeroth order, then at $t > t_*$, Eq. (6.23) reads $\ddot{\mathcal{R}}_k + 3H_*\dot{\mathcal{R}}_k \approx 0$. The general solution is

$$\mathcal{R}_k \approx c_1 + c_2\, e^{-3H_* t} \,, \tag{6.24}$$

implying that $\mathcal{R}_k$ settles to a constant, c_1. The value of c_1 can be deduced from Eq. (6.17), by pushing the initial time, t_i, close to t_*. The absolute value squared of Eq. (6.17) at this moment reproduces the power spectrum in Eq. (5.54). Even if these considerations are a bit sloppy, it is comforting that the correct result is recovered.

The predictions for the CMB observables from Eqs. (5.55) and (5.58) can now be expressed in terms of the slow-roll parameters. Straightforward but somewhat tedious substitutions lead to expressions widely used in the literature,

$$A_{\mathrm{s}} \overset{(5.54)}{\underset{(5.55)}{\approx}} \frac{H_*^4}{4\pi^2\dot{\varphi}_*^2} \overset{(6.18)}{\approx} \frac{9H_*^6}{4\pi^2 V_{,\varphi}^2} \overset{(6.19)}{\approx} \frac{9H_*^6}{64\pi^3 G V^2 \epsilon_V} \overset{(6.20)}{\approx} \frac{8G^2 V}{3\epsilon_V} \,, \tag{6.25}$$

$$n_{\mathrm{s}} \overset{(5.58)}{\underset{(6.21)}{\approx}} 1 + \frac{2}{H_*}\left(\frac{2\dot{H}_*}{H_*} - \frac{\ddot{\ddot{\varphi}}_*}{\dot{\varphi}_*} \right) \overset{(6.21)}{\underset{(6.22)}{\approx}} 1 - 6\epsilon_V + 2\eta_V \,. \tag{6.26}$$

For the tensor-to-scalar ratio r, mentioned in Eq. (2.17) and defined in Eq. (10.32), the corresponding prediction is worked out in Eq. (10.33).

As a final remark, the quasi de Sitter regime breaks down when $\epsilon_V \gtrsim 1$ (cf. Eq. (6.21)). From Eq. (6.20), this happens when $\dot{\varphi}^2/2 \gtrsim V$. The slow-roll approximation breaks down also if $|\eta_V|$ becomes large. In particular, in the *ultra-slow-roll regime*, discussed in a paragraph preceding Eq. (5.42), we still have $\epsilon_V \ll 1$ according to Eq. (6.19), given that $V_{,\varphi}$ had been set to zero. But otherwise the estimates above cannot be used, as the assumption introduced in Eq. (6.18) is not respected (the background equation reads $\ddot{\ddot{\varphi}} \approx -3H\dot{\varphi}$).

6.3   Stochastic Derivation with Quantum Noise

We now proceed to re-formulating the initial-value problem in Sect. 6.1 in a way that the information contained in the initial conditions at t_i (cf. Eqs. (6.15) and (6.17)), can instead be imposed at a later time, for instance at t_*. Even if this may feel a bit like a 'trick', the result has a nice physical interpretation, and has also paved the way for an approximation scheme that is used in non-linear numerical simulations (cf. Sect. 6.4). For the present section, we return to conformal time, and denote rescaled quantities with a wide hat, '$\widehat{\cdots}$'. The equation to solve is the original **x**-space version of Eq. (6.2), given in Eq. (5.30).

The starting point is to introduce a *window function*, or rather a distribution, $W_k(\tau)$, which splits the comoving momentum space into ultraviolet (UV) and infrared (IR) parts. The window is such that in the distant past, $\tau \to -\infty$, all modes are selected by W_k, i.e.

$$W_k(-\infty) \;=\; 1\,,\quad W_k'(-\infty) \;=\; 0 \;\;\forall k\,. \tag{6.27}$$

Later on, physical momenta redshift towards the IR. We consider a final moment, τ_f, at which the modes of interest are in the IR and fall out of the window, so that

$$W_k(\tau_f) \;\approx\; 0\,. \tag{6.28}$$

Often the UV part is identified as *'inside the Hubble horizon'* and the IR part as *'outside the Hubble horizon'*, which requires that τ_f is chosen inside a specific time window (cf. Fig. 1.1 (right) in Sect. 1.4).

Following Ref. [3], we now split the mode expansion in Eq. (5.31) into UV and IR parts,

$$\widehat{\mathcal{Q}}(\tau,\mathbf{x}) \;=\; \widehat{\mathcal{Q}}_<(\tau,\mathbf{x}) + \widehat{\mathcal{Q}}_>(\tau,\mathbf{x})\,, \tag{6.29}$$

with the short-distance (UV) part given by

$$\widehat{\mathcal{Q}}_<(\tau,\mathbf{x}) \equiv \int \frac{d^3k}{\sqrt{(2\pi)^3}}\, W_k(\tau)\left[\,w_\mathbf{k}\,\widehat{\mathcal{Q}}_k(\tau)\,e^{i\mathbf{k}\cdot\mathbf{x}} + w_\mathbf{k}^\dagger\,\widehat{\mathcal{Q}}_k^*(\tau)\,e^{-i\mathbf{k}\cdot\mathbf{x}}\,\right]. \tag{6.30}$$

Inserting Eq. (6.29) into Eq. (5.30), a *quantum noise* is defined as

$$\left[\partial_\tau^2 + \hat{\epsilon}_\mathbf{x}^2(\tau)\right]\widehat{\mathcal{Q}}_>(\tau,\mathbf{x}) = \widehat{\varrho}_Q(\tau,\mathbf{x})\,,\quad \widehat{\varrho}_Q(\tau,\mathbf{x}) \equiv -\left[\partial_\tau^2 + \hat{\epsilon}_\mathbf{x}^2(\tau)\right]\widehat{\mathcal{Q}}_<(\tau,\mathbf{x})\,. \tag{6.31}$$

The idea is that the long-distance modes are sourced by the quantum noise rather than by initial conditions, which is why the formalism is called 'stochastic'.

Operating on Eq. (6.30), and noting that several terms drop out, thanks to the equation satisfied by the mode functions, the noise can be expressed as

$$\widehat{\varrho}_Q(\tau,\mathbf{x}) = -\int \frac{d^3k}{\sqrt{(2\pi)^3}}\left[\,w_\mathbf{k}\,\widehat{f}_k(\tau)\,e^{i\mathbf{k}\cdot\mathbf{x}} + w_\mathbf{k}^\dagger\,\widehat{f}_k^*(\tau)\,e^{-i\mathbf{k}\cdot\mathbf{x}}\,\right], \tag{6.32}$$

$$\widehat{f}_k(\tau) \equiv W_k''(\tau)\widehat{\mathcal{Q}}_k(\tau) + 2\,W_k'(\tau)\widehat{\mathcal{Q}}_k'(\tau)\,. \tag{6.33}$$

The derivatives W_k'' and W_k' are to be understood in the sense of distributions, as we ultimately integrate over τ (see below).

Defining a *retarded Green's function*, as

$$\left[\, \partial_\tau^2 + \hat{\epsilon}_{\mathbf{x}}^2(\tau) \,\right] G_{|\mathbf{x}-\mathbf{z}|}(\tau; \tau_i) = \delta(\tau - \tau_i)\, \delta^{(3)}(\mathbf{x} - \mathbf{z}) \,, \tag{6.34}$$

$$G_{|\mathbf{x}-\mathbf{z}|}(\tau; \tau_i) \equiv 0 \quad \text{for} \quad \tau < \tau_i \,, \tag{6.35}$$

our long-distance IR solution from Eq. (6.31) is

$$\widehat{Q}_>(\tau, \mathbf{x}) = \int_{-\infty}^{\tau} \mathrm{d}\tau_i \int_{\mathbf{z}} G_{|\mathbf{x}-\mathbf{z}|}(\tau; \tau_i)\, \widehat{\varrho}_Q(\tau_i, \mathbf{z}) \,. \tag{6.36}$$

In order to transform the solution to momentum space, we define

$$G_k(\tau; \tau_i) \overset{(9)}{\equiv} \int_{\mathbf{x}} e^{-i\mathbf{k}\cdot(\mathbf{x}-\mathbf{z})}\, G_{|\mathbf{x}-\mathbf{z}|}(\tau; \tau_i) \,, \tag{6.37}$$

which fulfils

$$\left[\, \partial_\tau^2 + \hat{\epsilon}_k^2(\tau) \,\right] G_k(\tau; \tau_i) \overset{(6.34)}{\underset{(6.37)}{=}} \delta(\tau - \tau_i) \tag{6.38}$$

$$G_k(\tau; \tau_i) \equiv 0 \quad \text{for} \quad \tau < \tau_i \,. \tag{6.39}$$

Then we get

$$\begin{aligned}
\widehat{Q}_>(\tau, \mathbf{k}) &\overset{(9)}{\equiv} \int_{\mathbf{x}} e^{-i\mathbf{k}\cdot\mathbf{x}}\, \widehat{Q}_>(\tau, \mathbf{x}) \\
&\overset{(6.36)}{=} \int_{-\infty}^{\tau} \mathrm{d}\tau_i \int_{\mathbf{x},\mathbf{z}} e^{i\mathbf{k}\cdot(\mathbf{z}-\mathbf{x})}\, G_{|\mathbf{x}-\mathbf{z}|}(\tau; \tau_i)\, e^{-i\mathbf{k}\cdot\mathbf{z}}\, \widehat{\varrho}_Q(\tau_i, \mathbf{z}) \\
&\overset{(6.32)}{\underset{(6.37)}{=}} -\sqrt{(2\pi)^3} \int_{-\infty}^{\tau} \mathrm{d}\tau_i\, G_k(\tau; \tau_i)\left[\, w_{\mathbf{k}}\, \widehat{f}_k(\tau_i) + w_{-\mathbf{k}}^\dagger\, \widehat{f}_k^*(\tau_i) \,\right].
\end{aligned} \tag{6.40}$$

Now, like in the quantum-mechanical problem with a Dirac-δ potential, the initial conditions for G_k at $\tau = \tau_i^+$ can be obtained by imposing continuity at τ_i, and by figuring out the correct value of the first derivative by integrating over Eq. (6.38). This yields

$$\lim_{\tau \to \tau_i^+} G_k(\tau; \tau_i) = 0 \,, \qquad \lim_{\tau \to \tau_i^+} \partial_\tau G_k(\tau; \tau_i) = 1 \,. \tag{6.41}$$

Remarkably, the Green's function satisfying these initial conditions can be expressed in terms of the mode functions from Eq. (6.2), recalling that the Wronskian of a mode function and its complex conjugate obeys Eq. (6.3). The solution reads

$$G_k(\tau; \tau_i) = 2\, \mathrm{Im}\!\left[\, \widehat{Q}_k^*(\tau)\widehat{Q}_k(\tau_i) \,\right], \quad \tau > \tau_i \,, \tag{6.42}$$

because

$$2\,\partial_\tau \operatorname{Im}\left[\,\widehat{\mathcal{Q}}_k^*(\tau)\widehat{\mathcal{Q}}_k(\tau_i)\,\right] = \frac{1}{i}\left[\,\widehat{\mathcal{Q}}_k^{*\prime}(\tau)\widehat{\mathcal{Q}}_k(\tau_i) - \widehat{\mathcal{Q}}_k'(\tau)\widehat{\mathcal{Q}}_k^*(\tau_i)\,\right]$$

$$\overset{(5.34)}{=} \frac{1}{i}\,\mathcal{W}\left[\,\widehat{\mathcal{Q}}_k(\tau_i),\,\widehat{\mathcal{Q}}_k^*(\tau)\,\right] \overset{\tau\to\tau_i}{\underset{(6.3)}{\longrightarrow}} 1\,. \tag{6.43}$$

Let us then inspect the integral over τ_i in Eq. (6.40). Inserting Eq. (6.33), and carrying out a partial integration, we find

$$\int_{-\infty}^{\tau} d\tau_i\, G_k(\tau;\tau_i)\,\widehat{f}_k(\tau_i) \overset{(6.33)}{=} \int_{-\infty}^{\tau} d\tau_i\, G_k(\tau;\tau_i)\left[\,W_k''(\tau_i)\,\widehat{\mathcal{Q}}_k(\tau_i) + 2\,W_k'(\tau_i)\widehat{\mathcal{Q}}_k'(\tau_i)\,\right]$$

$$\overset{(6.27)}{\underset{(6.41)}{=}} \int_{-\infty}^{\tau} d\tau_i\, W_k'(\tau_i)\left[\,-\partial_{\tau_i} G_k(\tau;\tau_i)\,\widehat{\mathcal{Q}}_k(\tau_i) + G_k(\tau;\tau_i)\,\widehat{\mathcal{Q}}_k'(\tau_i)\,\right]$$

$$\overset{(6.42)}{\underset{(6.43)}{=}} \int_{-\infty}^{\tau} d\tau_i\, \frac{W_k'(\tau_i)}{i}\Big\{\left[-\widehat{\mathcal{Q}}_k^*(\tau)\widehat{\mathcal{Q}}_k'(\tau_i) + \widehat{\mathcal{Q}}_k(\tau)\widehat{\mathcal{Q}}_k^{*\prime}(\tau_i)\right]\widehat{\mathcal{Q}}_k(\tau_i)$$

$$+ \left[\,\widehat{\mathcal{Q}}_k^*(\tau)\widehat{\mathcal{Q}}_k(\tau_i) - \widehat{\mathcal{Q}}_k(\tau)\widehat{\mathcal{Q}}_k^*(\tau_i)\,\right]\widehat{\mathcal{Q}}_k'(\tau_i)\Big\}$$

$$= \int_{-\infty}^{\tau} d\tau_i\, \frac{W_k'(\tau_i)}{i}\,\mathcal{W}\left[\,\widehat{\mathcal{Q}}_k(\tau_i),\,\widehat{\mathcal{Q}}_k^*(\tau_i)\,\right]\widehat{\mathcal{Q}}_k(\tau)$$

$$\overset{(6.3)}{=} \left[\,W_k(\tau) - W_k(-\infty)\,\right]\widehat{\mathcal{Q}}_k(\tau) \overset{(6.27)}{=} -\left[1 - W_k(\tau)\right]\widehat{\mathcal{Q}}_k(\tau)\,. \tag{6.44}$$

If we choose a late time moment, according to Eq. (6.28), Eq. (6.40) becomes

$$\widehat{\mathcal{Q}}_>(\tau_f,\mathbf{k}) \overset{(6.40)}{\underset{(6.44),\,(6.28)}{=}} \sqrt{(2\pi)^3}\left[\,w_{\mathbf{k}}\,\widehat{\mathcal{Q}}_k(\tau_f) + w_{-\mathbf{k}}^{\dagger}\,\widehat{\mathcal{Q}}_k^*(\tau_f)\,\right]\,. \tag{6.45}$$

Actually, Eq. (6.45) could have been written down immediately, without a derivation, as it is nothing but a Fourier transform of the full field operator from Eq. (5.31). This should not be surprising, as we have made no approximation along the way. However, the derivation suggests a non-trivial interpretation. If we inspect the form in Eq. (6.32), and assume that $W_k(\tau_i)$ is a (perhaps smoothened) step function, so that $W_k'(\tau_i)$ is sharply localized, then $\widehat{\varrho}_Q(\tau_1,\mathbf{x}_1)$ and $\widehat{\varrho}_Q(\tau_2,\mathbf{x}_2)$ with $\tau_1 \neq \tau_2$ are generated from different momentum shells, $k_1 \neq k_2$. As the corresponding $w_{\mathbf{k}_1}$ and $w_{\mathbf{k}_2}^{\dagger}$ commute, the noise can be viewed as 'classical'. This is said to imply that initial quantum fluctuations have become classical ones, when they have redshifted into the IR domain. However, this is not strictly true: the full quantum information (both amplitude and phase) about the initial conditions is still stored in the quantum noise (technically, this is taken care of by the appearance of both W_k'' and W_k' in Eq. (6.33)).

Conventionally, the dividing line between UV and IR modes is adopted as the *crossing of the Hubble horizon*, i.e. at $k \sim \mathcal{H}$. So, according to Eq. (1.96),

$$W_k(\tau) \overset{\text{de Sitter}}{\underset{(1.96)}{\simeq}} \theta\left(k + \frac{\epsilon}{\tau}\right), \tag{6.46}$$

where $\epsilon \sim 1$ is a parameter which should drop out from physical results. We remark that with such a distribution, W_k' is Dirac-δ and W_k'' is singular, but this is not a problem, given that all derivatives of W_k are eliminated by the integrations in Eq. (6.44).

6.4   A Pragmatic Version of the Stochastic Formalism

The stochastic formalism described in Sect. 6.3 is equivalent to the full quantum-mechanical treatment of the problem, but does not offer any practical advantages for determining the solution numerically. There exists, however, a simplified version of the stochastic formalism, elaborated upon in Ref. [4], whose status is a bit different. In some sense, it is conceived to be an effective theory. This means that when we derive the formalism, approximations or restrictions of generality are made:

- We assume to find ourselves in the slow-roll regime (cf. Sect. 6.2).
- The window function is taken from Eq. (6.46), and written in physical time as

$$W_k(t) \equiv \theta\big(k - \epsilon\, a(t) H\big)\,, \tag{6.47}$$

  where H is assumed to be slowly varying (in the following, constant).
- We choose $\epsilon \ll 1$ in Eq. (6.47), so that modes fall out of the window only once they are far outside the Hubble horizon. Philosophically, this choice makes the construction similar in spirit to the so-called *separate universes* approach to cosmological perturbations [5,6].
- The simple mode function from Eq. (5.46) is used for describing the UV modes.
- A gauge is fixed, frequently by going over to the so-called δN formalism [7].

In the literature, there are attempts to relax some of these assumptions, nevertheless certain compromises appear necessary in order to obtain the desired simplifications.

On the other hand, the effective description is conjectured to have merits as well, notably that it could be valid beyond the linear order in perturbations for treating the IR modes. This is apparent already in the choice of variables: rather than splitting the inflaton field into three parts, $\bar{\varphi}$, $\delta\varphi_>$ and $\delta\varphi_<$, as would correspond to the approach of Sect. 6.3, the IR part $\bar{\varphi} + \delta\varphi_>$ is treated as a single variable. Simulations of this framework are being used for probing the likelihood of rare (non-Gaussian) large fluctuations, which might collapse into *primordial black holes*. It is fair to say, however, that the question of whether the formalism faithfully accounts for all non-linearities has an unclear answer at present.

In the following, we illustrate some basic ingredients of the simplified stochastic formalism. We remain, however, at linear order in perturbations, and use the gauge-invariant variable corresponding to $\delta\varphi$, $\mathcal{Q}_\varphi$ from Eq. (5.17). In Sect. 5.1, we saw that the ideal variable in conformal time is its rescaled version, $\widehat{\mathcal{Q}}_\varphi$ (cf. Eq. (5.16)), whereas in Sect. 6.1, for considerations in physical time, the best variable turned out to be the curvature perturbation $\mathcal{R}_\varphi$, from Eq. (6.1). In the present approach, it is advantageous to mix physical and conformal times (see below), and work with $\mathcal{Q}_\varphi$, rather than the previous $\widehat{\mathcal{Q}}_\varphi$ or $\mathcal{R}_\varphi$.

In order to obtain an evolution equation for $\mathcal{Q}_\varphi$, we start with the Mukhanov-Sasaki equation from Eqs. (5.28) and (5.29),

$$\left[\partial_\tau^2 - \nabla^2 + \widehat{m}^2(\tau)\right]\widehat{\mathcal{Q}}_\varphi \overset{(5.28)}{=} 0\,, \quad \widehat{m}^2(\tau) \overset{(5.29)}{=} -\frac{\mathcal{H}}{a\bar\varphi'}\left(\frac{a\bar\varphi'}{\mathcal{H}}\right)'' . \tag{6.48}$$

Similarly to Eqs. (6.5)–(6.7), we substitute

$$\widehat{\mathcal{Q}}_\varphi \overset{(5.16)}{=} a\mathcal{Q}_\varphi\,, \tag{6.49}$$

$$\widehat{\mathcal{Q}}_\varphi' = a'\mathcal{Q}_\varphi + a\mathcal{Q}_\varphi'\,, \tag{6.50}$$

$$\widehat{\mathcal{Q}}_\varphi'' = a''\mathcal{Q}_\varphi + 2a'\mathcal{Q}_\varphi' + a\mathcal{Q}_\varphi''\,, \tag{6.51}$$

and divide then the whole equation by a, leading to

$$\mathcal{Q}_\varphi'' + 2\mathcal{H}\mathcal{Q}_\varphi' + \left(\frac{a''}{a} - \nabla^2 + \widehat{m}^2\right)\mathcal{Q}_\varphi = 0\,. \tag{6.52}$$

Then we go to physical time, like in Eq. (6.10),

$$\mathcal{Q}_\varphi' \overset{(6.9)}{=} a\dot{\mathcal{Q}}_\varphi\,, \quad \mathcal{Q}_\varphi'' \overset{(6.10)}{=} a^2\left(\ddot{\mathcal{Q}}_\varphi + \frac{\dot a}{a}\dot{\mathcal{Q}}_\varphi\right) . \tag{6.53}$$

Dividing by a^2, Eq. (6.52) gets thereby converted into

$$\ddot{\mathcal{Q}}_\varphi + 3H\dot{\mathcal{Q}}_\varphi + \left(\frac{\ddot a}{a} + \frac{\dot a^2}{a^2} - \frac{\nabla^2}{a^2} + \frac{\widehat{m}^2}{a^2}\right)\mathcal{Q}_\varphi = 0\,. \tag{6.54}$$

For the mass parameter appearing here, we write

$$\frac{\widehat{m}^2}{a^2} \overset{(6.48)}{=} -\frac{H}{a^3\dot{\bar\varphi}} \times a\partial_t\left[\overbrace{a\partial_t\left(\frac{a\dot{\bar\varphi}}{H}\right)}^{a\dot a\frac{\dot{\bar\varphi}}{H} + a^2\left(\frac{\dot{\bar\varphi}}{H}\right)^{\cdot}}\right]$$

$$= -\frac{H}{a^2\dot{\bar\varphi}}\left[\dot a^2\frac{\dot{\bar\varphi}}{H} + a\ddot a\frac{\dot{\bar\varphi}}{H} + 3a\dot a\left(\frac{\dot{\bar\varphi}}{H}\right)^{\cdot} + a^2\left(\frac{\dot{\bar\varphi}}{H}\right)^{\cdot\cdot}\right]$$

$$= -\frac{\dot a^2}{a^2} - \frac{\ddot a}{a} - 3H \times \frac{H}{\dot{\bar\varphi}}\left(\frac{\dot{\bar\varphi}}{H}\right)^{\cdot} - \frac{H}{\dot{\bar\varphi}}\left(\frac{\dot{\bar\varphi}}{H}\right)^{\cdot\cdot} . \tag{6.55}$$

Recalling Eq. (6.12), we find

$$\mathcal{F} \stackrel{(6.12)}{=} \frac{H}{\dot{\varphi}}\left(\frac{\dot{\varphi}}{H}\right)^{\cdot} = \frac{\ddot{\varphi}}{\dot{\varphi}} - \frac{\dot{H}}{H} \tag{6.56}$$

$$\Rightarrow \quad \dot{\mathcal{F}} = \underbrace{\frac{\dot{H}}{\dot{\varphi}}\left(\frac{\dot{\varphi}}{H}\right)^{\cdot} - \frac{H\ddot{\varphi}}{(\dot{\varphi})^2}\left(\frac{\dot{\varphi}}{H}\right)^{\cdot}}_{-\mathcal{F}^2} + \frac{H}{\dot{\varphi}}\left(\frac{\dot{\varphi}}{H}\right)^{\cdot\cdot} . \tag{6.57}$$

Inserting Eqs. (6.55) and (6.57), and going to momentum space, Eq. (6.54) becomes

$$\boxed{\ddot{\mathcal{Q}}_k + 3H\dot{\mathcal{Q}}_k + \left(\frac{k^2}{a^2} - 3\mathcal{F}H - \dot{\mathcal{F}} - \mathcal{F}^2\right)\mathcal{Q}_k = 0 .} \tag{6.58}$$

We now recall from Eqs. (6.21) and (6.22) and the discussion below them that in the slow-roll regime, $\mathcal{F} \approx (2\epsilon_v - \eta_v)H \ll H$. Therefore, k^2/a^2 dominates the third term in Eq. (6.58) as long as $k/a \geq H$. Under this assumption, the previously determined conformal-time mode function from Eq. (5.46) indeed solves Eq. (6.58). To show this, we consider

$$\mathcal{Q}_k(\tau) \stackrel{(5.16)}{=} \frac{\widehat{\mathcal{Q}}_k(\tau)}{a} \stackrel{(5.46)}{=} \frac{e^{-ik\tau}}{\sqrt{2k}}\left(\tau - \frac{i}{k}\right)\frac{1}{a\tau} \stackrel{(5.40)}{\approx} -\frac{H}{\sqrt{2k}}\left(\tau - \frac{i}{k}\right)e^{-ik\tau} , \tag{6.59}$$

where we inserted $H \approx -1/(a\tau)$ from Eq. (5.40). Being in quasi de Sitter space-time, H is treated as approximately constant. Then

$$\partial_t \mathcal{Q}_k(\tau) \stackrel{(6.9)}{=} \frac{\mathcal{Q}_k'(\tau)}{a} \stackrel{(6.59)}{\approx} -\frac{H}{a\sqrt{2k}}\left[\cancel{1} - ik\left(\tau - \cancel{\frac{i}{k}}\right)\right]e^{-ik\tau}$$

$$\underset{H \approx -1/(a\tau)}{\stackrel{(5.40)}{\approx}} \frac{H^2}{\sqrt{2k}}\left(-ik\tau^2\right)e^{-ik\tau} , \tag{6.60}$$

$$\partial_t^2 \mathcal{Q}_k(\tau) \underset{(6.60)}{\stackrel{(6.9)}{=}} \frac{H^2}{a\sqrt{2k}}\left(-2ik\tau - k^2\tau^2\right)e^{-ik\tau}$$

$$\underset{H \approx -1/(a\tau)}{\stackrel{(5.40)}{\approx}} \frac{H^3}{\sqrt{2k}}\left(2ik\tau^2 + k^2\tau^3\right)e^{-ik\tau} . \tag{6.61}$$

It follows that

$$\left(\partial_t^2 + 3H\partial_t + \frac{k^2}{a^2}\right)\mathcal{Q}_k(\tau)$$

$$\approx \frac{H^3}{\sqrt{2k}}\left[\overbrace{2ik\tau^2 + k^2\tau^3}^{\text{from (6.61)}} \overbrace{- 3ik\tau^2}^{\text{from (6.60)}} - \overbrace{\underbrace{\frac{k^2}{a^2 H^2}}_{\approx k^2\tau^2}\left(\cancel{\tau} - \cancel{\frac{i}{k}}\right)}^{\text{from (6.59)}}\right]e^{-ik\tau} \approx 0 . \tag{6.62}$$

From Eq. (6.60), we also obtain the information that $\partial_t \mathcal{Q}_k(\tau) \approx 0$ when $\tau \to 0^-$, i.e. that $\mathcal{Q}_k$ is constant when the modes are far outside of the Hubble horizon.

Equipped with this knowledge, we take inspiration from Sect. 6.3 and define IR perturbations in analogy with the mode expansion in Eq. (6.30), i.e. by subtracting the UV part from the full mode expansion,

$$\mathcal{Q}_>(t, \mathbf{x}) \equiv \int \frac{d^3 \mathbf{k}}{\sqrt{(2\pi)^3}} \left[1 - W_k(t) \right] \left[w_{\mathbf{k}} \, \mathcal{Q}_k(\tau) \, e^{i\mathbf{k}\cdot\mathbf{x}} + w_{\mathbf{k}}^\dagger \, \mathcal{Q}_k^*(\tau) \, e^{-i\mathbf{k}\cdot\mathbf{x}} \right],$$

$$(6.63)$$

where the window function is from Eq. (6.47). Let us consider a time derivative of this operator. Because of the weight $1 - W_k$, each momentum k contributes only once the mode is far outside of the Hubble horizon, but as follows from Eq. (6.60) and was discussed below Eq. (6.62), $\mathcal{Q}_k(\tau)$ is approximately constant in this domain. Therefore the time derivative acts only on the window function, leading to

$$\partial_t \left[1 - W_k(t) \right] \overset{(6.47)}{\underset{H \approx \text{const}}{\approx}} \epsilon \dot{a}(t) H \delta\big(k - \epsilon a(t) H\big)$$

$$= \epsilon \dot{a}(t) H \, \delta\{ k - \epsilon[a(t_*) + \dot{a}(t_*)(t - t_*)] H \}$$

$$= \frac{\epsilon \dot{a}(t) H}{|\epsilon \dot{a}(t_*) H|} \delta\big(t - t_*(k) \big) \overset{\dot{a} \geq 0}{=} \delta\big(t - t_*(k) \big), \qquad (6.64)$$

where $t_*(k)$ is the moment at which a given momentum exits the Hubble horizon, i.e. $k \equiv \epsilon a(t_*) H$. All in all, the time derivative of the IR mode function reads

$$\dot{\mathcal{Q}}_>(t, \mathbf{x}) \approx \varrho_Q(t, \mathbf{x}), \qquad (6.65)$$

$$\varrho_Q(t, \mathbf{x}) \overset{(6.63)}{\underset{(6.64)}{\equiv}} \int \frac{d^3 \mathbf{k}}{\sqrt{(2\pi)^3}} \, \delta\big(t - t_*(k) \big) \left[w_{\mathbf{k}} \, \mathcal{Q}_k(\tau) \, e^{i\mathbf{k}\cdot\mathbf{x}} + w_{\mathbf{k}}^\dagger \, \mathcal{Q}_k^*(\tau) \, e^{-i\mathbf{k}\cdot\mathbf{x}} \right]. \quad (6.66)$$

Here, in analogy with Eq. (6.31), we have introduced the notion of a *quantum noise*.

In the literature, the conventional next step is to determine the autocorrelator of the quantum noise in coordinate space. We obtain [4]

$$\langle 0 | \varrho_Q(t_1, \mathbf{x}_1) \varrho_Q(t_2, \mathbf{x}_2) | 0 \rangle$$

$$\overset{(6.66)}{=} \int \frac{d^3 \mathbf{k} \, d^3 \mathbf{q}}{(2\pi)^3} \, \delta\big(t_1 - t_*(k) \big) \delta\big(t_2 - t_*(q) \big) \langle 0 | w_{\mathbf{k}} w_{\mathbf{q}}^\dagger | 0 \rangle \, \mathcal{Q}_k(\tau_1) \mathcal{Q}_q^*(\tau_2) \, e^{i(\mathbf{k}\cdot\mathbf{x}_1 - \mathbf{q}\cdot\mathbf{x}_2)}$$

$$\overset{(5.33)}{=} \int \frac{d^3 \mathbf{k}}{(2\pi)^3} \, \delta\big(t_1 - t_*(k) \big) \delta\big(t_2 - t_1 \big) \, |\mathcal{Q}_k(\tau_1)|^2 \, e^{i\mathbf{k}\cdot(\mathbf{x}_1 - \mathbf{x}_2)}$$

$$\overset{(6.59)}{\underset{(6.64)}{\approx}} \delta\big(t_1 - t_2 \big) \int_0^\infty \frac{dk \, k^2}{4\pi^2} \underbrace{\overbrace{\epsilon a(t_1) H^2 \, \delta\big(k - \epsilon a(t_1) H\big)}^{\delta(t_1 - t_*(k)) \text{ via } (6.64)}}_{kH} \overbrace{|\mathcal{Q}_k(\tau_1)|^2}^{(6.59):\, \frac{H^2}{2k^3}} \overbrace{\int_{-1}^{+1} dz \, e^{ik|\mathbf{x}_1 - \mathbf{x}_2|z}}^{\frac{2\sin(k|\mathbf{x}_1 - \mathbf{x}_2|)}{k|\mathbf{x}_1 - \mathbf{x}_2|}}$$

$$= \delta\big(t_1 - t_2\big) \frac{H^3}{4\pi^2} \frac{\sin[\,\epsilon a(t_1)H|\mathbf{x}_1 - \mathbf{x}_2|\,]}{\epsilon a(t_1)H|\mathbf{x}_1 - \mathbf{x}_2|} . \tag{6.67}$$

In the penultimate step, we made use of the asymptotics of Eq. (6.59) outside of the Hubble horizon. We thus find *white noise* in the time direction, similarly to Eq. (1.63). However, the oscillatory behaviour of Eq. (6.67) in the spatial directions, and the strong dependence on the arbitrary parameter ϵ, look artificial. Nevertheless, the $\epsilon \to 0$ limit (corresponding to $k \to 0$ according to Eq. (6.47)) exists, and is often used as a quantum noise for the evolution of the background field, $\bar{\varphi}$.

The meaning of the noise becomes more transparent if we go to momentum space. Fourier transforming Eq. (6.66), we find

$$\varrho_Q(t, \mathbf{k}) \stackrel{(9)}{=} \int \mathrm{d}^3\mathbf{x}\, \varrho_Q(t, \mathbf{x})\, e^{-i\mathbf{k}\cdot\mathbf{x}}$$

$$\stackrel{(6.66)}{=} \sqrt{(2\pi)^3}\, \delta\big(t - t_*(k)\big)\Big[\, w_{\mathbf{k}}\, \mathcal{Q}_k(\tau) + w^\dagger_{-\mathbf{k}}\, \mathcal{Q}_k^*(\tau)\,\Big]. \tag{6.68}$$

Then the autocorrelator reads

$$\langle 0|\varrho_Q(t_1, \mathbf{k})\varrho_Q(t_2, \mathbf{q})|0\rangle$$

$$\stackrel{(6.68)}{=} (2\pi)^3\delta\big(t_1 - t_*(k)\big)\delta\big(t_2 - t_*(q)\big)\,\langle 0|w_{\mathbf{k}}w^\dagger_{-\mathbf{q}}|0\rangle\, \mathcal{Q}_k(\tau_1)\, \mathcal{Q}_q^*(\tau_2)$$

$$\stackrel{(5.33)}{=} (2\pi)^3\delta^{(3)}(\mathbf{k} + \mathbf{q})\, \delta\big(t_1 - t_*(k)\big)\delta\big(t_2 - t_*(q)\big)\, \underbrace{|\mathcal{Q}_k(\tau_1)|^2}_{(6.59):\ \frac{H^2}{2k^3}} . \tag{6.69}$$

Here we again made use of the small-τ asymptotics of Eq. (6.59). What Eq. (6.69) signifies is that the noise autocorrelator is zero most of the time. It is non-vanishing only at the moment when a given mode exits the Hubble horizon.

More concretely, we may interpret Eq. (6.65) as an evolution equation, but viewing as the underlying variable not the IR field itself, $\mathcal{Q}_>(t, \mathbf{x})$, but rather the corresponding mode function, which we denote by $\mathcal{Q}_{k>}(t)$. Then Eq. (6.69) implies that

$$\dot{\mathcal{Q}}_{k>}(t) \underset{(6.69)}{\overset{(6.65)}{\simeq}} \delta\big(t - t_*(k)\big)\, \frac{H}{\sqrt{2k^3}} . \tag{6.70}$$

Integrating this equation in time with a vanishing initial condition, $\mathcal{Q}_{k>}$ jumps to its final value as the corresponding mode crosses into the IR domain. The power spectrum following from Eq. (6.70) reads

$$\mathcal{P}_{\mathcal{Q}_\varphi}(t_{\mathrm{out}}, k) \stackrel{t_{\mathrm{out}} \geq t_*}{=} \frac{k^3}{2\pi^2} \frac{H^2}{2k^3} = \frac{H^2}{(2\pi)^2} , \tag{6.71}$$

which agrees with the second term of Eq. (5.53).

To summarize, we have shown that, staying at linear order and restricting ourselves to the slow-roll regime, the simplified version of the stochastic formalism actually corresponds to a *non-stochastic* evolution equation, Eq. (6.70). The strength of the stochastic formalism lies in its presumed generalization beyond the linear order, however this property is not easy to prove rigorously. In the non-linear case, different momenta are not independent, and the evolution equation is typically solved in coordinate rather than momentum space, with a noise autocorrelator reminiscent of Eq. (6.67). In practice, the equation becomes non-linear if the Hubble rate appearing on the right-hand side is made a function of $\mathcal{Q}_\varphi$, as would be natural in the separate universes picture [5,6].

6.5   Numerical Solution for Curvature Power Spectrum

In this section we show how the evolution Eq. (6.11), with initial conditions fixed according to Eqs. (6.15) and (6.17), can be solved numerically, in order to obtain the curvature power spectrum from Eq. (6.13). The numerical result is plotted in Fig. 6.1.

It is worth stressing that the numerical solution of Eq. (6.11) poses a number of challenges, if we want to integrate over a long period of time (a discussion of dedicated softwares can be found in Ref. [8]). At early times, $\mathcal{R}_k$ undergoes rapid phase oscillations. These are expensive to treat, and therefore we can start the integration of $\mathcal{R}_k$ only when $k/(aH)$ is moderate, say $k/(aH) \sim 10^3$. Once the modes exit the Hubble horizon, the rapid phase oscillations cease, and $\mathcal{R}_k$ freezes out. Since its frozen value is the main variable of our interest, this dynamics has to be determined precisely. After a while, the background field, $\bar{\varphi}$, may start to oscillate. Then the coefficient $\mathcal{F}$ in Eq. (6.11) can become singular (cf. Eq. (6.12)). The singular value needs to be regularized, but care is required, as we want to render the equation solvable without changing the nature of the solution. A procedure for this, making use of complexified variables, has been discussed in Ref. [9]. Finally, at late times, it is costly to treat every single background oscillation individually. Then we can go over to an averaged solution, as explained in Sect. 1.7. Below we show a `python` script which implements these ingredients (the evolution has been split into 4 periods, as explained in the script, so that the accuracy of each can be tuned separately if necessary).

```python
# numerical solution for curvature perturbations in the cold regime [numerics_curvature.py]
#
# import basic tools and integration routines
import numpy as np
from scipy.integrate import solve_ivp

# parameters [mpl]
fa = 1.25              # inflaton decay constant
m = 1.09e-6            # inflaton mass
koa_ref = 7.35e52      # momentum of perturbations at initial time
delta = 1.e-15         # regulator for circumventing poles [mpl^2]
Pi = np.pi

# potential of inflaton (phi) [mpl^4]
def V(phi): return m*m*fa*fa*( 1 - np.cos(phi/fa) )

# derivative of inflaton potential by phi [mpl^3]
def Vd(phi): return m*m*fa*np.sin(phi/fa)

# total energy density of inflaton field [mpl^4]
def etot(phi,phid): return phid*phid/2 + V(phi)
```

```python
# Hubble rate [mpl]
def H(energy_density): return np.sqrt(8*Pi*np.abs(energy_density)/3)

# time derivative of Hubble rate [mpl^2]
def dotH(phid): return - 4*Pi*phid*phid

# solve for [dot phi, ddot phi, number of efolds] for the background evolution
# derivatives are taken with respect to t = Href*time
def derivatives1(t,y):
    Phi, Phid, Nfolds = y
    Hubble = H(etot(Phi, Phid));          dN_dt = Hubble/Href
    dphi_dt = Phid/Href;                  ddphi_ddt = ( -3*Hubble*Phid - Vd(Phi) )/Href
    dy_dt = [dphi_dt, ddphi_ddt, dN_dt]
    return dy_dt

# solve for [dot phi, ddot phi, number of efolds, dot Rk, ddot Rk] including curvature perturbations
def derivatives2(t,y):
    Phi, Phid, efolds, reRk, imRk, reRkd, imRkd = y
    Hubble = H(etot(Phi, Phid));          dHubble = dotH(Phid);          dN_dt = Hubble/Href
    dphi_dt = Phid/Href;                  ddphi_ddt = ( -3*Hubble*Phid - Vd(Phi) )/Href
    recalF = ddphi_ddt*Href*np.real(1/(Phid + 1j* delta)) - dHubble/Hubble
    imcalF = ddphi_ddt*Href*np.imag(1/(Phid + 1j* delta))
    dreRk_dt = reRkd/Href;                dimRk_dt = imRkd/Href
    ddreRk_ddt = - ( 2*(recalF*reRkd-imcalF*imRkd) + 3*Hubble*reRkd
                    + koa_ref*koa_ref*np.exp(-2*efolds)*reRk )/Href
    ddimRk_ddt = - ( 2*(recalF*imRkd+imcalF*reRkd) + 3*Hubble*imRkd
                    + koa_ref*koa_ref*np.exp(-2*efolds)*imRk )/Href
    dy_dt = [dphi_dt, ddphi_ddt, dN_dt, dreRk_dt, dimRk_dt, ddreRk_ddt, ddimRk_ddt]
    return dy_dt

# solve for [e_phi, number of efolds, dot Rk, ddot Rk] after averaging over oscillations
def derivatives3(t,y):
    e_phi, efolds, reRk, imRk, reRkd, imRkd = y
    Hubble = H(e_phi);                    dHubble = -4*Pi*e_phi;          dN_dt = Hubble/Href
    recalF =  - dHubble/Hubble;           imcalF = 0
    dreRk_dt = reRkd/Href;                dimRk_dt = imRkd/Href
    ddreRk_ddt = - ( 2*(recalF*reRkd-imcalF*imRkd) + 3*Hubble*reRkd
                    + koa_ref*koa_ref*np.exp(-2*efolds)*reRk )/Href
    ddimRk_ddt = - ( 2*(recalF*imRkd+imcalF*reRkd) + 3*Hubble*imRkd
                    + koa_ref*koa_ref*np.exp(-2*efolds)*imRk )/Href
    dy_dt = [-3*Hubble*e_phi/Href, dN_dt, dreRk_dt, dimRk_dt, ddreRk_ddt, ddimRk_ddt]
    return dy_dt

# initial conditions and reference values
phi_0 = 3.5                              # mpl
Href = np.sqrt(4*Pi/3)*m*phi_0          # mpl
phid_0 = - Vd(phi_0)/(3*Href)           # approximate slow-roll value for initial derivative [mpl^2]
N_0 = 0                                  # initial e-folds
koaH_start = 1e3                         # when to start solving for perturbations
time_end = 250                           # choose large enough that desired koaH is reached

# event function monitoring k/aH (here _not_ the correct H, rather the initial approximation)
def koaH(t, y):  return koa_ref*np.exp(-y[2])/Href - koaH_start

# integrate without curvature perturbations until k/aH obtains prescribed value koaH_start
sol1 = solve_ivp(derivatives1, [1, time_end], [phi_0, phid_0, N_0], events=koaH, rtol=1e-10)
t_match1 = sol1.t_events[0][0]          # point at which desired value was reached
time1 = sol1.t[sol1.t <= t_match1];     n_match1 = len(time1)
phi1 = sol1.y[0][:n_match1];            phid1 = sol1.y[1][:n_match1]
ephi1 = etot(phi1, phid1);             efolds1 = sol1.y[2][:n_match1]
reRk1 = np.ones(n_match1);              imRk1 = np.zeros(n_match1)
slowroll1 = H(ephi1)**4/(2*Pi*phid1)**2

# define initial conditions for curvature perturbations at moment t_match1
time_0 = t_match1
phi_0 = sol1.y_events[0][0][0]; phid_0 = sol1.y_events[0][0][1]; N_0 = sol1.y_events[0][0][2]
reRk_0 = -H(etot(phi_0,phid_0))/(2*Pi*phid_0)*koa_ref*np.exp(-N_0);  imRk_0 = 0.
dreRk_0 = 0.;                           dimRk_0 = - reRk_0* koa_ref*np.exp(-N_0)
print("start Rk-evolution at t/tref = ",time_0,", with k/aH approx ",koaH_start)
koaH_start = 10                         # re-adjust target k/aH
time_end = 250                          # choose large enough that desired koaH is reached

# integrate with curvature perturbations until k/aH obtains prescribed value koaH_start
sol2 = solve_ivp(derivatives2, [time_0,time_end],
                [phi_0, phid_0, N_0, reRk_0, imRk_0, dreRk_0, dimRk_0],
                events=koaH,t_eval=np.linspace(time_0,time_end,1000),
                atol=1e-12,rtol=1e-13,method='DOP853')
```

```python
t_match2 = sol2.t_events[0][0]            # point at which desired value was reached
time2 = sol2.t[sol2.t <= t_match2];       n_match2 = len(time2)
phi2 = sol2.y[0][:n_match2];              phid2 = sol2.y[1][:n_match2]
ephi2 = etot(phi2,phid2);                 efolds2 = sol2.y[2][:n_match2]
reRk2 = sol2.y[3][:n_match2];             imRk2 = sol2.y[4][:n_match2]
slowroll2 = H(ephi2)**4/(2*Pi*phid2)**2


# define initial conditions for next range (crossing outside of Hubble horizon)
time_0 = t_match2
phi_0 = sol2.y_events[0][0][0] ;  phid_0 = sol2.y_events[0][0][1]; N_0 = sol2.y_events[0][0][2]
reRk_0 = sol2.y_events[0][0][3];      imRk_0 = sol2.y_events[0][0][4]
dreRk_0 = sol2.y_events[0][0][5];     dimRk_0 = sol2.y_events[0][0][6]
print("start horizon crossing at t/tref = ",time_0,", with k/aH approx ",koaH_start)
time_end = 1e3                            # choose large enough that 10 full oscillations take place


# define the event function counting the zeros of phi
def phi_crossings(t, y):  return y[0]


# integrate up to 10 oscillations (21 crossings of zero)
sol3 = solve_ivp(derivatives2, [time_0,time_end],
                 [phi_0, phid_0, N_0, reRk_0, imRk_0, dreRk_0, dimRk_0],
                 events=phi_crossings,t_eval=np.linspace(time_0,time_end,1000),
                 atol=1e-12,rtol=1e-13,method='DOP853')
t_match3 = sol3.t_events[0][20]           # point at which desired value was reached
time3 = sol3.t[sol3.t <= t_match3];       n_match3 = len(time3)
phi3 = sol3.y[0][:n_match3];              phid3 = sol3.y[1][:n_match3]
ephi3 = etot(phi3,phid3);                 efolds3 = sol3.y[2][:n_match3]
reRk3 = sol3.y[3][:n_match3];             imRk3 = sol3.y[4][:n_match3]
slowroll3 = H(ephi3)**4/(2*Pi*phid3)**2


# define initial conditions for averaged regime
time_0 = t_match3;                        time_end = 1e8
phi_0 = sol3.y_events[0][20][0];          phid_0 = sol3.y_events[0][20][1]
etot_0 = etot(phi_0,phid_0);              N_0 = sol3.y_events[0][20][2]
reRk_0 = sol3.y_events[0][20][3];         imRk_0 = sol3.y_events[0][20][4]
dreRk_0 = sol3.y_events[0][20][5];        dimRk_0 = sol3.y_events[0][20][6]
print("start averaged regime at t/tref = ",time_0)


# integrate in averaged regime (matter-dominated era)
sol4 = solve_ivp(derivatives3, [time_0, time_end],
                 [etot_0, N_0, reRk_0, imRk_0, dreRk_0, dimRk_0],
                 atol=1e-12,rtol=1e-13,method='DOP853')
time4 = sol4.t;                           n_match4 = len(time4)
ephi4 = sol4.y[0];                        efolds4 = sol4.y[1]
reRk4 = sol4.y[2];                        imRk4 = sol4.y[3];              slowroll4 = np.zeros(n_match4)


# assemble together the complete solution
time = np.concatenate((time1,time2,time3,time4)); e_phi = np.concatenate((ephi1,ephi2,ephi3,ephi4))
efolds = np.concatenate((efolds1,efolds2,efolds3,efolds4)); koa = koa_ref*np.exp(-efolds)
reRk = np.concatenate((reRk1,reRk2,reRk3,reRk4)); imRk = np.concatenate((imRk1,imRk2,imRk3,imRk4))
slowroll = np.concatenate((slowroll1,slowroll2,slowroll3,slowroll4))


# print to file
np.savetxt('numerics_curvature.dat',
           np.c_[time, e_phi, H(e_phi), efolds, koa, reRk**2+imRk**2, slowroll],fmt='%.6e',newline='\n',
           header='columns: t*H_ref, e_phi/mpl**4, H/mpl, efolds, k_ref/a/mpl, P_R, slowroll')
```

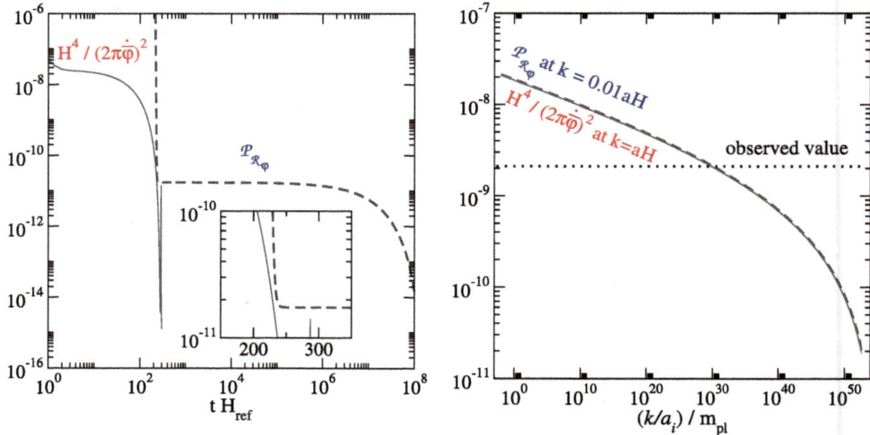

Fig. 6.1 Left: The dashed blue line shows a numerical solution of Eq. (6.11), with initial conditions fixed through Eqs. (6.15) and (6.17), and the result normalized according to Eq. (6.13). The parameters are like in Fig. 1.2 in Sect. 1.7. The solid red curve shows the result from Eq. (5.54) but *without* fixing to the moment H_* of horizon crossing. We see how $\mathcal{P}_{\mathcal{R}_\varphi}$ is normally much larger than $H^4/(2\pi\dot{\varphi})^2$, due to the term k^2/a^2 in Eq. (5.53). However, when k/a becomes smaller than H, $\mathcal{P}_{\mathcal{R}_\varphi}$ freezes out to the value that it had at that moment. Afterwards, $H^4/(2\pi\dot{\varphi})^2$ continues evolving, becoming oscillatory when $\dot{\varphi}$ starts crossing zeros. In contrast, $\mathcal{P}_{\mathcal{R}_\varphi}$ stays fixed, until the mode crosses back inside the Hubble horizon. The initial decrease of $H^4/(2\pi\dot{\varphi})^2$ is due to the fact that the initial condition for $\dot{\varphi}$ from Eq. (1.103) is approximate, because H_{ref} is not the correct Hubble rate, and it takes a few Hubble times to adjust to the attractor trajectory. Right: The dependence of $\mathcal{P}_{\mathcal{R}_\varphi}$, evaluated at $k = 0.01\,aH$, and $H^4/(2\pi\dot{\varphi})^2$, evaluated at $k = aH$, on the initial momentum, fixed at $t = t_i = H_{\text{ref}}^{-1}$ (cf. Eq. (1.104)). We find that $\mathcal{P}_{\mathcal{R}_\varphi}$ at $k = 0.01\,aH$ is $(2-5)\%$ larger than $H^4/(2\pi\dot{\varphi})^2$ at $k = aH$. The results are compared with the observed $\mathcal{P}_{\mathcal{R}_\varphi}(k_*)$, given in Eq. (2.8). Whether the k at which the observed value is reached indeed corresponds to k_*, depends on the reheating history after inflation, and we return to this in Fig. 7.2 in Sect. 7.7

References

1. Liddle, A.R., Parsons, P., Barrow, J.D.: Formalizing the slow-roll approximation in inflation. Phys. Rev. D **50**, 7222 (1994). https://doi.org/10.1103/PhysRevD.50.7222, astro-ph/9408015
2. Auclair, P., Ringeval, C.: Slow-roll inflation at N3LO. Phys. Rev. D **106**, 063512 (2022). https://doi.org/10.1103/PhysRevD.106.063512, 2205.12608
3. Sasaki, M., Nambu, Y., Nakao, K.-i.: Classical behavior of a scalar field in the inflationary universe. Nucl. Phys. B **308**, 868 (1988). https://doi.org/10.1016/0550-3213(88)90132-0
4. Starobinsky, A.A., Yokoyama, J.: Equilibrium state of a self-interacting scalar field in the de Sitter background. Phys. Rev. D **50**, 6357 (1994). https://doi.org/10.1103/PhysRevD.50.6357, astro-ph/9407016
5. Salopek, D.S., Bond, J.R.: Nonlinear evolution of long-wavelength metric fluctuations in inflationary models. Phys. Rev. D **42**, 3936 (1990). https://doi.org/10.1103/PhysRevD.42.3936
6. Wands, D., Malik, K.A., Lyth, D.H., Liddle, A.R.: New approach to the evolution of cosmological perturbations on large scales. Phys. Rev. D **62**, 043527 (2000). https://doi.org/10.1103/PhysRevD.62.043527, astro-ph/0003278

7. Sasaki, M., Tanaka, T.: Super-Horizon Scale Dynamics of Multi-Scalar Inflation. Prog. Theor. Phys. **99**, 763 (1998). https://doi.org/10.1143/PTP.99.763, gr-qc/9801017
8. Haddadin, W.I.J., Handley, W.J.: Rapid numerical solutions for the Mukhanov-Sasaki equation. Phys. Rev. D **103**, 123513 (2021). https://doi.org/10.1103/PhysRevD.103.123513, 1809.11095
9. Laine, M., Procacci, S., Rogelj, A.: Evolution of coupled scalar perturbations through smooth reheating. Part I. Dissipative regime. JCAP **10**, 040 (2024). https://doi.org/10.1088/1475-7516/2024/10/040, 2407.17074

Evolution Equations in the Presence of a Thermalizing Plasma

7

Abstract

As time goes by, other matter components than the inflaton field play an increasingly important role. The expectation is that some of them should be Standard Model particles, interacting fairly strongly with each other, and eventually thermalizing, setting up the required environment for big-bang nucleosynthesis. The equilibrated system is called a primordial plasma, while the equilibration process, culminating in a radiation-dominated universe, is known as 'reheating'. Introducing generic couplings between the inflaton and the plasma, we write down the corresponding background and perturbed equations. We show how a 'seed' temperature may emerge as a fixed point of the background solution already during the slow-roll stage of inflation. We demonstrate how reheating influences inflationary predictions, through the overall redshift between horizon crossing and the late universe. We indicate how interactions damp initial quantum fluctuations, but also generate new thermal fluctuations, via thermal noise.

Keywords

Temperature · Langevin equation · Equipartition · Smooth reheating · Redshift · Inflaton and plasma equilibration rates · Thermal noise · Fluctuation-dissipation relation · Rayleigh-Jeans divergence · Quantum statistical physics · Coupled curvature perturbations

7.1 What Is Temperature?

When we discuss thermal effects, which ultimately lead to the notions of a *hot big bang* and the generation of a *primordial plasma*, the question of how temperature is defined is a central one. In text-book statistical physics, a starting point is offered by the *microcanonical ensemble*. Let us inspect a closed system of a total energy E. The *microcanonical partition function* $\Omega(E)$ counts the number of states in an energy interval $(E - \Delta E, E]$. The Boltzmann entropy is defined as $S(E) \equiv \ln \Omega(E)$, and

M. Laine and S. Procacci, *From Inflation to Hot Big Bang*, Lecture Notes in Physics 1047,
https://doi.org/10.1007/978-3-032-09893-1_7

the temperature as

$$\frac{1}{T} \equiv \frac{\partial S(E)}{\partial E} . \tag{7.1}$$

Multiplying by $T \, \mathrm{d}E$, this yields the first law of thermodynamics, $\mathrm{d}E = T \, \mathrm{d}S + \cdots$.

There is, however, an implicit assumption in this definition of the temperature. The relations above make no reference to interactions, only to the counting of states in the vicinity of a given energy (this could be done even in a non-interacting system). The implicit assumption is that of *equipartition*: all those states should be filled with equal probability, $1/\Omega(E)$. This implies that the thermal system contains minimal information: if we carry out physical computations, which require reactions between states, all possibilities allowed by energy conservation are treated on equal footing, and averaged over.

In the real world, establishing equipartition from a given initial state takes time. For instance, let us think of a system made of unstable heavy particles (such as inflatons) and their decay products (perhaps the much lighter Standard Model particles). In an initial decay, the large energy released is carried mostly by the momenta of the decay products. But this is only one of the possible states of the same energy. Another would be that the energy is redistributed evenly between, say, a hundred light particles. Combinatorially, counting the directions of the momenta, there are many more states of the latter type. In thermal equilibrium, *all possible final states* should be filled with equal probability.

It should be clear from the picture described that in any realistic setting, it is impossible to carry out an exact computation of how the initial energy released in the decay gets redistributed to all possible final states. Rather, we try to capture the efficiency of this process by defining an *equilibration rate*, Γ. We could compute Γ by first estimating a would-be T from the thermodynamic definition. Then, we can ask how efficiently the processes proceed within a *Hubble time*, $\Delta t \equiv H^{-1}$. If $\Gamma \Delta t \gg 1$, we can be confident that equipartition is established, and make use of the usual tools of equilibrium statistical physics. If $\Gamma \Delta t \ll 1$, there is no time for equipartition, and we say that the multiparticle system is 'out of equilibrium'. The latter type of systems are in general hard to handle.

Even if out-of-equilibrium systems are somewhat intractable, there is one generic idea that helps to classify them, and that is that we may view them as a collection of separate *subsystems*. In the example above, inflatons constitute one subsystem, the decay products another one. In full equilibrium, every reaction, including the initial decay, can also go in the opposite direction: many low-energy Standard Model particles could merge together, to form one inflaton particle. The rate of the inflaton equilibration processes, which we denote by Υ, is often much smaller than Γ. One reason is that, for simple model building, it is beneficial if inflaton-plasma interactions do not induce large radiative corrections to the inflaton potential. This can be achieved if the inflaton couples weakly to the Standard Model, for instance only via a higher-dimensional operator, suppressed by powers of m_{pl}. Such models possess the hierarchy $\Upsilon \ll \Gamma$. As illustrated in Fig. 7.1, this implies the presence of different epochs in the history of the universe, as the Hubble rate H decreases.

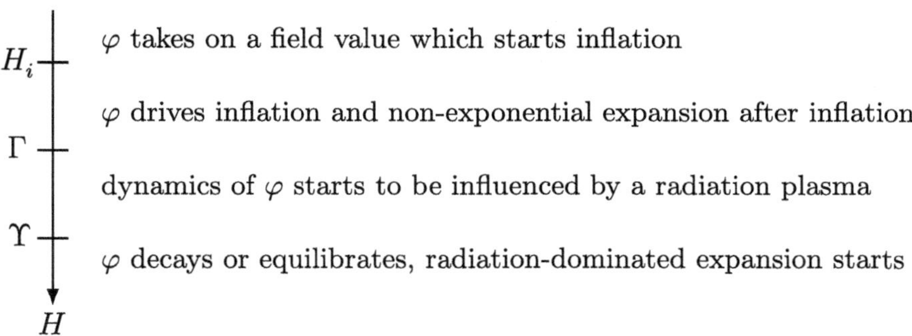

Fig. 7.1 A sketch of a typical timeline of how inflation ends, if $\Upsilon \ll \Gamma$. Here Υ is the equilibration rate of the low-momentum modes of the inflaton, and Γ is the equilibration rate of the plasma particles, with physical momenta $p \sim T$. As the Hubble rate decreases, any interacting particles that are present tend to equilibrate when $\Gamma > H$, and form a plasma. In the end, when $\Upsilon > H$, the inflaton background decays and the inflaton particles may equilibrate with the plasma

The remainder of this chapter is organized as follows. In Sect. 7.2, we show how Υ can be computed from a certain Green's function, given a coupling of φ to a radiation plasma. In Sect. 7.3, we discuss what can be said about the other plasma-induced ingredient in Eq. (1.62), the thermal noise ϱ. The background evolution equations satisfied by the radiation plasma are elaborated upon in Sect. 7.4, whereas in Sect. 7.5, we derive the evolution equations for curvature perturbations in the presence of Υ and ϱ. Section 7.6 illustrates what happens when $\Upsilon \gg H$, so that the inflaton itself may equilibrate, and Sect. 7.7 shows a numerical solution in the presence of a radiation plasma, from the inflationary epoch through reheating until a late universe. We aim to keep the discussion model independent, which means that it is to be understood in an effective theory sense.

The details of how the universe heated up after inflation [1,2] are, however, model dependent. Let us mention a few notions that appear frequently in the literature:

- An example of a radiation plasma equilibration rate, Γ, is that of a non-Abelian gauge theory. It has been studied extensively, motivated by heavy-ion collision experiments (but also cosmology, cf., e.g., Ref. [3]). In terms of microscopic processes, a necessary ingredient is to estimate the ensemble average of a cross section for 'large-angle' scattering, so that momenta are efficiently redistributed, leading to *kinetic equilibration*. In the limit of a weak self-coupling of the plasma, $\alpha \equiv g^2/(4\pi) \ll 1$ (implying that its temperature is well above the confinement scale), and considering physical momenta around the *thermal energy scale, $p \sim T$*, the result is $\Gamma \sim \alpha^2 T / \ln(\alpha^{-1})$ [4]. The rate of *helicity equilibration* has been studied in less detail, but is believed to be of order $\alpha^3 T$ at weak coupling (cf., e.g., Ref. [5]).

- The process of heating up is often called *reheating*. The name originates from the fact that early models of inflation [6–8] involved a phase transition taking place in the hot early universe at the corresponding critical temperature. After inflation, the universe must *heat up again*, to produce the hot environment observed in the CMB. However, as an alternative, an exponential expansion can also be driven by the energy density of a slowly evolving scalar field (cf., e.g., Ref. [9]), or other effects (cf., e.g., Ref. [10]), and no pre-existing thermal state is needed.

- The original stages of how the inflaton loses energy to other degrees of freedom, before a proper thermal state is established, are often referred to as *preheating* [11,12]. Nowadays preheating is typically studied by solving numerically a coupled set of non-linear classical field equations, for the inflaton field and for some 'spectators', which model the constituents of a radiation plasma. However, classical dynamics is a good approximation only for large occupation numbers, relevant for low-energy bosons, thanks to their Bose enhancement. Quantum mechanics was discovered by Planck to explain the spectrum of blackbody radiation, and correspondingly, quantum mechanics is essential for the main stage of the heating-up process.

- The normal tool for studying equilibration, once the occupation numbers are already close to unity, is *kinetic theory* (*Boltzmann equations*). However, this framework is not equivalent to the full quantum field theory, either. In fact, even for weakly coupled systems, it must often be combined with various 'resummations', in order to account for the emergence of 'quasiparticle' states, whereby the effective masses carried by different excitations are modified by interactions with the thermal ensemble (see, e.g., Ref. [4]). These effects can have a large qualitative effect, because modified masses may permit to open or close scattering channels, changing the ways in which equipartition can be established.

- A famous concept from quantum field theory in a curved background, notably in de Sitter space-time, is that of a *Gibbons-Hawking temperature* [13]. A Gibbons-Hawking temperature is obtained for any observer, independent of their velocity, and of the interactions that are present. In contrast, as discussed at the beginning of this section, for the physics that we are interested in, a temperature is a meaningful notion when interactions have time to re-distribute energy insertions. This leads to the corresponding equilibration rates, Γ and Υ. Once interactions are present, they also establish a *preferred frame*, the plasma rest frame: any moving excitation loses energy until it comes to rest with respect to this frame. The absence of these properties indicates that the Gibbons-Hawking temperature, though it can be defined, does *not* capture the plasma physics that is of relevance to us here.

- A notion somewhat related to the Gibbons-Hawking temperature is that of cosmological *gravitational particle production* (for reviews see, e.g., Refs. [14,15]). When the background metric is time-dependent, 'instantaneous' vacuum states can be defined at different times, and they are not equivalent to each other. What looked to be a vacuum state at an initial time seems to contain many particles if considered from the viewpoint of another vacuum. The semiclassical transition from one to another vacuum is called a *Bogolyubov transformation*. The particles produced when mode functions initialized in a Bunch-Davies vacuum are projected onto a post-reheating vacuum state might be interpreted as dark matter, for instance. In the particle-physics language, the same physics can be captured by Feynman diagrams of $n \to 2$ processes, where $n \geq 2$. Here n refers to inflaton particles, annihilating into an off-shell 's-channel' gravitational perturbation ($\succ\!\!\prec$), which in turn decays into dark matter particles (and in general Standard Model particles as well). The particles produced have a non-thermal spectrum, so that this mechanism does *not* explain reheating, though it could produce an initial state from which it can start.

We end by anticipating that the idea of many subsystems, mentioned above, applies not only to stages of the heating-up process, but also to a much later universe. For instance, a precise understanding of large-scale structure formation requires treating dark matter, neutrinos, baryons, and photons, as decoupled degrees of freedom.

7.2 How to Estimate the Inflaton Equilibration Rate, Υ?

The dynamics of a *scalar field interacting with a plasma* can under certain circumstances be described by a *Langevin equation*, given in Eqs. (1.62) and (1.63), *viz.*

$$\varphi^{;\mu}{}_{;\mu} - \Upsilon u^\mu \varphi_{,\mu} - V_{,\varphi} + \varrho \overset{(1.62)}{=} 0 \,, \quad \left\langle \varrho(\mathcal{X})\varrho(\mathcal{Y}) \right\rangle \overset{(1.63)}{\approx} \frac{\Omega\, \delta^{(4)}(\mathcal{X} - \mathcal{Y})}{\sqrt{-g}} \,. \tag{7.2}$$

Here the matching coefficients Υ and Ω describe the interactions of φ with the plasma, in particular Υ can be referred to as its *equilibration rate* (to be precise, it is the equilibration rate of the momentum modes $k/a \ll T$; that for $k/a \geq T$ is in general different, cf., e.g., Ref. [16]). Equation (7.2) should be viewed as a (dissipative) *low-energy effective theory*, in which the details of the plasma decomposition do not matter. However, we may backtrack to a more fundamental description, notably an action such as in Eq. (1.45), and ask how Υ and ϱ originate in that context?

Specifically, we consider a Lagrangian of the form

$$\mathcal{L} = -\frac{1}{2} g^{\mu\nu} \partial_\mu \varphi \, \partial_\nu \varphi - V_0(\varphi) - \varphi J + \mathcal{L}_{\mathrm{bath}} \,, \tag{7.3}$$

where V_0 is the inflaton self-interaction potential, J is an operator made of plasma fields, and $\mathcal{L}_{\mathrm{bath}}$ is a Lagrangian describing their kinetic terms and self-interactions. As discussed above, we assume that the coupling between φ and the plasma is weak; for instance, J could be suppressed by the Planck mass, m_{pl}. This then leads to a hierarchy $\Upsilon \ll \Gamma$.

If we have simultaneously $\Upsilon \ll \Gamma$, $k/a \ll \Gamma$, and $\widehat{m}/a \ll \Gamma$, where $\widehat{m}$ is from Eq. (6.2), then the evolution of φ, as described by Eq. (7.2), is slow compared with the equilibration dynamics of the plasma. The background field, $\bar{\varphi}$, then follows the evolution equation

$$\ddot{\bar{\varphi}} + 3H\dot{\bar{\varphi}} + V_{,\varphi}(\bar{\varphi}) \overset{(7.2)}{\underset{(1.67)}{=}} -\Upsilon\dot{\bar{\varphi}} \,. \tag{7.4}$$

We now want to determine the values of V and Υ by *matching*, i.e. by looking at the equation of motion following from Eq. (7.3), and setting it in the same form as Eq. (7.4). In order to simplify the matching, we work in local Minkowskian coordinates, in the plasma rest frame. By the equivalence principle, the resulting evolution equation can be written in a covariant form, and finally evaluated in an expanding FLRW universe.

In a Minkowskian frame, the Euler-Lagrange equation for $\bar\varphi$ from Eq. (7.3) reads

$$\ddot{\bar\varphi} + V_{0,\varphi}(\bar\varphi) = -\langle J(t)\rangle_{\bar\varphi} \,. \tag{7.5}$$

The right-hand side is non-trivial, because the average value of J depends on the slow variables of the problem, notably $\bar\varphi$ and T. To find this dependence, we can invoke a *linear response* argument. Let us assume that a heat bath is present from a time $t = 0$ onwards. The heat bath dynamics is considered in the canonical formalism, so that the Hamiltonian reads

$$\hat{H} = \hat{H}_{\mathrm{bath}} + \bar\varphi\,\hat{J} \,. \tag{7.6}$$

With the help of the *density matrix* of the heat bath, $\hat\rho(t)$, the average value of $\hat{J}$ can be written as

$$\langle \hat{J}(t)\rangle_{\bar\varphi} \;\equiv\; \mathrm{tr}[\,\hat\rho(t)\hat{J}(t)\,] \,. \tag{7.7}$$

The density matrix satisfies the *Liouville-von Neumann equation*,

$$i\partial_t\hat\rho(t) = [\hat{H}(t), \hat\rho(t)] \,. \tag{7.8}$$

It is helpful to go over to an *interaction picture*, in which the time evolution of the operators is determined by an unperturbed Hamiltonian ($\hat{H}_{\mathrm{bath}}$ in Eq. (7.6)), whereas that of states originates from a perturbation ($\bar\varphi\,\hat{J}$ in Eq. (7.6)). The corresponding density matrix $\hat\rho_I$ and operator $\hat{J}_I$ are obtained by substituting

$$\hat\rho(t) \;\equiv\; e^{-i\hat{H}_{\mathrm{bath}}t}\,\hat\rho_I(t)\,e^{i\hat{H}_{\mathrm{bath}}t} \,, \quad \hat{J}(t) \;\equiv\; e^{-i\hat{H}_{\mathrm{bath}}t}\,\hat{J}_I(t)\,e^{i\hat{H}_{\mathrm{bath}}t} \tag{7.9}$$

in Eqs. (7.6) and (7.8). Then the Liouville-von Neumann equation becomes

$$i\partial_t\hat\rho_I(t) \overset{(7.8)}{\underset{(7.9)}{=}} \bar\varphi(t)\,[\hat{J}_I(t), \hat\rho_I(t)] \,, \tag{7.10}$$

which can be solved for $\hat\rho_I$ iteratively in the perturbation,

$$\hat\rho_I(t) \overset{(7.10)}{=} \hat\rho_I(0) - i\int_0^t \mathrm{d}t'\,\bar\varphi(t')\,[\hat{J}_I(t'), \hat\rho_I(0)] + \mathcal{O}(\bar\varphi^2 J_I^2) \,, \tag{7.11}$$

where the initial time has been chosen as $t = 0$. Inserting the result in Eq. (7.7), where $\mathrm{tr}[\hat{\rho}(t)\hat{J}(t)] = \mathrm{tr}[\hat{\rho}_I(t)\hat{J}_I(t)]$ by the cyclic property of the trace, we find

$$
\langle \hat{J}(t) \rangle_{\bar{\varphi}} = \mathrm{tr}[\hat{\rho}_I(0)\hat{J}_I(t)] - i \int_0^t dt'\, \bar{\varphi}(t')\, \mathrm{tr}\{[\hat{J}_I(t'), \hat{\rho}_I(0)]\hat{J}_I(t)\} + \mathcal{O}(\bar{\varphi}^2 J_I^3)
$$

$$
= \mathrm{tr}[\hat{\rho}_I(0)\hat{J}_I(t)] - i \int_0^t dt'\, \bar{\varphi}(t')\, \mathrm{tr}\{\hat{\rho}_I(0)[\hat{J}_I(t), \hat{J}_I(t')]\} + \mathcal{O}(\bar{\varphi}^2 J_I^3) \ . \quad (7.12)
$$

According to Eq. (7.11), at time $t = 0$ the density matrix $\hat{\rho}_I$ does not depend on $\bar{\varphi}$. Given that the equilibration dynamics of the heat bath has been assumed faster than the evolution of φ, we may assume its initial density matrix to be a thermal one,

$$
\hat{\rho}_I(0) \overset{(7.9)}{=} \hat{\rho}(0) \equiv \frac{e^{-\hat{H}_{\mathrm{bath}}/T}}{\mathcal{Z}} \ , \quad \mathcal{Z} \equiv \mathrm{tr}\left(e^{-\hat{H}_{\mathrm{bath}}/T} \right) , \qquad (7.13)
$$

where $\mathcal{Z}$ is the *canonical partition function*. We define

$$
\langle \hat{J}(t) \rangle_0 \overset{(7.12)}{\equiv} \mathrm{tr}\left[\hat{\rho}_I(0)\hat{J}_I(t)\right] \overset{(7.9)}{\underset{(7.13)}{=}} \mathrm{tr}\left[\hat{\rho}(0)\hat{J}(t)\right] , \qquad (7.14)
$$

$$
G_{\mathrm{R}}(t - t') \overset{(7.12)}{\equiv} \theta(t - t')\, \mathrm{tr}\left\{ \hat{\rho}_I(0)\, i[\hat{J}_I(t), \hat{J}_I(t')] \right\}
$$

$$
\overset{(7.9)}{\underset{(7.13)}{=}} \theta(t - t')\, \mathrm{tr}\left\{ \hat{\rho}(0)\, i[\hat{J}(t), \hat{J}(t')] \right\} , \qquad (7.15)
$$

where G_{R} is a *retarded Green's function*, of the type that we have met before but now in the context of quantum statistical physics. Then Eq. (7.12) becomes

$$
\langle \hat{J}(t) \rangle_{\bar{\varphi}} \overset{(7.12)}{\underset{(7.14),(7.15)}{=}} \langle \hat{J}(t) \rangle_0 - \int_0^t dt'\,\bar{\varphi}(t')\, G_{\mathrm{R}}(t - t') + \mathcal{O}(\bar{\varphi}^2 J_I^3) \ . \qquad (7.16)
$$

This is called a *linear response* relation, given that $\bar{\varphi}$ is expanded to linear order.

Returning now to Eq. (7.5), let us assume that we can define a global symmetry $\bar{\varphi} \to -\bar{\varphi}$, which guarantees that in a stable vacuum, we have $\bar{\varphi} = 0$. In order to keep the full Lagrangian invariant under this symmetry, so that the symmetry is respected by quantum corrections, J needs to be odd. This then implies that $\langle \hat{J}(t) \rangle_0 = 0$. Thereby Eq. (7.5) combined with Eq. (7.16) can be turned into an effective equation of motion,

$$
\ddot{\bar{\varphi}} + V_{0,\varphi}(\bar{\varphi}) - \int_0^t dt'\,\bar{\varphi}(t')\, G_{\mathrm{R}}(t - t') \overset{(7.5)}{\underset{(7.16)}{\approx}} 0 \ . \qquad (7.17)
$$

In order to match Eqs. (7.4) and (7.17), let us transform them to frequency space, recalling that the time evolution starts at $t = 0$,

$$
\bar{\varphi}(\omega) \equiv \int_0^\infty dt\, e^{i\omega t}\, \bar{\varphi}(t) \ , \qquad \bar{\varphi}(t)\theta(t) = \int_{-\infty}^\infty \frac{d\omega}{2\pi} e^{-i\omega t}\, \bar{\varphi}(\omega) \ , \qquad (7.18)
$$

$$
G_{\mathrm{R}}(\omega) = \int_0^\infty dt\, e^{i\omega t}\, G_{\mathrm{R}}(t) \ , \qquad G_{\mathrm{R}}(t) = \int_{-\infty}^\infty \frac{d\omega}{2\pi} e^{-i\omega t}\, G_{\mathrm{R}}(\omega) \ . \qquad (7.19)
$$

The one-sided Fourier transforms of $\dot{\bar\varphi}$ and $\ddot{\bar\varphi}$ yield

$$\int_0^\infty \mathrm{d}t\,\dot{\bar\varphi}(t)\,e^{i\omega t} \;=\; -\bar\varphi(0) - i\omega\bar\varphi(\omega)\,, \tag{7.20}$$

$$\int_0^\infty \mathrm{d}t\,\ddot{\bar\varphi}(t)\,e^{i\omega t} \;=\; -\dot{\bar\varphi}(0) + i\omega\bar\varphi(0) - \omega^2\bar\varphi(\omega)\,, \tag{7.21}$$

and for a convolution we find

$$\int_0^\infty \mathrm{d}t\,e^{i\omega t}\int_0^t \mathrm{d}t'\,\bar\varphi(t')\,G_{\mathrm R}(t-t') \;=\; \overbrace{\int_0^\infty \mathrm{d}t\int_0^t \mathrm{d}t'}^{\int_0^\infty \mathrm{d}t'\,\int_{t'}^\infty \mathrm{d}t}\; e^{i\omega t'}\bar\varphi(t')\,e^{i\omega(t-t')}G_{\mathrm R}(t-t')$$

$$\overset{t=t'+\tilde{t}}{=\!=\!=}\; \underbrace{\int_0^\infty \mathrm{d}t'\,e^{i\omega t'}\bar\varphi(t')}_{\bar\varphi(\omega)}\; \underbrace{\int_0^\infty \mathrm{d}\tilde{t}\,e^{i\omega\tilde{t}}\,G_{\mathrm R}(\tilde{t})}_{G_{\mathrm R}(\omega)}\,. \tag{7.22}$$

Furthermore, we denote the Fourier transform of the self-interaction 'force' by

$$\mathcal{F}_\omega[V_{0,\varphi}] \;\equiv\; \int_0^\infty \mathrm{d}t\,V_{0,\varphi}\,e^{i\omega t}\,. \tag{7.23}$$

Then we find (in a Minkowskian frame, so that the Hubble rate vanishes)

$$(7.4): \quad -\dot{\bar\varphi}(0) + i\omega\bar\varphi(0) - \omega^2\bar\varphi(\omega) - \Upsilon\,[\,\bar\varphi(0) + i\omega\bar\varphi(\omega)\,] + \mathcal{F}_\omega[V_{,\varphi}(\bar\varphi)] \;\approx\; 0\,, \tag{7.24}$$

$$(7.17): \quad -\dot{\bar\varphi}(0) + i\omega\bar\varphi(0) - \omega^2\bar\varphi(\omega) - \bar\varphi(\omega)\,G_{\mathrm R}(\omega) + \mathcal{F}_\omega[V_{0,\varphi}(\bar\varphi)] \;\approx\; 0\,. \tag{7.25}$$

The terms related to initial conditions should play no role for the late-time dynamics. With $G_{\mathrm R}(\omega) = \mathrm{Re}\,G_{\mathrm R}(\omega) + i\,\mathrm{Im}\,G_{\mathrm R}(\omega)$, a comparison of Eqs. (7.24) and (7.25) yields

$$\mathcal{F}_\omega[V_{,\varphi}(\bar\varphi)] \;\overset{(7.24)}{\underset{(7.25)}{\approx}}\; \mathcal{F}_\omega[V_{0,\varphi}(\bar\varphi)] - \bar\varphi(\omega)\,\mathrm{Re}\,G_{\mathrm R}(\omega)\,, \tag{7.26}$$

$$\Upsilon \;\overset{(7.24)}{\underset{(7.25)}{\approx}}\; \frac{\mathrm{Im}\,G_{\mathrm R}(\omega)}{\omega}\,. \tag{7.27}$$

To be precise, Eq. (7.26) is only valid at linear order in $\bar\varphi$, as otherwise the ω-dependence of $\mathcal{F}_\omega$ could contain both a real and an imaginary part. In principle, effects non-linear in $\bar\varphi$ can also be incorporated [17].

The friction coefficient in Eq. (7.27) has an important property, namely that

$$\Upsilon \;\geq\; 0\,. \tag{7.28}$$

On a general level, this can be related to the 'optical theorem', according to which imaginary parts of scattering amplitudes are related to their absolute values squared. More concretely, we could evaluate the Green's function from Eq. (7.15) in the eigenbasis of $\hat{H}_{\mathrm{bath}}$, $\hat{H}_{\mathrm{bath}}|n\rangle = E_n|n\rangle$, resulting in a weighted sum over transition matrix elements like $|\langle m|\hat{J}|n\rangle|^2$. Even if this is a nice exercise, the details play no role in practical computations, and thus we do not show them here.

The interpretation of Eqs. (7.26) and (7.27) is not entirely trivial. Equation (7.26) suggests that $\mathrm{Re}\,G_{\mathrm{R}} \equiv -\delta m_T^2$ plays the role of a thermal mass correction, and Eq. (7.27) shows how the friction coefficient Υ arises. However, these relations are functions of ω, and it must be asked how ω is chosen. If the inflaton oscillates around the minimum of its potential, so that $V_{0,\varphi} \approx m^2\,\bar{\varphi}$, then it is natural to identify the frequency with the one of the oscillations, $\omega \to m$, and the thermally corrected mass reads $m_T^2 = m^2 + \delta m_T^2(m)$. On the other hand, before the oscillation period, if the inflaton dynamics is slow compared with plasma equilibration rate, as indicated above Eq. (7.4), then we can approximate the plasma Green's function by the leading term of its expansion around $\omega = 0$. Generically, both $\lim_{\omega \to 0} \mathrm{Re}\,G_{\mathrm{R}}$ and $\lim_{\omega \to 0} \mathrm{Im}\,G_{\mathrm{R}}/\omega$ are non-zero.

Equations (7.26) and (7.27) underline an important challenge for coupling the inflaton field to other degrees of freedom. If we introduce an operator like in Eq. (7.3), we see that the retarded correlator associated with J yields simultaneously a mass correction and a friction coefficient. If a plasma is present already during the inflationary period, this may be regarded as a problem [18], given that large mass corrections may spoil the desired inflationary predictions, based on V_0. At the same time, we would like to have a non-zero Υ, in order to efficiently heat up the universe after inflation. These conflicting requirements pose (sometimes ignored) constraints on viable inflationary models.

7.3   Which Role Is Played by the Thermal Noise, ϱ?

We now turn to the other part of the effective Langevin description (cf. Eq. (7.2)) that is related to the interactions between the inflaton and the radiation plasma, namely the *thermal noise*, ϱ. Before describing the influence of the noise within the Langevin equation, let us specify the 'target' for what the noise should achieve.

We place ourselves in a Minkowskian frame like in Sect. 7.2, and furthermore consider a thought experiment leading to $\Upsilon \Delta t \gg 1$, so that the inflaton has equilibrated. It can still experience fluctuations around its global minimum, $\bar{\varphi} = 0$. We assume that the curvature of the potential is $m^2 \equiv V_{,\varphi\varphi}$ (we have redefined $m_T^2 \to m^2$ for notational simplicity). Then the inflaton fluctuations have a 2-point correlator related to that determined in Eq. (5.51), but now modified by thermal corrections (cf., e.g., Ref. [19, Eq. (3.39)]),

$$\langle \delta\widehat{\varphi}(t, \mathbf{x})\, \delta\widehat{\varphi}(t, \mathbf{y})\rangle \overset{\text{Gaussianity}}{=} \int \frac{\mathrm{d}^3\mathbf{k}}{(2\pi)^3}\, \frac{e^{i\mathbf{k}\cdot(\mathbf{x}-\mathbf{y})}}{2\hat{\epsilon}_k}\left[\underbrace{1}_{\Leftrightarrow (5.52) \atop \text{at } |k\tau| \gg 1} + \underbrace{\frac{2}{e^{\hat{\epsilon}_k/(aT)} - 1}}_{\equiv\, 2n_{\mathrm{B}}(\hat{\epsilon}_k/a)} \right],$$

$$(7.29)$$

where the Gaussian approximation refers to the perturbative approach (cf. the discussion in Sect. 2.5). In Eq. (7.29), $\hat{\epsilon}_k \equiv \sqrt{k^2 + a^2 m^2}$, and n_{B} is the *Bose distribution*.

In order to simplify the notation further, we go over to non-conformal (physical) coordinates in the following, denoted by

$$\mathbf{r} \equiv a\mathbf{x}\,, \quad \mathbf{s} \equiv a\mathbf{y}\,, \quad \mathbf{p} \equiv \frac{\mathbf{k}}{a}\,, \quad \epsilon_p \equiv \sqrt{p^2 + m^2}\,. \tag{7.30}$$

Recalling also $\delta\widehat{\varphi} = a\delta\varphi$, Eq. (7.29) becomes

$$\langle \delta\varphi(t, \mathbf{r})\, \delta\varphi(t, \mathbf{s}) \rangle \overset{(7.29)}{\underset{(7.30)}{=}} \int \frac{\mathrm{d}^3\mathbf{p}}{(2\pi)^3} \frac{e^{i\mathbf{p}\cdot(\mathbf{r}-\mathbf{s})}}{2\epsilon_p} \left[1 + 2n_{\mathrm{B}}(\epsilon_p) \right]\,. \tag{7.31}$$

Next, we would like to compute the same correlator as in Eq. (7.31) from the Langevin equation, Eq. (7.2). In a local Minkowskian coordinate system, around the global minimum, and boosting to the plasma rest frame, the perturbations satisfy

$$\left(\partial_t^2 - \nabla_{\mathbf{r}}^2 + \Upsilon\partial_t + m^2 \right)\delta\varphi(t, \mathbf{r}) \overset{(7.2)}{\underset{\varphi=\bar{\varphi}+\delta\varphi}{=}} \varrho(t, \mathbf{r})\,. \tag{7.32}$$

As this is a linear partial differential equation, its general solution is a sum of the general solution of the homogeneous equation and a special solution of the inhomogeneous one. We assume the solution of the homogeneous equation to be represented in momentum space, and its integration constants to be fixed like for a quantum-mechanical mode function, from Eqs. (5.37) and (5.38), though it now gradually decays, due to the dissipative coefficient Υ. Let us denote this solution by $\delta\varphi_{\mathrm{vac}}$. The special solution originates from the noise, ϱ, and we call it the classical one, $\delta\varphi_{\mathrm{cl}}$. The general solution is then

$$\delta\varphi = \delta\varphi_{\mathrm{vac}} + \delta\varphi_{\mathrm{cl}}\,. \tag{7.33}$$

Let us stress that we treat both parts of $\delta\varphi$ as complex numbers (not operators), in accordance with the classical nature of the Langevin equation. However, the initial value of $\delta\varphi_{\mathrm{vac}}$ is normalized like a quantum-mechanical mode function, from Eqs. (5.37) and (5.38). In quantum mechanics, the power spectrum is obtained from the absolute value squared of the mode function, cf. Eq. (5.52), whereas in the classical description, there are no creation and annihilation operators, and the power spectrum originates like in Eq. (2.49). Either way, in the absence of Υ, $\delta\varphi_{\mathrm{vac}}$ accounts for the vacuum part of Eq. (7.31), $1/(2\epsilon_p)$.

On the other hand, the special solution, $\delta\varphi_{\mathrm{cl}}$, can be obtained with a Green's function. To streamline the notation, we denote

$$\mathcal{R} \equiv (t, \mathbf{r})\,, \quad \int_{\mathcal{R}} \overset{(11)}{\underset{(7.30)}{=}} \int \mathrm{d}^4\mathcal{R}\,, \quad \mathcal{P} \equiv (\omega, \mathbf{p})\,, \quad \int_{\mathcal{P}} \overset{(10)}{\underset{(7.30)}{=}} \int_{-\infty}^{\infty} \frac{\mathrm{d}\omega}{2\pi} \int \frac{\mathrm{d}^3\mathbf{p}}{(2\pi)^3}\,. \tag{7.34}$$

With different letters we denote coordinate ($\mathcal{R}, \mathcal{S}, \mathcal{U}, \mathcal{V}$) or momentum space ($\mathcal{P}, \mathcal{Q}$).

With this notation, the special solution is given by

$$\delta\varphi_{\rm cl}(\mathcal{R}) \;=\; \int_{\mathcal{U}} G_{\rm R}(\mathcal{R}-\mathcal{U})\varrho(\mathcal{U})\,, \tag{7.35}$$

where the *retarded Green's function*,

$$\big(\partial_t^2 - \nabla_{\mathbf r}^2 + \Upsilon\partial_t + m^2\big)G_{\rm R}(\mathcal{R}) = \delta^{(4)}(\mathcal{R})\,, \tag{7.36}$$

can be represented in momentum space,

$$G_{\rm R}(\mathcal{R}) \overset{(10)}{=} \int_{\mathcal{P}} e^{-i\omega t + i\mathbf{p}\cdot\mathbf{r}}\,\widetilde{G}_{\rm R}(\mathcal{P})\,,\quad \widetilde{G}_{\rm R}(\mathcal{P}) \overset{(7.36)}{=} \frac{1}{-\omega^2 - i\omega\Upsilon + \epsilon_p^2}\,. \tag{7.37}$$

With the Green's function, the contribution of the classical fluctuations to the 2-point correlator becomes

$$
\begin{aligned}
\langle\,\delta\varphi_{\rm cl}(\mathcal{R})\,\delta\varphi_{\rm cl}(\mathcal{S})\,\rangle &\overset{(7.35)}{=} \int_{\mathcal{U},\mathcal{V}} G_{\rm R}(\mathcal{R}-\mathcal{U})\,G_{\rm R}(\mathcal{S}-\mathcal{V})\,\langle\,\varrho(\mathcal{U})\varrho(\mathcal{V})\,\rangle \\[4pt]
&\overset{(7.2)}{=} \Omega\int_{\mathcal{U}} G_{\rm R}(\mathcal{R}-\mathcal{U})\,G_{\rm R}(\mathcal{S}-\mathcal{U}) \\[4pt]
&\overset{(7.37)}{=} \Omega\int_{\mathcal{P},\mathcal{Q}} \underbrace{\int_{\mathcal{U}} e^{i\mathcal{P}\cdot(\mathcal{R}-\mathcal{U})+i\mathcal{Q}\cdot(\mathcal{S}-\mathcal{U})}}_{(2\pi)^4\delta^{(4)}(\mathcal{P}+\mathcal{Q})\,e^{i\mathcal{P}\cdot(\mathcal{R}-\mathcal{S})}}\,\widetilde{G}_{\rm R}(\mathcal{P})\widetilde{G}_{\rm R}(\mathcal{Q}) \\[4pt]
&\xrightarrow[\text{time}]{\text{equal}} \Omega\int_{-\infty}^{\infty}\frac{d\omega}{2\pi}\int_{\mathbf p} e^{i\mathbf{p}\cdot(\mathbf{r}-\mathbf{s})}\,\widetilde{G}_{\rm R}(\mathcal{P})\widetilde{G}_{\rm R}(-\mathcal{P})\,. \tag{7.38}
\end{aligned}
$$

We can search for the poles of the Green's functions,

$$[-\widetilde{G}_{\rm R}(\mathcal{P})]^{-1} = \omega^2 + i\omega\Upsilon - \epsilon_p^2 = \prod_{\sigma=\pm}\left(\omega + \tfrac{i\Upsilon}{2} + \sigma\sqrt{\epsilon_p^2 - \tfrac{\Upsilon^2}{4}}\right)$$

$$\Rightarrow \text{poles in the lower half-plane if } \epsilon_p^2 > 0\,, \tag{7.39}$$

$$[-\widetilde{G}_{\rm R}(-\mathcal{P})]^{-1} = \omega^2 - i\omega\Upsilon - \epsilon_p^2 = \prod_{\sigma=\pm}\left(\omega - \tfrac{i\Upsilon}{2} + \sigma\sqrt{\epsilon_p^2 - \tfrac{\Upsilon^2}{4}}\right)$$

$$\Rightarrow \text{poles in the upper half-plane if } \epsilon_p^2 > 0\,. \tag{7.40}$$

We integrate over ω in Eq. (7.38) by closing the contour in the upper half-plane, obtaining

$$
\begin{aligned}
&\langle \delta\varphi_{\mathrm{cl}}(\mathcal{R})\, \delta\varphi_{\mathrm{cl}}(\mathcal{S}) \rangle \\
&\xrightarrow[\text{time}]{\text{equal}} \Omega \int_{\mathbf{p}} e^{i\mathbf{p}\cdot(\mathbf{r}-\mathbf{s})} \frac{2\pi i}{2\pi} \left[\underbrace{\frac{1}{i\Upsilon(i\Upsilon+2\sqrt{\ldots})2\sqrt{\ldots}}}_{\text{from pole at } \omega=\frac{i\Upsilon}{2}+\sqrt{\ldots}} + \underbrace{\frac{1}{(i\Upsilon-2\sqrt{\ldots})i\Upsilon(-2\sqrt{\ldots})}}_{\text{from pole at } \omega=\frac{i\Upsilon}{2}-\sqrt{\ldots}} \right] \\
&= \frac{\Omega}{\Upsilon} \int_{\mathbf{p}} e^{i\mathbf{p}\cdot(\mathbf{r}-\mathbf{s})} \left[\frac{\cancel{i\Upsilon}-2\sqrt{\ldots}}{-\cancel{\Upsilon^2}-4(\epsilon_p^2-\cancel{\tfrac{\Upsilon^2}{4}})} - \frac{\cancel{i\Upsilon}+2\sqrt{\ldots}}{-\cancel{\Upsilon^2}-4(\epsilon_p^2-\cancel{\tfrac{\Upsilon^2}{4}})} \right] \frac{1}{2\sqrt{\ldots}} \\
&= \frac{\Omega}{\Upsilon} \int_{\mathbf{p}} e^{i\mathbf{p}\cdot(\mathbf{r}-\mathbf{s})} \frac{1}{2\epsilon_p^2} \, .
\end{aligned}
\tag{7.41}
$$

Let us compare Eq. (7.41) with the target expression in Eq. (7.31). We recall that $\delta\varphi_{\mathrm{vac}}$ is responsible for $1/(2\epsilon_p)$. Obviously, Eq. (7.41) does *not* agree with the other part, containing the Bose distribution. However, the low-energy expansion of the Bose distribution reads

$$
n_{\mathrm{B}}(\epsilon_p) = \frac{1}{e^{\epsilon_p/T}-1} \overset{\epsilon_p \ll T}{\approx} \frac{1}{\frac{\epsilon_p}{T}+\frac{\epsilon_p^2}{2T^2}+\ldots} \overset{\epsilon_p \ll T}{\approx} \frac{T}{\epsilon_p}-\frac{1}{2}+\ldots \, .
\tag{7.42}
$$

It follows that

$$
\langle \delta\varphi(t,\mathbf{r})\, \delta\varphi(t,\mathbf{s}) \rangle \overset{(7.31)}{\underset{(7.42)}{\supset}} \int_{\mathbf{p}} e^{i\mathbf{p}\cdot(\mathbf{r}-\mathbf{s})} \left(\frac{T}{\epsilon_p^2}+\ldots \right) .
\tag{7.43}
$$

To summarize, Eqs. (7.41) and (7.43) match, if the *noise autocorrelator* is

$$
\Omega = 2T\Upsilon \, .
\tag{7.44}
$$

This relation between Ω and Γ is known as the *fluctuation-dissipation theorem*. Making use of Eq. (7.44) guarantees that the noise ϱ drives φ_{cl} towards the thermal state. At the same time, the approximate nature of the matching underlines that the Langevin description is 'only' an effective theory, valid for inflaton excitations with energies $\epsilon_p \ll T$. We elaborate on this point around the end of Sect. 7.6. The possibility to keep the full Bose distribution in the noise autocorrelator, in order to extend the validity of the Langevin description, has been discussed in Ref. [20].

7.4 Temperature Evolution and Its Influence on CMB Observables

Having discussed what temperature means in Sect. 7.1, and how thermal effects can be incorporated via the friction coefficient, Υ, in Sect. 7.2, and the noise, ϱ, in Sect. 7.3, we now ask which value the temperature can take, assuming that it is a meaningful notion ($\Gamma > H$), and which role it plays for CMB observables (the last point is discussed at the end of this section). To proceed, we need an evolution equation for the radiation plasma.

The key input for determining the temperature evolution comes from overall energy conservation via Eq. (1.73). At the background level, the conservation implies

$$\dot{\bar{e}} + 3H(\bar{e} + \bar{p}) \overset{(1.73)}{=} -(\ddot{\bar{\varphi}} + 3H\dot{\bar{\varphi}})\,\dot{\bar{\varphi}}\,, \tag{7.45}$$

where, according to Eq. (1.65),

$$\bar{e} \overset{(1.65)}{\equiv} \bar{e}_r + V - TV_{,T}\,, \qquad \bar{p} \overset{(1.65)}{\equiv} \bar{p}_r - V\,, \tag{7.46}$$

$$\dot{\bar{e}} \overset{(7.46)}{=} \dot{\bar{e}}_r + V_{,\varphi}\dot{\bar{\varphi}} + V_{,T}\dot{T} - \dot{T}V_{,T} - T\dot{V}_{,T}\,. \tag{7.47}$$

We can simplify Eq. (7.45) by combining it with the evolution equation for $\bar{\varphi}$. Concretely, let us multiply Eq. (7.4), *viz.*

$$\boxed{\ddot{\bar{\varphi}} + (3H + \Upsilon)\dot{\bar{\varphi}} + V_{,\varphi} \overset{(7.4)}{=} 0\,,} \tag{7.48}$$

by $\dot{\bar{\varphi}}$, and add it to Eq. (7.45). Then, with the help of Eq. (7.47), we get

$$\boxed{\dot{\bar{e}}_r + V_{,\varphi}\dot{\bar{\varphi}} - T\dot{V}_{,T} + 3H(\bar{e}_r + \bar{p}_r - TV_{,T}) \overset{(7.45)-(7.48)}{=} V_{,\varphi}\dot{\bar{\varphi}} + \Upsilon\dot{\bar{\varphi}}^2\,.} \tag{7.49}$$

Here, given that the radiation energy density is a function of T only, we could furthermore write $\dot{\bar{e}}_r = \dot{T}\bar{c}_r$, where $\bar{c}_r \equiv \bar{e}_{r,T}$ is the *heat capacity*. Therefore, Eq. (7.49) determines the time evolution of the temperature. To have the complete set of equations in one place, let us also repeat the Hubble rate (cf. Eq. (1.69)) once again,

$$\boxed{\frac{8\pi}{3m_{\mathrm{pl}}^2}\left(\frac{\dot{\bar{\varphi}}^2}{2} + \bar{e}\right) \overset{\substack{(1.69),(1) \\ \kappa=0}}{=} H^2\,.} \tag{7.50}$$

We recall from Eq. (2.32) that if the system has no chemical potentials related to conserved charges, then

$$\bar{e}_r + \bar{p}_r \overset{(2.32)}{=} T\bar{s}_r\,, \tag{7.51}$$

where $s_r = \bar{p}_{r,T}$ is the *entropy density*. In Eq. (7.49), the term $-V_{,T}$ represents a contribution to the entropy density. The negative sign means that storing free energy density in $\bar{\varphi}$, whose value carries definite information, decreases the entropy density.

However, as mentioned in the last paragraph of Sect. 7.2, inflationary models typically have $V_{,T} \approx 0$.

Let us now ask whether Eq. (7.49) could have a stationary solution, $\dot{\bar{e}}_r \approx 0$? We also set $V_{,T} \approx 0$ for simplicity. Furthermore, could this happen already during the slow-roll stage of inflation, when $\ddot{\varphi} \approx 0$? Solving for $\dot{\varphi}$ from Eq. (7.48), substituting the answer in Eq. (7.49), and inserting Eq. (7.51), we find

$$T\bar{s}_r \overset{\text{stationary}}{\underset{(7.48)-(7.51)}{\approx}} \frac{\Upsilon}{3H}\left(\frac{V_{,\varphi}}{3H + \Upsilon}\right)^2 . \tag{7.52}$$

The left-hand side of this equation is a rapidly varying function of T, while the right-hand side is normally slowly varying. Therefore, a solution can be found (cf., e.g., Refs. [21–23]), and it turns out to be a stable *fixed point*, in the sense that if the initial temperature is above or below the stationary value, it adjusts to it. An example of a numerical solution that displays these features is shown in Sect. 7.7.

However, whenever discussing T, we have to keep in mind the discussion from Sect. 7.1: temperature is a self-consistent notion only if we can also show that the plasma self-interaction rate satisfies $\Gamma \gg H$ during the period considered. Here Γ should be evaluated at the thermal energy scale, $p \sim T$, because such momenta contribute most to the thermal energy density and pressure (here we have in mind ultrarelativistic particles, with masses $m \ll T$). Given that the parameters entering Γ are different from those entering Υ and H, it is a model-dependent question when $\Gamma \gg H$ is satisfied.

An example of a simple estimate of Γ and H can be obtained by considering a radiation-dominated universe, in which the Hubble rate is parametrically

$$H \overset{(7.50)}{\sim} \frac{\sqrt{\bar{e}}}{m_{\rm pl}} \overset{(7.55)}{\sim} \frac{T^2}{m_{\rm pl}} . \tag{7.53}$$

At the same time, for kinetic equilibration, $\Gamma \sim \alpha^2 T / \ln(\alpha^{-1})$ according to the discussion in Sect. 7.1, with $\alpha < 1$ referring to a fine-structure constant of non-Abelian gauge interactions [4]. So, we see that $\Gamma \gg H$ is satisfied if $T \ll \alpha^2 m_{\rm pl}/\ln(\alpha^{-1})$. For a numerical value, we could insert $\alpha \sim 0.01$ for QCD, obtaining $T \ll 10^{15}$ GeV.

The consideration below Eq. (7.53) is, however, not a strict criterion for when we can talk about a temperature. In principle, the notion of a temperature may be meaningful already before the universe entered a radiation-dominated epoch, as the inflaton loses energy to the other particles, which attain a would-be temperature according to Eq. (7.1). We may then ask what the *maximal temperature* of the universe could be. For this, we may look for a solution of Eq. (7.49), with $\dot{T} = 0$ at $T = T_{\rm max}$ (together with $\ddot{T} < 0$). However, now we *cannot* use the slow-roll approximation on the right-hand side, given that $T_{\rm max}$ could be reached after the end of inflation. Therefore Eq. (7.52) should be rephrased as

$$\left(T\bar{s}_r\right)_{\rm max} \simeq \left(\Upsilon \frac{\dot{\varphi}^2}{3H}\right)_{\rm max} . \tag{7.54}$$

The solution is model-dependent, through the properties of the radiation plasma (via s_r), through its interactions with the inflaton (via Υ), and through inflaton dynamics (via $\dot{\bar{\varphi}}^2$). In some models, the solution coincides with the stationary value from Eq. (7.52). If the plasma is strongly self-interacting ('confining'), $T_{\max}$ could be particularly high, as its entropy density is otherwise exponentially suppressed [24].

After having reached $T_{\max}$, the temperature starts to decrease. For a while, $\dot{\bar{\varphi}}^2$ and V can dominate the total energy density. If V can be approximated as quadratic around its minimum, this leads to a matter-dominated epoch, as discussed in Sect. 1.7. The matter-dominated epoch stops when $\Upsilon \Delta t \approx \Upsilon/H \gg 1$. Then we enter the radiation-dominated era, where Eq. (7.53) applies. This whole process may be called *smooth reheating*. With this concept we underline the distinction to *instantaneous reheating*, which is often employed as a simplified model (cf. Fig. 1.1 in Sect. 1.4).

Three further remarks are in order. First, if $\Upsilon = 0$, like in cold inflation, there is no source term for the radiation plasma, and any possible initial temperature just redshifts away. However, if Υ is proportional to a positive power of T, the solution $T = 0$ corresponds to an unstable fixed point. Mathematically, just a small perturbation drives the system to the solution of Eq. (7.52).

Second, we have not yet specified the radiation *thermodynamic functions*. In general, they can be parameterized as

$$e_r = \frac{g_* \pi^2 T^4}{30}, \quad s_r = \frac{2 h_* \pi^2 T^3}{45}, \quad c_r = \frac{2 i_* \pi^2 T^3}{15}, \tag{7.55}$$

with the pressure obtained as $p_r = T s_r - e_r$ (for small chemical potentials). The functions g_*, h_* and i_* are referred to as *effective numbers of massless degrees of freedom*. If the plasma were non-interacting, and its constituents were massless bosons, these functions would be constant integers, with $g_* = h_* = i_*$. In realistic systems, the functions vary slowly and are close to each other, unless the system becomes strongly interacting, like QCD at temperatures $(0.1...1.0)$ GeV, in which case they evolve fast. The determination of these functions for the Standard Model is a topic of its own (cf., e.g., Ref. [25]).

Third, reheating dynamics plays an important role for the inflationary curvature power spectrum, because it influences the *redshift* between early universe and present-day momenta. We illustrate this with a concrete example in Fig. 7.2 in Sect. 7.7. We note that in the literature, if reheating dynamics is not addressed, it is conventional to vary the time at which the pivot scale exited the Hubble horizon between 50 and 60 e-folds before inflation ends, and display this variation as an error band. Here the *end of inflation* is defined as the moment at which the slow-roll parameter ϵ_V from Eq. (6.19) grows to be of order unity or, in physical terms, when the change of the Hubble rate during a Hubble time is as large as the Hubble rate itself, $|\Delta H| \equiv |\Delta t \, \dot{H}| \equiv |\dot{H}/H| \sim H$ (cf. Eq. (6.21)).

7.5   Evolution Equations for Curvature Perturbations

Having determined the background equations in the presence of a plasma and discussed their solution (cf. Sect. 7.4), the next challenge is to work out the evolution equations for the perturbations. Restricting to scalar perturbations, this amounts to generalizing the derivation in Sect. 5.1 to include the effects from Υ, ϱ, δe, δp, and $v \equiv \delta v$. As a starting point, we take Eqs. (3.92), (3.93), (3.94), (3.96) and (3.114).

To be concrete, a closed set for inflaton and metric perturbations is constituted by

$$
\begin{aligned}
a^2 \varrho \stackrel{(3.114)}{=} &\; \delta\varphi'' + (2\mathcal{H} + a\Upsilon)\,\delta\varphi' - \nabla^2\delta\varphi - (h_0' + 3h_{\mathrm{D}}' + \nabla^2 h)\,\bar{\varphi}' \\
&+ a(\delta\Upsilon + h_0\Upsilon)\bar{\varphi}' + a^2(\delta V_{,\varphi} + 2\,h_0 V_{,\varphi})\,,
\end{aligned}
\tag{7.56}
$$

$$
\begin{aligned}
-(h_0' + 3h_{\mathrm{D}}' + \nabla^2 h) \stackrel{(3.96)}{=} &\; 2\left(2\mathcal{H} + \frac{\mathcal{H}'}{\mathcal{H}}\right)h_0 + \frac{1}{\mathcal{H}}(\partial_\tau^2 + 2\mathcal{H}\partial_\tau - \nabla^2)Y \\
&- \frac{4\pi G a^2}{\mathcal{H}}\left(\delta p - \delta e + \frac{2}{3}\nabla^2\Pi\right),
\end{aligned}
\tag{7.57}
$$

$$
h_0 \stackrel{(3.93)}{=} -\frac{Y'}{\mathcal{H}} + \frac{4\pi G}{\mathcal{H}}\left[a^2(\bar{e} + \bar{p})(v - h) + \bar{\varphi}'\delta\varphi\right].
\tag{7.58}
$$

Here, gauge dependence, which will be demonstrated to cancel, is put into one of the quantities introduced in Eq. (5.12), related to spatial curvature (cf. Eq. (3.61)),

$$
Y \stackrel{(5.12)}{\equiv} h_{\mathrm{D}} + \frac{\nabla^2\vartheta}{3}.
\tag{7.59}
$$

As a basic variable suitable for a treatment in conformal time, we adopt the gauge-invariant field perturbation from Eqs. (5.16) and (5.17),

$$
\widehat{\mathcal{Q}}_\varphi \stackrel[(5.17)]{(5.16)}{\equiv} a\left(\delta\varphi + \frac{\bar{\varphi}'}{\mathcal{H}}\,Y\right) \stackrel{(4.60)}{=} -\frac{a\bar{\varphi}'}{\mathcal{H}}\mathcal{R}_\varphi.
\tag{7.60}
$$

The velocity and temperature perturbations are written in terms of the gauge-invariant curvature perturbations, $\mathcal{R}_v$ and $\mathcal{R}_T$, defined in Eqs. (4.61) and (4.62), respectively. However, it is helpful to express the curvature part in terms of $\mathcal{R}_\varphi$, and represent the velocity and temperature in terms of *isocurvature perturbations*, i.e. differences of curvature perturbations in which the spatial curvature term, Y, drops out,

$$
\mathcal{S}_v \stackrel[(4.61)]{(4.60)}{\equiv} (\bar{e} + \bar{p})(\mathcal{R}_v - \mathcal{R}_\varphi)\,,
\tag{7.61}
$$

$$
\mathcal{S}_T \stackrel[(4.62)]{(4.60)}{\equiv} \bar{e}_{,T}\,\dot{T}\,(\mathcal{R}_T - \mathcal{R}_\varphi) \stackrel[(1.7)]{(1.6)}{=} \frac{\bar{e}_{,T}\,T'}{a}(\mathcal{R}_T - \mathcal{R}_\varphi)\,.
\tag{7.62}
$$

We now express the right-hand sides of Eqs. (7.56)–(7.58) in terms of the variables in Eqs. (7.60)–(7.62). For $\delta\varphi$ and its derivatives, Eq. (7.60) yields

$$\delta\varphi \overset{(7.60)}{=} \frac{1}{a}\,\widehat{\mathcal{Q}}_\varphi - \frac{\bar\varphi'}{\mathcal{H}}\,Y\,, \tag{7.63}$$

$$\delta\varphi' = \frac{1}{a}\big(\widehat{\mathcal{Q}}'_\varphi - \mathcal{H}\widehat{\mathcal{Q}}_\varphi\big) - \left(\frac{\bar\varphi'}{\mathcal{H}}\right)' Y - \left(\frac{\bar\varphi'}{\mathcal{H}}\right) Y'\,, \tag{7.64}$$

$$\delta\varphi'' = \frac{1}{a}\Big[\,\widehat{\mathcal{Q}}''_\varphi - 2\mathcal{H}\widehat{\mathcal{Q}}'_\varphi + (\mathcal{H}^2 - \mathcal{H}')\widehat{\mathcal{Q}}_\varphi\,\Big]$$
$$- \left(\frac{\bar\varphi'}{\mathcal{H}}\right)'' Y - 2\left(\frac{\bar\varphi'}{\mathcal{H}}\right)' Y' - \left(\frac{\bar\varphi'}{\mathcal{H}}\right) Y''\,. \tag{7.65}$$

For $\delta\Upsilon$, which may be a function of φ and T, we find

$$\delta\Upsilon = \Upsilon_{,\varphi}\,\delta\varphi + \Upsilon_{,T}\,\delta T$$
$$\overset{(7.63)}{\underset{(4.62)}{=}} \Upsilon_{,\varphi}\left(\frac{1}{a}\widehat{\mathcal{Q}}_\varphi - \frac{\bar\varphi'}{\mathcal{H}}Y\right) - \Upsilon_{,T}\frac{T'}{\mathcal{H}}\big(\mathcal{R}_T - \mathcal{R}_\varphi + \mathcal{R}_\varphi + Y\big)$$
$$\overset{(7.60)}{\underset{(7.62)}{=}} \frac{\Upsilon_{,\varphi}}{a}\left(\widehat{\mathcal{Q}}_\varphi - \frac{a\bar\varphi'}{\mathcal{H}}Y\right) - \frac{a\Upsilon_{,T}}{\mathcal{H}\bar{e}_{,T}}\mathcal{S}_T + \frac{\Upsilon_{,T}T'}{a\bar\varphi'}\left(\widehat{\mathcal{Q}}_\varphi - \frac{a\bar\varphi'}{\mathcal{H}}Y\right)$$
$$= \frac{\Upsilon'}{a\bar\varphi'}\left(\widehat{\mathcal{Q}}_\varphi - \frac{a\bar\varphi'}{\mathcal{H}}Y\right) - \frac{a\Upsilon_{,T}}{\mathcal{H}\bar{e}_{,T}}\mathcal{S}_T\,, \tag{7.66}$$

where in the last step we made use of $\Upsilon' = \Upsilon_{,\varphi}\bar\varphi' + \Upsilon_{,T}T'$. In complete analogy,

$$\delta V_{,\varphi} = \frac{V'_{,\varphi}}{a\bar\varphi'}\left(\widehat{\mathcal{Q}}_\varphi - \frac{a\bar\varphi'}{\mathcal{H}}Y\right) - \frac{aV_{,\varphi T}}{\mathcal{H}\bar{e}_{,T}}\mathcal{S}_T\,, \tag{7.67}$$

$$\delta p - \delta e = \frac{\bar{p}' - \bar{e}'}{a\bar\varphi'}\left(\widehat{\mathcal{Q}}_\varphi - \frac{a\bar\varphi'}{\mathcal{H}}Y\right) + \frac{a}{\mathcal{H}}\left(1 - \frac{\bar{p}_{,T}}{\bar{e}_{,T}}\right)\mathcal{S}_T\,. \tag{7.68}$$

Finally, for the velocity perturbation,

$$(\bar{e} + \bar{p})(v - h) \overset{(4.61)}{\underset{(7.59)}{=}} -\frac{\bar{e} + \bar{p}}{\mathcal{H}}\big(\mathcal{R}_v - \mathcal{R}_\varphi + \mathcal{R}_\varphi + Y\big)$$
$$\overset{(7.60)}{\underset{(7.61)}{=}} \frac{\bar{e} + \bar{p}}{a\bar\varphi'}\left(\widehat{\mathcal{Q}}_\varphi - \frac{a\bar\varphi'}{\mathcal{H}}Y\right) - \frac{1}{\mathcal{H}}\mathcal{S}_v\,. \tag{7.69}$$

Inserting the results in Eqs. (7.63)–(7.69) into (7.56)–(7.58), the expressions become lengthy. However, they can be simplified by making use of the background identities (1.66)–(1.73). As an example, h_0 from Eq. (7.58) becomes

$$
h_0 \stackrel[(7.63),(7.69)]{(7.58)}{=} -\frac{Y'}{\mathcal{H}} + \overbrace{\frac{4\pi G[a^2(\bar{e}+\bar{p})+(\bar{\varphi}')^2]}{\mathcal{H}a\bar{\varphi}'}}^{\text{from (1.70): } (\mathcal{H}-\frac{\mathcal{H}'}{\mathcal{H}})/(a\bar{\varphi}')} \left(\widehat{\mathcal{Q}}_\varphi - \frac{a\bar{\varphi}'}{\mathcal{H}} Y \right) - \frac{4\pi G a^2}{\mathcal{H}^2} \mathcal{S}_v .
$$
(7.70)

Another frequent simplification is obtained by collecting the terms multiplying h_0 from Eqs. (7.56) and (7.57), whereby they can be expressed as

$$
a^2 \varrho \stackrel[(7.57)]{(7.56)}{\supset} \left[2\bar{\varphi}'\left(2\mathcal{H}+\frac{\mathcal{H}'}{\mathcal{H}}\right) + a\bar{\varphi}'\Upsilon + 2a^2 V_{,\varphi} \right] h_0
$$

$$
\stackrel{(7.72)}{=} -\left[2\mathcal{H}\left(\frac{\bar{\varphi}'}{\mathcal{H}}\right)' + a\bar{\varphi}'\Upsilon \right] h_0 .
$$
(7.71)

Here we made use of

$$
-\left(\frac{\bar{\varphi}'}{\mathcal{H}}\right)' = -\frac{\bar{\varphi}''}{\mathcal{H}} + \frac{\mathcal{H}'\bar{\varphi}'}{\mathcal{H}^2} \stackrel{(1.66)}{=} \left(a\Upsilon + 2\mathcal{H} + \frac{\mathcal{H}'}{\mathcal{H}}\right)\frac{\bar{\varphi}'}{\mathcal{H}} + \frac{a^2 V_{,\varphi}}{\mathcal{H}} ,
$$
(7.72)

which is a generalization of Eq. (5.21) to $\Upsilon > 0$.

Given the complicated expressions, gauge invariance once again offers for a valuable crosscheck. Almost immediately, it can be seen that Y'' from Eqs. (7.57) and (7.65) cancel against each other, and the same happens for $\nabla^2 Y$. A less trivial crosscheck originates by inspecting the coefficient of Y',

$$
a^2 \varrho \stackrel{(7.56)}{\supset} \left[\overbrace{-2\left(\frac{\bar{\varphi}'}{\mathcal{H}}\right)}^{\delta\varphi'' \text{ from (7.65)}} \underbrace{-(2\mathcal{H}+a\Upsilon)\left(\frac{\bar{\varphi}'}{\mathcal{H}}\right)}_{\delta\varphi' \text{ from (7.64)}} \overbrace{+2\bar{\varphi}'}^{\text{from (7.57)}} \overbrace{+2\left(\frac{\bar{\varphi}'}{\mathcal{H}}\right) + \frac{a\bar{\varphi}'}{\mathcal{H}}\Upsilon}^{\text{from (7.70), (7.71)}} \right] Y' .
$$
(7.73)

The most non-trivial part is the cancellation of the coefficient of Y. To verify it, we generalize Eq. (5.25) to include Υ. For this, we take a derivative of Eq. (7.72),

$$
-\left(\frac{\bar{\varphi}'}{\mathcal{H}}\right)'' \stackrel{(7.72)}{=} \left(a\Upsilon + 2\mathcal{H} + \frac{\mathcal{H}'}{\mathcal{H}}\right)\left(\frac{\bar{\varphi}'}{\mathcal{H}}\right)'
$$

$$
+ \left(a\mathcal{H}\Upsilon + a\Upsilon' + 2\mathcal{H}' + \frac{\mathcal{H}''}{\mathcal{H}} - \frac{\mathcal{H}'^2}{\mathcal{H}^2}\right)\frac{\bar{\varphi}'}{\mathcal{H}}
$$

$$
+ \left(2 - \frac{\mathcal{H}'}{\mathcal{H}^2}\right)a^2 V_{,\varphi} + \frac{a^2 V'_{,\varphi}}{\mathcal{H}} .
$$
(7.74)

In addition we need a new background identity, relating $\mathcal{H}''$ to the difference $\bar{p}' - \bar{e}'$ appearing in Eq. (7.68). Subtracting Eqs. (1.68) and (1.70); omitting once again κ; and taking a time derivative, we find

$$2\mathcal{H}^2 + \mathcal{H}' \overset{(1.68)}{\underset{(1.70)}{=}} 4\pi G a^2(\bar{e} - \bar{p}) \overset{\partial_\tau}{\Rightarrow} 4\mathcal{H}\mathcal{H}' + \mathcal{H}'' = 2\mathcal{H}\overbrace{4\pi G a^2(\bar{e} - \bar{p})}^{2\mathcal{H}^2+\mathcal{H}'} + 4\pi G a^2(\bar{e}' - \bar{p}')$$

$$\Rightarrow \mathcal{H}'' + 2\mathcal{H}\mathcal{H}' - 4\mathcal{H}^3 = 4\pi G a^2(\bar{e}' - \bar{p}') . \tag{7.75}$$

Through a repeated use of Eqs. (7.72), (7.74) and (7.75), the complete cancellation of the coefficient of Y can indeed be verified.

It remains to work out the physical terms. Given that these steps offer no new insight beyond what was met in the context of deriving the vacuum Mukhanov-Sasaki equation, leading to Eq. (5.28), we refrain from showing them explicitly, but provide a brief outline. The isocurvature $\mathcal{S}_T$ originates from the last terms of Eqs. (7.66)–(7.68), and $\mathcal{S}_v$ from Eqs. (7.70) and (7.71). The anisotropic stress, $\nabla^2\Pi$, only appears in Eq. (7.57), so it does not cancel. The operator acting on $\widehat{\mathcal{Q}}_\varphi$ is fairly complicated. It has similarities with the operator acting on Y, except that there is no cancellation. However, Eqs. (7.72), (7.74) and (7.75) can again be employed in order to combine terms. Appearances of $\mathcal{H}$ and $\mathcal{H}'$ can be hidden by letting derivatives act on $(a\bar{\varphi}'/\mathcal{H})$ rather than $(\bar{\varphi}'/\mathcal{H})$, like in Eq. (5.29).

Proceeding along these lines, and putting the stochastic noise on the right-hand side as is conventionally done, the final gauge-invariant relation in conformal time reads

$$\overbrace{\left\{ \partial_\tau^2 - \nabla^2 - \frac{\mathcal{H}}{a\bar{\varphi}'}\left(\frac{a\bar{\varphi}'}{\mathcal{H}}\right)'' + a\Upsilon\left[\partial_\tau - \frac{\mathcal{H}}{a\bar{\varphi}'}\left(\frac{a\bar{\varphi}'}{\mathcal{H}}\right)' \right] \right\}}^{\equiv L_\varphi} \widehat{\mathcal{Q}}_\varphi$$

$$- \left\{ \frac{4\pi G a}{\mathcal{H}}\left(1 - \frac{\bar{p}_{,T}}{\bar{e}_{,T}}\right) + \frac{\Upsilon_{,T}\,\bar{\varphi}' + a V_{,\varphi T}}{\bar{e}_{,T}\,\bar{\varphi}'} \right\} \frac{a^3\bar{\varphi}'}{\mathcal{H}} \mathcal{S}_T$$

$$+ \left\{ \Upsilon + \frac{2\mathcal{H}}{a\bar{\varphi}'}\left(\frac{\bar{\varphi}'}{\mathcal{H}}\right)' \right\} \frac{4\pi G a^4\bar{\varphi}'}{\mathcal{H}^2} \mathcal{S}_v$$

$$- \frac{8\pi G a^3\bar{\varphi}'}{3\mathcal{H}} \nabla^2\Pi = a^3\varrho . \tag{7.76}$$

The first line has been given a special name for later convenience.

Next, we transcribe the equation to physical time, and simultaneously from $\widehat{\mathcal{Q}}_\varphi$ to the curvature perturbation $\mathcal{R}_\varphi$, according to Eq. (7.60). First, substituting $\widehat{\mathcal{Q}}_\varphi = -(a\bar{\varphi}'/\mathcal{H})\mathcal{R}_\varphi$ and following Eqs. (6.5)–(6.7),

$$\left\{ \partial_\tau^2 - \frac{\mathcal{H}}{a\bar{\varphi}'}\left(\frac{a\bar{\varphi}'}{\mathcal{H}}\right)'' \right\} \widehat{\mathcal{Q}}_\varphi \overset{(6.5)}{\underset{(6.7)}{=}} -\frac{a\bar{\varphi}'}{\mathcal{H}}\left\{ \mathcal{R}_\varphi'' + \frac{2\mathcal{H}}{a\bar{\varphi}'}\left(\frac{a\bar{\varphi}'}{\mathcal{H}}\right)' \mathcal{R}_\varphi' \right\} , \tag{7.77}$$

$$\left\{ \partial_\tau - \frac{\mathcal{H}}{a\bar{\varphi}'}\left(\frac{a\bar{\varphi}'}{\mathcal{H}}\right)' \right\} \widehat{\mathcal{Q}}_\varphi \overset{(6.5)}{\underset{(6.6)}{=}} -\frac{a\bar{\varphi}'}{\mathcal{H}}\{ \mathcal{R}_\varphi' \} . \tag{7.78}$$

This implies that

$$
L_\varphi \overset{(7.76)}{\underset{(7.77),(7.78)}{=}} -\frac{a\bar\varphi'}{\mathcal{H}}\left\{\partial_\tau^2 - \nabla^2 + \left[a\Upsilon + \frac{2\mathcal{H}}{a\bar\varphi'}\left(\frac{a\bar\varphi'}{\mathcal{H}}\right)'\right]\partial_\tau\right\}\mathcal{R}_\varphi .
\tag{7.79}
$$

Second, going to physical time by following Eqs. (6.9) and (6.10),

$$
L_\varphi \overset{(7.79)}{\underset{(6.9),(6.10)}{=}} -\frac{a^3\dot{\bar\varphi}}{H}\left\{\partial_t^2 - \frac{\nabla^2}{a^2} + \left[\frac{\dot a}{a} + \Upsilon + \frac{2H}{a\dot{\bar\varphi}}\left(\frac{a\dot{\bar\varphi}}{H}\right)^{\cdot}\right]\partial_t\right\}\mathcal{R}_\varphi .
\tag{7.80}
$$

Subsequently, multiplying the whole with $-H/(a^3\dot{\bar\varphi})$, and recalling the definition $\mathcal{F} = \ddot{\bar\varphi}/\dot{\bar\varphi} - \dot H/H$ from Eq. (6.12), the equation becomes

$$
\boxed{
\begin{aligned}
&\left\{\partial_t^2 + (\Upsilon + 2\mathcal{F} + 3H)\partial_t - \frac{\nabla^2}{a^2}\right\}\mathcal{R}_\varphi \\
&+ \left\{\frac{4\pi G}{H}\left(1 - \frac{\bar p_{,T}}{\bar e_{,T}}\right) + \frac{\Upsilon_{,T}\dot{\bar\varphi} + V_{,\varphi T}}{\bar e_{,T}\dot{\bar\varphi}}\right\}\mathcal{S}_T \\
&- \left\{\frac{4\pi G(\Upsilon + 2\mathcal{F})}{H}\right\}\mathcal{S}_v + \frac{8\pi G}{3}\nabla^2\Pi \overset{(7.76)}{\underset{(7.80)}{=}} -\frac{\varrho H}{\dot{\bar\varphi}} .
\end{aligned}
}
\tag{7.81}
$$

This generalizes Eq. (6.11) to an environment in which a thermal plasma is present, and the evolution of curvature perturbations is influenced by Υ and ϱ.

As for the anisotropic stress, Π, we recall from Eq. (3.134) that it contains

$$
\Pi \overset{(3.134)}{\supset} \frac{2\eta(v - \vartheta')}{a} \overset{(4.31)}{\underset{(4.61)}{=}} -\frac{2\eta}{a\mathcal{H}}(\psi + \mathcal{R}_v) \overset{(1.6)}{=} -\frac{2\eta}{a^2 H}(\psi + \mathcal{R}_\varphi + \mathcal{R}_v - \mathcal{R}_\varphi)
$$

$$
\overset{(7.61)}{=} -\frac{2\eta}{a^2 H}\left(\psi + \mathcal{R}_\varphi + \frac{\mathcal{S}_v}{\bar e + \bar p}\right) ,
\tag{7.82}
$$

where η is the shear viscosity. In addition, Π contains a hydrodynamic noise term (cf. Eqs. (3.127) and (3.139)), which could be moved to the right-hand side of the equation, where it plays a role similar to ϱ. All in all, Eq. (7.81) contains four dynamical variables ($\mathcal{R}_\varphi$, $\mathcal{S}_T$, $\mathcal{S}_v$ and ψ), and further equations are needed for specifying the solution.

To obtain equations for the isocurvature perturbations, $\mathcal{S}_v$ and $\mathcal{S}_T$, we start from generalizations of the energy-momentum conservation in Eqs. (3.119) and (3.120), including now both inflaton and fluid perturbations,

$$0 \;=\; \delta T_{\mu 0}{}^{;\mu} \overset{\text{code in}}{\underset{\text{Sect. 3.8}}{=}} \frac{1}{a^2}\Big\{\, \bar\varphi'(\nabla^2 - \partial_\tau^2)\delta\varphi - (\bar\varphi'' + 4\mathcal{H}\bar\varphi')\,\delta\varphi'$$

$$+ (\bar\varphi')^2(h_0' + 3h_{\mathrm{D}}' + \nabla^2 h) + 2\bar\varphi'(\bar\varphi'' + 2\mathcal{H}\bar\varphi')\,h_0 \Big\}$$

$$- \delta e' - 3\mathcal{H}(\delta e + \delta p) + (\bar e + \bar p)(3h_{\mathrm{D}}' + \nabla^2 v)\,, \tag{7.83}$$

$$0 \;=\; \delta T_{\mu i}{}^{;\mu} \overset{\text{code in}}{\underset{\text{Sect. 3.8}}{=}} -\frac{1}{a^2}(\bar\varphi'' + 2\mathcal{H}\bar\varphi')\,\delta\varphi_{,i}$$

$$+ \delta p_{,i} + (\bar e + \bar p)h_{0,i} + (\partial_\tau + 4\mathcal{H})[(\bar e + \bar p)(v_i - h_i)] + \Pi_{ik,k}\,. \tag{7.84}$$

We can eliminate second time derivatives from Eq. (7.83) by inserting $\delta\varphi''$ from Eq. (7.56). As for Eq. (7.84), it contains both a scalar and a vector part. The vector part retains the form in Eq. (3.122) in the presence of $\delta\varphi$, whereas the scalar part from Eq. (3.121) gets modified. Collecting together, the equations reduce to

$$0 \overset{(7.83)}{\underset{(7.56)}{=}} \frac{1}{a^2}\Big\{ \overbrace{(-\bar\varphi'' - 2\mathcal{H}\bar\varphi' + a\Upsilon\bar\varphi')}^{\text{from (7.87): } 2a\Upsilon\bar\varphi' + a^2 V_{,\varphi}}\,\delta\varphi' + a(\bar\varphi')^2\,\overbrace{\delta\Upsilon}^{\text{insert (7.66)}} + a^2\bar\varphi'(\overbrace{\delta V_{,\varphi}}^{\text{insert (7.67)}} - \varrho)$$

$$+ \bar\varphi'\,\overbrace{[\,2\bar\varphi'' + (4\mathcal{H} + a\Upsilon)\bar\varphi' + 2a^2 V_{,\varphi}\,]}^{\text{from (7.87): } -a\Upsilon\bar\varphi'}\;\overbrace{h_0}^{\text{insert (7.70)}} \Big\}$$

$$- \delta e' - 3\mathcal{H}\overbrace{(\delta e + \delta p)}^{\text{insert (7.68)}} + (\bar e + \bar p)\Big[\, h_0' + 3h_{\mathrm{D}}' + \overbrace{\nabla^2 h}^{\text{insert (7.57)}} + \overbrace{\nabla^2(v - h)}^{\text{insert (7.69)}} - h_0'\,\Big]\,, \tag{7.85}$$

$$0 \overset{(7.84)}{\underset{(3.77)}{=}} -\frac{1}{a^2}(\bar\varphi'' + 2\mathcal{H}\bar\varphi')\;\overbrace{\delta\varphi}^{\text{insert (7.63)}}$$

$$+ \overbrace{\delta p}^{\text{insert (7.68)}} + \overbrace{[(\bar e + \bar p)(h - v)]'}^{\text{insert (7.69)}} + (\bar e + \bar p)[\,\overbrace{h_0}^{\text{insert (7.70)}} + 4\mathcal{H}\,\overbrace{(h - v)}^{\text{insert (7.69)}}\,] + \frac{2}{3}\nabla^2\Pi\,. \tag{7.86}$$

Here at several points we employed the background identity from Eq. (1.66),

$$\bar\varphi'' + (2\mathcal{H} + a\Upsilon)\bar\varphi' + a^2 V_{,\varphi} \overset{(1.66)}{=} 0\,. \tag{7.87}$$

Let us first tackle Eq. (7.86), which leads to a simpler analysis than Eq. (7.85). We start with the coefficient of Y (cf. Eq. (7.59)), which should cancel. We find

$$(7.86)_{\rm R} \supset \left[\overbrace{\frac{(\bar{\varphi}'' + 2\mathcal{H}\bar{\varphi}')\,\bar{\varphi}'}{a^2\mathcal{H}}}^{\text{from }\delta\varphi} \overbrace{- \frac{\cancel{\bar{R}'}}{\mathcal{H}} + \frac{\bar{e}' + \cancel{\bar{R}'}}{\mathcal{H}}}^{\text{from }\delta p} \overbrace{- \frac{\cancel{(\bar{e}+\bar{p})\mathcal{H}'}}{\mathcal{H}^2} + \frac{\bar{e}+\bar{p}}{\cancel{\mathcal{H}}}\cancel{\partial_\tau} + 4(\bar{e}+\bar{p})}^{\text{from }(\bar{e}+\bar{p})(h-v)} \right.$$

$$\left. \overbrace{- \frac{\bar{e}+\bar{p}}{\cancel{\mathcal{H}}}\cancel{\partial_\tau} - \left(1 - \frac{\cancel{\mathcal{H}'}}{\cancel{\mathcal{H}^2}}\right)(\bar{e}+\bar{p})}^{\text{from }h_0} \right] Y$$

$$= \left[\frac{(\bar{\varphi}'' + 2\mathcal{H}\bar{\varphi}')\,\bar{\varphi}'}{a^2\mathcal{H}} + \frac{\bar{e}' + 3\mathcal{H}(\bar{e}+\bar{p})}{\mathcal{H}} \right] Y \overset{(1.72)}{=} 0 \,. \tag{7.88}$$

Encouraged by the cancellation of gauge dependence, we can proceed to the simplification of the physical terms. They can be combined into

$$0 \overset{(7.86)}{\underset{Y\to 0}{=}} \overbrace{\left[-\frac{a\bar{p}_{,T}}{\mathcal{H}\bar{e}_{,T}} \right]}^{\text{from }\delta p} S_T + \frac{1}{\mathcal{H}}\left[\overbrace{\partial_\tau - \frac{\mathcal{H}'}{\mathcal{H}} + 4\mathcal{H}}^{\text{from }(\bar{e}+\bar{p})(h-v)} \overbrace{- \frac{4\pi G a^2(\bar{e}+\bar{p})}{\mathcal{H}}}^{\text{from }h_0\,;\,\text{insert }(1.70)} \right] S_v$$

$$+ \left[\overbrace{-\frac{\bar{\varphi}'' + 2\mathcal{H}\bar{\varphi}'}{a^3}}^{\text{from }\delta\varphi} \overbrace{+ \frac{\bar{R}'}{a\bar{\varphi}'}}^{\text{from }\delta p} \overbrace{- \frac{\bar{e}+\bar{p}}{a\bar{\varphi}'}\partial_\tau - \frac{\bar{e}' + \bar{R}'}{a\bar{\varphi}'} + \frac{(\bar{e}+\bar{p})\bar{\varphi}''}{a(\bar{\varphi}')^2} - 3\mathcal{H}\frac{\bar{e}+\bar{p}}{a\bar{\varphi}'}}^{\text{from }(\bar{e}+\bar{p})(h-v)} \right.$$

$$\left. \overbrace{+ \left(\mathcal{H} - \frac{\mathcal{H}'}{\mathcal{H}}\right)\frac{\bar{e}+\bar{p}}{a\bar{\varphi}'}}^{\text{from }(\bar{e}+\bar{p})\,h_0} \right] \widehat{\mathcal{Q}}_\varphi + \frac{2}{3}\nabla^2\Pi$$

$$\overset{(1.72)}{=} \left[-\frac{a\bar{p}_{,T}}{\mathcal{H}\bar{e}_{,T}} \right] S_T + \frac{1}{\mathcal{H}}\left[\partial_\tau + 3\mathcal{H} + \frac{4\pi G(\bar{\varphi}')^2}{\mathcal{H}} \right] S_v$$

$$+ \frac{\bar{e}+\bar{p}}{a\bar{\varphi}'}\underbrace{\left[-\partial_\tau + \frac{\bar{\varphi}''}{\bar{\varphi}'} + \mathcal{H} - \frac{\mathcal{H}'}{\mathcal{H}} \right]}_{\frac{\mathcal{H}}{a\bar{\varphi}'}\left(\frac{a\bar{\varphi}'}{\mathcal{H}}\right)'} \widehat{\mathcal{Q}}_\varphi + \frac{2}{3}\nabla^2\Pi \,. \tag{7.89}$$

As a final step, we go over to the variable $\mathcal{R}_\varphi$ and to physical time, by inserting Eqs. (7.78) and (6.9), whereby we find

$$\boxed{\begin{aligned} &\left\{ \partial_t + 3H + \frac{4\pi G\dot{\bar{\varphi}}^2}{H} \right\} S_v - \left\{ \frac{\bar{p}_{,T}}{\bar{e}_{,T}} \right\} S_T \\ &+ \left\{ (\bar{e}+\bar{p})\partial_t \right\} \mathcal{R}_\varphi + \frac{2H}{3}\nabla^2\Pi \overset{H\times(7.89)}{\underset{(7.78),(6.9)}{=}} 0 \,. \end{aligned}} \tag{7.90}$$

This is an equation for the same variables as Eq. (7.81), however the two equations are linearly independent. Given that $\dot{\mathcal{S}}_v$ appears, Eq. (7.90) can be said to determine the time evolution of $\mathcal{S}_v$.

Next, we attack the most complicated relation, from Eq. (7.85). Two new ingredients are needed for it, $\delta e'$ and h_0'. From Eq. (7.68),

$$\delta e \stackrel{(7.68)}{=} \frac{\bar{e}'}{a\bar{\varphi}'}\left(\widehat{\mathcal{Q}}_\varphi - \frac{a\bar{\varphi}'}{\mathcal{H}}Y\right) - \frac{a}{\mathcal{H}}\mathcal{S}_T \tag{7.91}$$

$$\Rightarrow\ \delta e' = \left[\frac{\bar{e}''}{a\bar{\varphi}'} + \frac{\bar{e}'}{a\bar{\varphi}'}\left(\partial_\tau - \mathcal{H} - \frac{\bar{\varphi}''}{\bar{\varphi}'}\right)\right]\widehat{\mathcal{Q}}_\varphi$$
$$-\left[\frac{\bar{e}''}{\mathcal{H}} + \frac{\bar{e}'}{\mathcal{H}}\left(\partial_\tau - \frac{\mathcal{H}'}{\mathcal{H}}\right)\right]Y - \frac{a}{\mathcal{H}}\left(\partial_\tau + \mathcal{H} - \frac{\mathcal{H}'}{\mathcal{H}}\right)\mathcal{S}_T\ . \tag{7.92}$$

From Eq. (7.70),

$$h_0 \stackrel{(7.70)}{=} \left(\mathcal{H} - \frac{\mathcal{H}'}{\mathcal{H}}\right)\frac{\widehat{\mathcal{Q}}_\varphi}{a\bar{\varphi}'} - \left(\frac{1}{\mathcal{H}}\partial_\tau + 1 - \frac{\mathcal{H}'}{\mathcal{H}^2}\right)Y - \frac{4\pi G a^2}{\mathcal{H}^2}\mathcal{S}_v \tag{7.93}$$

$$\Rightarrow\ h_0' = \frac{1}{a\bar{\varphi}'}\left[\left(\mathcal{H} - \frac{\mathcal{H}'}{\mathcal{H}}\right)\partial_\tau + 2\mathcal{H}' - \mathcal{H}^2 - \frac{\mathcal{H}''}{\mathcal{H}} + \frac{\mathcal{H}'^2}{\mathcal{H}^2} - \left(\mathcal{H} - \frac{\mathcal{H}'}{\mathcal{H}}\right)\frac{\bar{\varphi}''}{\bar{\varphi}'}\right]\widehat{\mathcal{Q}}_\varphi$$
$$-\left[\frac{1}{\mathcal{H}}\partial_\tau^2 + \left(1 - \frac{2\mathcal{H}'}{\mathcal{H}^2}\right)\partial_\tau + \frac{2\mathcal{H}'^2}{\mathcal{H}^3} - \frac{\mathcal{H}''}{\mathcal{H}^2}\right]Y$$
$$-\frac{4\pi G a^2}{\mathcal{H}^2}\left[\partial_\tau + 2\left(\mathcal{H} - \frac{\mathcal{H}'}{\mathcal{H}}\right)\right]\mathcal{S}_v\ . \tag{7.94}$$

In order to eliminate $\bar{e}''$ from Eq. (7.92), we combine Eqs. (1.72) and (7.87) into

$$\bar{e}' \stackrel{(1.72)}{=} -3\mathcal{H}(\bar{e} + \bar{p}) - \frac{\bar{\varphi}'}{a^2}\overbrace{\left(\bar{\varphi}'' + 2\mathcal{H}\bar{\varphi}'\right)}^{\text{from (7.87): } -a\Upsilon\bar{\varphi}' - a^2 V_{,\varphi}}$$

$$= -3\mathcal{H}(\bar{e} + \bar{p}) + \frac{\Upsilon(\bar{\varphi}')^2}{a} + \bar{\varphi}'\,V_{,\varphi} \tag{7.95}$$

$$\Rightarrow\ \bar{e}'' = -3\mathcal{H}'(\bar{e} + \bar{p}) - 3\mathcal{H}(\bar{e}' + \bar{p}') + \frac{(\Upsilon' - \mathcal{H}\Upsilon)(\bar{\varphi}')^2}{a} + \frac{2\Upsilon\bar{\varphi}'\bar{\varphi}''}{a}$$
$$+ \bar{\varphi}''\,V_{,\varphi} + \bar{\varphi}'\,V_{,\varphi}'\ . \tag{7.96}$$

From Eq. (7.94), it is advantageous to eliminate $\mathcal{H}''$, by making use of Eq. (7.75), and $\mathcal{S}_v'$, by making use of Eq. (7.89). In addition, Eqs. (1.68) and (1.70) are needed.

It is inevitable that when we insert Eqs. (7.92) and (7.94) into (7.85), the expressions become lengthy. It is then all the more important that we have an efficient crosscheck available, from the cancellation of Y. Most simply, $(\partial_\tau^2 - \nabla^2)Y$ cancels between its appearances in $h_0' + 3h_D' + \nabla^2 h$ (cf. Eq. (7.57)), $\nabla^2(v - h)$ (cf. Eq. (7.69)), and $-h_0'$ (cf. Eq. (7.94)). The cancellation of Y' requires a bit more effort, making use of Eq. (7.95), whereas the coefficient of Y contains $\bar{e}''$ and $\mathcal{H}''$, so that Eqs. (7.96) and (7.75) need to be employed.

There are also many cancellations between the physical terms. Notably, after having eliminated $\mathcal{S}'_v$, by making use of Eq. (7.89), the anisotropic stress, $\nabla^2 \Pi$, drops out. By setting $\widehat{\mathcal{Q}}'_\varphi$ into the combination $\widehat{\mathcal{Q}}'_\varphi - \frac{\mathcal{H}}{a\bar{\varphi}'}\left(\frac{a\bar{\varphi}'}{\mathcal{H}}\right)'\widehat{\mathcal{Q}}_\varphi$, all other non-derivative appearances of $\widehat{\mathcal{Q}}_\varphi$ cancel. By a repeated use of background identities, the coefficients of $\widehat{\mathcal{Q}}'_\varphi$ and $\mathcal{S}_v$ are seen to be proportional to the same combination, $\Upsilon(\bar{\varphi}')^2/a^2 + 8\pi G a (\bar{e} + \bar{p})\bar{e}/\mathcal{H}$.

Finally, we go once again to physical time, by inserting Eqs. (7.78) and (6.9), as well as Eq. (1.6), in the form $\mathcal{H}' - \mathcal{H}^2 = a^2 \dot{H}$. This then leads to our third basic equation,

$$
\left\{ \partial_t + 3H\left(1 + \frac{\bar{p}_{,T}}{\bar{e}_{,T}}\right) + \frac{4\pi G(\bar{e} + \bar{p}) - \dot{H}}{H} - \frac{\dot{\bar{\varphi}}\,(\Upsilon_{,T}\,\dot{\bar{\varphi}} + V_{,\varphi T})}{\bar{e}_{,T}} \right\} \mathcal{S}_T
$$
$$
- \left\{ \Upsilon\dot{\bar{\varphi}}^2 + \frac{8\pi G(\bar{e} + \bar{p})\bar{e}}{H} \right\}\left\{ \dot{\mathcal{R}}_\varphi - \frac{4\pi G}{H}\mathcal{S}_v \right\} - \frac{\nabla^2}{a^2}\left\{ (\bar{e} + \bar{p})\mathcal{R}_\varphi + \mathcal{S}_v \right\} = \varrho\,\dot{\bar{\varphi}}H \, .
$$

$$(7.97)$$

This is linearly independent of Eqs. (7.81) and (7.90), and given the appearance of $\dot{\mathcal{S}}_T$, can be said to determine the time evolution of $\mathcal{S}_T$.

Let us summarize the situation so far. In Eqs. (7.81), (7.90) and (7.97), we have three independent equations for three variables, $\mathcal{R}_\varphi$, $\mathcal{S}_v$ and $\mathcal{S}_T$. If the anisotropic stress, Π, can be omitted, i.e. if shear viscous corrections are small (cf. the discussion below Eq. (3.143)), then this is sufficient for specifying the time evolution of the three physical perturbations. However, if the anisotropic stress plays a role, then, as shown in Eq. (7.82), one of the Bardeen potentials, ψ (cf. Eq. (4.31)), also makes an appearance. In Eq. (4.56), we already obtained one equation for the Bardeen potentials,

$$
\phi - \psi + 8\pi G a^2 \Pi \overset{(4.56)}{=} 0 \, .
$$

$$(7.98)$$

However, since this contains the other Bardeen potential, ϕ (cf. Eq. (4.30)), we also need a fifth equation, in order to fully specify the time evolution of the perturbations of a non-perfect fluid coupled to an inflaton field.

A possible starting point for deriving the remaining equation is given by Eq. (7.93). Even if we have already employed this equation as a part of other computations, we can still extract novel information from it. Notably, we express the left-hand side of Eq. (7.93) in terms of the Bardeen potential from Eq. (4.30),

$$
h_0 \overset{(4.30)}{=} \phi - (\partial_\tau + \mathcal{H})X \, , \quad X \overset{(5.12)}{\equiv} h - \vartheta' \, .
$$

$$(7.99)$$

On the right-hand side of Eq. (7.93), we express the gauge-variant combination Y from Eq. (7.59) in terms of ψ via Eq. (4.31),

$$
Y \overset{(4.31)}{\underset{(5.12)}{=}} \psi + \mathcal{H}X \, .
$$

$$(7.100)$$

Substituting Eqs. (7.99) and (7.100) into Eq. (7.93), where we also insert $\widehat{Q}_\varphi = -(a\bar\varphi'/\mathcal{H})\mathcal{R}_\varphi$ according to Eq. (7.60), it can be verified that all appearances of X and X' drop out. The remaining terms give

$$\phi \overset{(7.93),(7.60)}{\underset{(7.99),(7.100)}{=}} \left(\frac{\mathcal{H}'}{\mathcal{H}^2} - 1\right)\mathcal{R}_\varphi - \left(\frac{1}{\mathcal{H}}\partial_\tau + 1 - \frac{\mathcal{H}'}{\mathcal{H}^2}\right)\psi - \frac{4\pi G a^2}{\mathcal{H}^2}\mathcal{S}_v . \quad (7.101)$$

In this equation, all the quantities appearing are gauge-invariant. Moving the terms to the left-hand side, and going over to physical time, with $\mathcal{H}'/\mathcal{H}^2 = 1 + \dot{H}/H^2$ from Eq. (1.6), we obtain our fifth and final relation,

$$\boxed{\left(\partial_t - \frac{\dot{H}}{H}\right)\psi + H\phi - \frac{\dot{H}}{H}\mathcal{R}_\varphi + \frac{4\pi G}{H}\mathcal{S}_v = 0 .} \quad (7.102)$$

Given that $\dot\psi$ appears, Eq. (7.102) can be said to determine the time evolution of ψ.

We now have the full set of equations needed for determining the time evolution of the coupled system at linear order. The solutions following from Eqs. (7.81), (7.90), (7.97), (7.98) and (7.102) will be discussed in Chap. 8 (for Hubble horizon exit) and Chap. 9 (for times after Hubble horizon re-entry).

7.6 Perturbative Thermodynamics for a Thermalized Inflaton

If $\Upsilon \gg H$, the inflaton field thermalizes (cf. Fig. 7.1). In this regime (if it gets realized for all k), it is possible to adopt a 'complementary' viewpoint, in which the inflaton is actually part of the radiation plasma. In the present section, we elaborate on this from a few different perspectives. We first show two ways to determine the contributions of a thermalized inflaton to the energy density and pressure. Finally, we recall why the effective theory that we have used, based on the Langevin equation, ultimately fails in this regime.

The first method goes through the effective potential. In Eq. (7.3), we met V_0, the 'bare' self-interaction potential appearing in a Lagrangian. In the effective-theory description of Eq. (7.4), this had been replaced by V. Through a matching computation, we saw that V obtains a thermal mass correction from interactions with the medium (cf. Eq. (7.26)). However, through the formalism of thermal field theory, we can also determine the influence of the medium on V to all orders in $\bar\varphi$, not only to linear order, as we had done to arrive at Eq. (7.26). In fact, important effects already arise at *zeroth order* in $\bar\varphi$, through an overall modification of the energy density and pressure. To be specific, let us consider a time at which the inflaton is settled around its global minimum, $\bar\varphi = 0$, but still experiences thermal fluctuations. In statistical physics, the effect of fluctuations is obtained by computing the *canonical partition function*, $\mathcal{Z}$. From $\mathcal{Z}$, we can extract the *free energy density*, and this is exactly what is called the *effective potential*, V_{eff}.

The leading non-trivial contribution to the effective potential is referred to as a '1-loop' contribution. To compute it in the perturbative approach, we treat the inflaton as a weakly coupled massive scalar field, $V_0 \approx m^2\varphi^2/2$. Then the 1-loop correction reads (cf., e.g. Ref. [19, Eq. (9.26)])

$$V_{\text{eff}}^{(1)} = -\lim_{L\to\infty}\frac{T}{L^3}\ln\mathcal{Z}^{(1)} = \int\frac{\mathrm{d}^3\mathbf{p}}{(2\pi)^3}\left[\frac{\epsilon_p}{2} + T\ln\left(1 - e^{-\epsilon_p/T}\right)\right]_{\epsilon_p = \sqrt{p^2+m^2}} .$$

(7.103)

The first, temperature-independent term, requires regularization and renormalization; let us assume that it modifies the parameters appearing in V_0. The second term gives the thermal effects that we are interested in.

Changing integration variables as

$$\mathrm{d}^3\mathbf{p} = 4\pi\mathrm{d}p\, p^2 = 4\pi\mathrm{d}\epsilon_p\,\epsilon_p\sqrt{\epsilon_p^2 - m^2}\,,$$

(7.104)

the contribution to the pressure according to Eq. (1.65) is then

$$p_\varphi \overset{(1.65)}{\underset{(7.103),(7.104)}{\supset}} -\frac{T}{2\pi^2}\int_m^\infty\mathrm{d}\epsilon_p\,\epsilon_p\sqrt{\epsilon_p^2 - m^2}\,\ln\left(1 - e^{-\epsilon_p/T}\right) .$$

(7.105)

For the energy density from Eq. (1.65), we need a Legendre transform. Noting that

$$\left(1 - T\partial_T\right)\left[T\ln\left(1 - e^{-\epsilon_p/T}\right)\right] = -T^2\partial_T\ln\left(1 - e^{-\epsilon_p/T}\right) = -T^2\frac{-\epsilon_p\, e^{-\epsilon_p/T}}{1 - e^{-\epsilon_p/T}}\frac{1}{T^2}$$

$$= \frac{\epsilon_p}{e^{\epsilon_p/T} - 1} \equiv \epsilon_p\, n_{\text{B}}(\epsilon_p)\,,$$

(7.106)

we find

$$e_\varphi \overset{(7.105)}{\underset{(7.106)}{\supset}} \frac{1}{2\pi^2}\int_m^\infty\mathrm{d}\epsilon_p\,\epsilon_p^2\sqrt{\epsilon_p^2 - m^2}\, n_{\text{B}}(\epsilon_p) .$$

(7.107)

Even if the integral representations of p_φ and e_φ in Eqs. (7.105) and (7.107) are rapidly convergent, neither can be evaluated analytically. However, by expanding the integrands in $e^{-\epsilon_p/T}$, sum representations in terms of modified Bessel functions can be found. Analytic results become available by sending $m/T \to 0$, and then Eqs. (7.105) nor (7.107) amount just to adding one light degree of freedom to the radiation plasma, i.e. changing $g_* \to g_* + 1$ and $h_* \to h_* + 1$ in Eq. (7.55).

Let us now inspect the same results from a different perspective. In a flat Minkowskian frame, the energy-momentum tensor of the scalar field from Eq. (1.47), after the replacement $V \to V_0 \to \frac{1}{2}m^2\varphi^2$, has the components

$$T_{00} \overset{(1.47)}{=} \frac{1}{2}\left(\dot\varphi^2 + \sum_l\varphi_{,l}\varphi_{,l} + m^2\varphi^2\right) ,$$

(7.108)

$$T_{ij} \overset{(1.47)}{=} \varphi_{,i}\varphi_{,j} + \delta_{ij}\frac{1}{2}\left(\dot\varphi^2 - \sum_l\varphi_{,l}\varphi_{,l} - m^2\varphi^2\right) .$$

(7.109)

Placing ourselves in the same situation as before, with $\bar{\varphi} \to 0$, the φ here is $\delta\varphi$. We can average over the thermal noise influencing $\delta\varphi$, denoting the results by $\langle T_{00} \rangle$ and $\langle T_{ij} \rangle$. Employing the 'target' expression from Eq. (7.31) for the 2-point correlator of $\delta\varphi$; inserting a plane-wave time dependence; and only including the convergent thermal parts, we obtain

$$\langle T_{00} \rangle \overset{(7.108)}{\underset{(7.31)}{\supset}} \int \frac{d^3\mathbf{p}}{(2\pi)^3} \frac{n_{\rm B}(\epsilon_p)}{\epsilon_p} \frac{1}{2}\left(\epsilon_p^2 + p^2 + m^2\right) = \int \frac{d^3\mathbf{p}}{(2\pi)^3} \epsilon_p\, n_{\rm B}(\epsilon_p) \ . \qquad (7.110)$$

This agrees with Eq. (7.107) when we make use of Eq. (7.104).

We have to work a bit harder for the spatial components, $\langle T_{ij} \rangle$. Inserting the thermal part of Eq. (7.31), and making use of isotropy, we get

$$\langle T_{ij} \rangle \overset{(7.109)}{\underset{(7.31)}{\supset}} \int \frac{d^3\mathbf{p}}{(2\pi)^3} \frac{n_{\rm B}(\epsilon_p)}{\epsilon_p}\left[p_i p_j + \delta_{ij} \frac{1}{2}\left(\epsilon_p^2 - p^2 - m^2\right)\right]$$

$$\overset{\text{isotropy}}{=} \int \frac{d^3\mathbf{p}}{(2\pi)^3} \frac{p^2 \delta_{ij}}{3\epsilon_p}\, n_{\rm B}(\epsilon_p) \ . \qquad (7.111)$$

Similarly to Eq. (7.106), we now note that

$$\partial_p\left[T \ln\left(1 - e^{-\epsilon_p/T}\right)\right] = T \frac{-e^{-\epsilon_p/T}}{1 - e^{-\epsilon_p/T}} \frac{(-\partial_p \epsilon_p)}{T} = \frac{p}{\epsilon_p} n_{\rm B}(\epsilon_p) \ . \qquad (7.112)$$

Therefore

$$\langle T_{ij} \rangle \overset{(7.111)}{\underset{(7.112)}{\supset}} \frac{\delta_{ij}}{2\pi^2} \int_0^\infty dp\, \frac{p^3}{3} \partial_p\left[T \ln\left(1 - e^{-\epsilon_p/T}\right)\right]$$

$$= -\frac{\delta_{ij}}{2\pi^2} \int_0^\infty dp\, p^2\, T \ln\left(1 - e^{-\epsilon_p/T}\right) \ , \qquad (7.113)$$

where we carried out a partial integration. This agrees with Eq. (7.105).

To summarize, if we go to *second order in perturbations*, and assume that the inflaton field has equilibrated, its fluctuations contribute to the energy density and pressure just like a massive scalar field that is part of the radiation plasma.

Now, however, we come to an issue. Our actual effective-theory framework is not based on Eq. (7.31), but rather on the Langevin equation, which yields Eq. (7.43). This corresponds to replacing $n_{\rm B}(\epsilon_p) \to T/\epsilon_p$ in Eqs. (7.110) and (7.111). Then the integrals over the momenta are UV-divergent. This is a reflection of the famous *Rayleigh-Jeans divergence* of classical field theory.

The analysis in this book stays mostly at linear order in perturbations. Then, if we restrict ourselves to small enough comoving momenta, the Rayleigh-Jeans divergence is of no concern. However, it is important to keep in mind that equilibration implies a transfer of energy, from the initial quantum state towards thermal fluctuations, which carry a typical momentum $p \sim T$ for a massless particle ($m \ll T$), or $p \sim \sqrt{2mT}$ for a massive one ($m \gg T$). If we encounter a phenomenon which is sensitive to the thermal energy scales, $\epsilon_p \sim T$, we have to employ quantum-statistical tools, in order to obtain physically meaningful results. Notably, this can be important if we consider thermally induced *non-Gaussianities* in power spectra (cf., e.g. Ref. [26]), or *second-order effects* producing gravitational waves (cf. the end of Sect. 10.4). In principle, the issue is also relevant for the autocorrelator of the thermal noise (cf. Eq. (7.44)), because at early times, any comoving momentum lies in the domain $p = k/a \gg T$. At linear order, the last issue can be rectified by modifying the noise autocorrelator, as mentioned below Eq. (7.44).

7.7   Numerical Solution for Smooth Reheating

We show here how the background equations (7.48)–(7.50) can be solved numerically. The functions e_r and p_r are parametrized according to Eq. (7.55), with

$$g_* = h_* = i_* = 106.75 \, . \tag{7.114}$$

The number 106.75 corresponds to the Standard Model in the limit that all interactions are weak enough to be insignificant, and the temperature is high enough for masses to be negligible (the fractional number originates from fermionic contributions). For the friction we take the ansatz

$$\Upsilon \equiv \frac{\kappa_T \, (\pi T)^3 + \kappa_m \, m^3}{(4\pi)^3 \, f_a^2} \, . \tag{7.115}$$

Otherwise the setup is identical to that described in Sect. 1.7. A `python` script is displayed below, and it produces the data as shown in Fig. 7.2.

Having available a background solution from inflation until reheating, we can determine how momenta *redshift across different epochs*. Let us denote by a_e a moment when the inflationary dynamics has ended and we can employ an extrapolation of Standard Model thermodynamics to describe the dominant plasma component (we choose $T_e = 10^{-12} m_{\rm pl}$ for the matching point). Then we can write a current momentum mode (in astrophysical units) as

$$\frac{(k/a_0)}{\mathrm{Mpc}^{-1}} = \underbrace{\frac{(k/a_i)}{m_{\rm pl}}}_{\text{like fig. 6.1}} \underbrace{\left(\frac{a_i}{a_e}\right)}_{e^{-N_e}} \underbrace{\left(\frac{a_e}{a_0}\right)}_{\text{like (2.35)}} \underbrace{m_{\rm pl} \, \mathrm{Mpc}}_{1.9092 \times 10^{57}} \, . \tag{7.116}$$

For the chosen $T_e = 10^{-12} m_{\rm pl}$, a computation like in Eq. (2.35) yields $a_e/a_0 \approx e^{-46.5}$. The factor $a_i/a_e \equiv e^{-N_e}$ needs to be determined by integrating the $H(t)$ given by a numerical computation, like that in Fig. 7.2 (left). Putting everything together, we obtain Fig. 7.2 (right), which shows the initial momentum that corresponds to a given physical value today. The corresponding data file is generated at the end of the `python` script.

```python
# numerical solution for reheating and/or warm inflation [numerics_bg_warm.py]
#
# import basic tools and integration routines
import numpy as np
from scipy.integrate import solve_ivp

# parameters [mpl]
fa = 1.25;        m = 1.09e-6;    Pi = np.pi     # inflaton decay constant and mass, shorthand for pi
gstar = 106.75;  hstar = gstar;  istar = gstar  # energy density, entropy density, heat capacity

# parameters for the cases considered
time_end = 1e20        # time sufficiently large that the system has reheated
case = 'thin' # 'thick'
if case == 'thin':
    kappaT = 1e0;        kappam = 1e0;     T_0 = 1.e-9        # initial seed temperature [mpl]
else:
    kappaT = 1e7;        kappam = 1e9;     T_0 = 1.e-8

# potential of inflaton (phi) [mpl^4]
def V(phi): return m*m*fa*fa*( 1 - np.cos(phi/fa) )

# derivative of inflaton potential by phi [mpl^3]
def Vd(phi): return m*m*fa*np.sin(phi/fa)

# friction coefficient [mpl]
def Ups(T): return (kappaT*Pi*Pi*T*T*T + kappam*m*m*m)/(4*4*4*Pi*Pi*Pi)/(fa*fa)

# radiation energy density [mpl^4]
def er(T): return gstar*Pi*Pi*T*T*T*T/30

# radiation entropy density [mpl^3]
def sr(T): return 2*hstar*Pi*Pi*T*T*T/45

# radiation heat capacity [mpl^3]
def cr(T): return 2*istar*Pi*Pi*T*T*T/15

# energy density of inflaton field [mpl^4]
def efield(phi,phid): return phid*phid/2 + V(phi)

# Hubble rate [mpl]
def H(energy_density): return np.sqrt(8*Pi*np.abs(energy_density)/3)

# evolution at early times: solve for [phi, dot phi, efolds, T]
# derivatives are taken with respect to t=Href*time
def derivatives1(t,y):
    Phi, Phid, Nfolds, T = y
    Hubble = H(efield(Phi, Phid)+er(T))
    dphi_dt = Phid/Href;          ddphi_ddt = (-(3*Hubble+Ups(T))*Phid - Vd(Phi))/Href
    dN_dt = Hubble/Href;          dT_dt = (Ups(T)*Phid*Phid-3*Hubble*T*sr(T))/cr(T)/Href
    dy_dt = [dphi_dt, ddphi_ddt, dN_dt, dT_dt]; return dy_dt

# evolution in averaged regime: solve for [e_phi, efolds, T]
def derivatives2(t,y):
    e_phi, Nfolds, T = y
    Hubble = H(e_phi+er(T))
    dephi_dt = (-(3*Hubble+Ups(T))*e_phi)/Href
    dN_dt = Hubble/Href;          dT_dt = (Ups(T)*e_phi-3*Hubble*T*sr(T))/cr(T)/Href
    dy_dt = [dephi_dt, dN_dt, dT_dt];   return dy_dt

# evolution after e_phi has become insignificant: solve for [efolds, T]
def derivatives3(t,y):
    Nfolds, T = y
    dN_dt = H(er(T))/Href;     dT_dt = (-3*H(er(T))*T*sr(T))/cr(T)/Href
    dy_dt = [dN_dt, dT_dt];  return dy_dt

# initial conditions and reference values [mpl]
phi_0 = 3.5;   Href = np.sqrt(4*Pi/3)*m*phi_0;    phid_0 = - Vd(phi_0)/(3*Href);    N_0 = 0

# define the event function counting the zeros of phi
def phi_crossings(t, y): return y[0]

# integrate up to 10 oscillations (21 crossings of zero)
sol1 = solve_ivp(derivatives1, [1, 1e4], [phi_0, phid_0, N_0, T_0],
                 events=phi_crossings,method='DOP853',atol=1e-12,rtol=1e-13)
t_match = sol1.t_events[0][20]          # time at which condition was met
time1 = sol1.t[sol1.t <= t_match];       n_match1 = len(time1)
ephi1 = efield(sol1.y[0][:n_match1], sol1.y[1][:n_match1]);
```

```python
efolds1 = sol1.y[2][:n_match1];          T1 = sol1.y[3][:n_match1]

# initial conditions for averaged regime
time_0 = t_match;   ephi_0 = efield(sol1.y_events[0][20][0], sol1.y_events[0][20][1])
efolds_0 = sol1.y_events[0][20][2];   T_0 = sol1.y_events[0][20][3]
print("go over to averaged evolution at t/t_ref = ",time_0)

# define the event function locating when e_field = e_r/500
def phi_decays(t, y): return y[0]-er(y[2])/500
setattr(phi_decays,'terminal',True)     # for stopping immediately when condition is met

# integrate in averaged regime until phi decays
if(ephi_0 > er(T_0)/500):
    sol2 = solve_ivp(derivatives2, [time_0, time_end], [ephi_0, efolds_0, T_0],
                     events=phi_decays,method='DOP853',atol=1e-12,rtol=1e-13)
    t_match = sol2.t_events[0][0]       # time at which condition was met
    time2 = sol2.t[sol2.t <= t_match]; n_match2 = len(time2)
    ephi2 = sol2.y[0][:n_match2];         efolds2 = sol2.y[1][:n_match2];    T2 = sol2.y[2][:n_match2]
    time_0 = t_match;                     efolds_0 = sol2.y_events[0][0][1]; T_0 = sol2.y_events[0][0][2]
else:
    t_match = time_0;                     time2 = np.array([])
    ephi2 = np.array([]);                 efolds2 = np.array([]);            T2 = np.array([])
print("go over to evolution without e_phi at t/t_ref = ",time_0)

# parameters for matching to Standard Model
mplMpc = 1.9092e57                        # conversion between microscopic and macroscopic scales
Tswitchmpl = 1.e-12                       # temperature at which we switch to Standard Model [mpl]
efolds_after_switch = 46.5                # efolds from Tswitch until today

# define the event function locating the switching temperature
def Tswitch(t, y): return y[1]-Tswitchmpl

# integrate until temperature is reached
sol3 = solve_ivp(derivatives3, [time_0, time_end], [efolds_0, T_0],
                 events=Tswitch,method='DOP853',atol=1e-12,rtol=1e-13)
t_match = sol3.t_events[0][0]            # time at which condition was met
time3 = sol3.t[sol3.t <= t_match];       n_match3 = len(time3)
ephi3 = np.zeros(n_match3);              efolds3 = sol3.y[0][:n_match3];    T3 = sol3.y[1][:n_match3]

# collect information about switching point
efolds_till_switch = sol3.y_events[0][0][0]
print("Tswitch reached at t/tref = ",t_match," after ",efolds_till_switch," e-folds")
coeff = np.exp(-efolds_till_switch)*np.exp(-efolds_after_switch)*mplMpc

# initial conditions for the last domain
time_0 = t_match;                        efolds_0 = sol3.y_events[0][0][0]; T_0 = sol3.y_events[0][0][1]

# integrate till the very end
sol4 = solve_ivp(derivatives3, [time_0, time_end], [efolds_0, T_0],
                 method='DOP853',atol=1e-12,rtol=1e-13)
time4 = sol4.t;                          n_match4 = len(time4)
t_match = time4[n_match4-1]              # time reached at the end
ephi4 = np.zeros(n_match4);              efolds4 = sol4.y[0];              T4 = sol4.y[1]
print("integration terminated at t/tref = ",t_match," after ",efolds4[n_match4-1]," e-folds")

# assemble together the complete solution
time = np.concatenate((time1,time2,time3,time4)); e_phi = np.concatenate((ephi1,ephi2,ephi3,ephi4))
efolds = np.concatenate((efolds1,efolds2,efolds3,efolds4)); T = np.concatenate((T1,T2,T3,T4))
e_r = er(T)

# print out background solution
np.savetxt('numerics_bg_warm.dat',
           np.c_[time, e_phi, H(e_phi+e_r), T, Ups(T), efolds],fmt='%.6e',newline='\n',
           header='columns: t*H_ref, e_phi/mpl**4, H/mpl, T/mpl, Ups/mpl, efolds')

# print out relation of microscopic and macroscopic momentum scales
koampl = np.geomspace(1e-10,1e60,100); koaMpc = coeff*koampl
np.savetxt('numerics_redshift.dat',
           np.c_[koaMpc,koampl],fmt='%.6e',newline='\n',
           header='columns: k/a_0*Mpc, k/a_i/mpl')
```

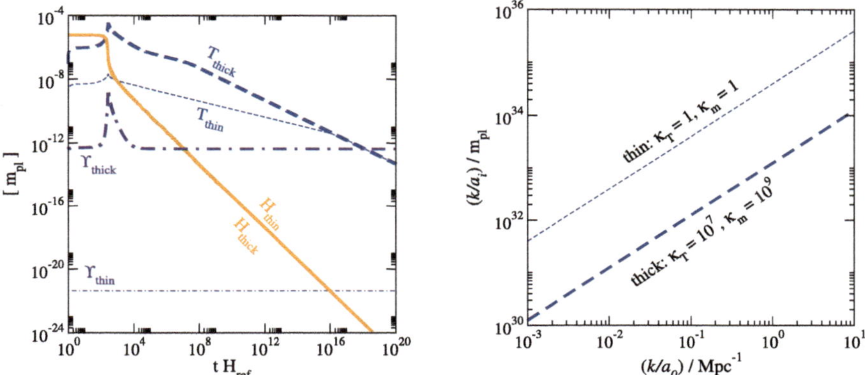

Fig. 7.2 Left: A numerical solution for the reheating dynamics described by Eqs. (7.48)–(7.50). The vacuum potential is like in Eqs. (1.101) and (1.102), and H_{ref} is from Eq. (1.104), whereas the equation of state of the plasma is parametrized according to Eq. (7.114), and the friction according to Eq. (7.115). If the friction is small (thin lines), the universe is matter-dominated until $t\,H_{\mathrm{ref}} \approx 10^{16}$ (cf. discussion around Eq. (1.107)), when $\Upsilon \approx H$ and T starts to decrease faster. If the friction is larger (thick lines), a higher maximal temperature is reached, and radiation domination starts already at $t\,H_{\mathrm{ref}} \approx 10^7$. Right: mapping from current momenta in the CMB range, $(k/a_0)/\mathrm{Mpc}^{-1}$, onto initial momenta at $t = H_{\mathrm{ref}}^{-1}$, $(k/a_i)/m_{\mathrm{pl}}$, as determined from Eq. (7.116). With a larger friction (thick lines), the matter-domination period is shorter, and less e-folds separate inflation and present day. Consequently, we should consider smaller initial momenta. Given that the curvature power spectrum depends on the momentum mode considered (cf. Fig. 6.1 in Sect. 6.5), this influences inflationary predictions. In the literature, reheating dynamics is often not addressed quantitatively, and the results rather contain an error band, obtained by requiring that the pivot scale (cf. Eq. (2.7)) exited the Hubble horizon 50–60 e-folds before inflation ended

References

1. Albrecht, A., Steinhardt, P.J., Turner, M.S., Wilczek, F.: Reheating an Inflationary Universe. Phys. Rev. Lett. **48**, 1437 (1982). https://doi.org/10.1103/PhysRevLett.48.1437
2. Abbott, L.F., Farhi, E., Wise, M.B.: Particle production in the new inflationary cosmology. Phys. Lett. B **117**, 29 (1982). https://doi.org/10.1016/0370-2693(82)90867-X
3. Mukaida, K., Yamada, M.: Perturbative reheating and thermalization of pure Yang-Mills plasma. JHEP **05**, 174 (2024). https://doi.org/10.1007/JHEP05(2024)174, 2402.14054
4. Fu, Y., Ghiglieri, J., Iqbal , S., Kurkela, A.: Thermalization of non-Abelian gauge theories at next-to-leading order. Phys. Rev. D **105**, 054031 (2022). https://doi.org/10.1103/PhysRevD.105.054031, 2110.01540
5. Broadberry, E., Hook, A., Mondal, S.: Warm Inflation with Pseudo-scalar couplings. 2505.07943
6. Guth, A.H.: Inflationary universe: A possible solution to the horizon and flatness problems. Phys. Rev. D **23**, 347 (1981). https://doi.org/10.1103/PhysRevD.23.347
7. Linde, A.D.: A new inflationary universe scenario: A possible solution of the horizon, flatness, homogeneity, isotropy and primordial monopole problems. Phys. Lett. B **108**, 389 (1982). https://doi.org/10.1016/0370-2693(82)91219-9
8. Albrecht, A., Steinhardt, P.J.: Cosmology for Grand Unified Theories with Radiatively Induced Symmetry Breaking. Phys. Rev. Lett. **48**, 1220 (1982). https://doi.org/10.1103/PhysRevLett.48.1220

9. Linde, A.D.: Chaotic inflation. Phys. Lett. B **129**, 177 (1983). https://doi.org/10.1016/0370-2693(83)90837-7
10. Starobinsky, A.A.: A new type of isotropic cosmological models without singularity. Phys. Lett. B **91**, 99 (1980). https://doi.org/10.1016/0370-2693(80)90670-X
11. Kofman, L., Linde, A.D., Starobinsky, A.A.: Towards the theory of reheating after inflation. Phys. Rev. D **56**, 3258 (1997). https://doi.org/10.1103/PhysRevD.56.3258, hep-ph/9704452
12. Barman, B., Bernal, N., Rubio, J.: Two or three things particle physicists (mis)understand about (pre)heating. Nucl. Phys. B **1018**, 116996 (2025). https://doi.org/10.1016/j.nuclphysb.2025.116996, 2503.19980
13. Gibbons, G.W., Hawking, S.W.: Cosmological event horizons, thermodynamics, and particle creation. Phys. Rev. D **15**, 2738 (1977). https://doi.org/10.1103/PhysRevD.15.2738
14. Ford, L.H.: Gravitational particle creation and inflation. Phys. Rev. D **35**, 2955 (1987). https://doi.org/10.1103/PhysRevD.35.2955
15. Kolb, E.W., Long, A.J.: Cosmological gravitational particle production and its implications for cosmological relics. Rev. Mod. Phys. **96**, 045005 (2024). https://doi.org/10.1103/RevModPhys.96.045005, 2312.09042
16. Bouzoud, K., Ghiglieri, J.: Thermal axion production at hard and soft momenta. JHEP **01**, 163 (2025). https://doi.org/10.1007/JHEP01(2025)163, 2404.06113
17. Bödeker, D., Nienaber, J.: Scalar field damping at high temperatures. Phys. Rev. D **106**, 056016 (2022). https://doi.org/10.1103/PhysRevD.106.056016, 2205.14166
18. Yokoyama, J., Linde, A.D.: Is warm inflation possible? Phys. Rev. D **60**, 083509 (1999). https://doi.org/10.1103/PhysRevD.60.083509, hep-ph/9809409
19. Laine, M., Vuorinen, A.: Basics of Thermal Field Theory. Lect. Notes Phys. **925**, 1 (2016). https://doi.org/10.1007/978-3-319-31933-9, 1701.01554
20. Laine, M., Procacci, S., Rogelj, A.: Evolution of coupled scalar perturbations through smooth reheating. Part II. Thermal fluctuation regime. 2507.12849
21. DeRocco, W., Graham, P.W., Kalia, S.: Warming up cold inflation. JCAP **11**, 011 (2021). https://doi.org/10.1088/1475-7516/2021/11/011, 2107.07517
22. Berghaus, K.V., Drewes, M., Zell, S.: Warm Inflation with the Standard Model. 2503.18829
23. Ramos, R.O., Rodrigues, G.S.: Viability of warm inflation with standard model interactions. Phys. Rev. D **111**, 123527 (2025). https://doi.org/10.1103/wn1m-19gt, 2504.20943
24. Kolesova, H., Laine, M., Procacci, S.: Maximal temperature of strongly-coupled dark sectors. JHEP **05**, 239 (2023). https://doi.org/10.1007/JHEP05(2023)239, 2303.17973
25. Laine, M., Meyer, M., Schröder, Y.: Data for the Standard Model equation of state. http://laine.itp.unibe.ch/eos15/
26. Mirbabayi, M.: Loosely coupled particles in warm inflation. JCAP **05**, 067 (2025). https://doi.org/10.1088/1475-7516/2025/05/067, 2409.17927

What Happens When Scalar Modes Exit the Hubble Horizon?   8

Abstract

The equations for scalar perturbations from Chap. 7 have a fairly complicated structure. We explain how their solution gets simplified once the wavelength redshifts enough that the mode 'exits the Hubble horizon', whereby it effectively decouples from local dynamics. Subsequently, the nature of the background solution may change, from vacuum-energy dominated expansion, via a possible intermediate period, ultimately into a radiation-dominated universe. A key feature of the inflationary paradigm is that the cosmological perturbations visible in the CMB are outside of the Hubble horizon throughout these processes, and are almost unaffected by them, or 'frozen'. We also explain how the frozen solution may have a richer structure in multi-field inflationary models, leading to the appearance of so-called isocurvature perturbations.

Keywords

Horizon exit · Freezing out of curvature or adiabatic perturbations · Isocurvature perturbations · Non-adiabatic modes · Entropy perturbations · Multi-field inflation · Spectator fields · Dark matter isocurvature

8.1   Overview

It is an important property of the inflationary paradigm that curvature perturbations 'freeze out' when they exit the Hubble horizon. For a system characterized by a single curvature perturbation, we have already seen this in Chap. 6 (cf. the discussion between Eqs. (6.22) and (6.25)). If several 'flavours' of curvature perturbations are present simultaneously (like in Eqs. (4.60)–(4.63)), they freeze out to the same value. As a consequence, differences between various flavours of curvature perturbations, which are sometimes called *isocurvature perturbations*, vanish outside of the Hubble horizon. The purpose of the present section is to enlarge the setup considered in Chap. 6 by the addition of a radiation plasma, and establish these properties.

M. Laine and S. Procacci, *From Inflation to Hot Big Bang*, Lecture Notes in Physics 1047,
https://doi.org/10.1007/978-3-032-09893-1_8

At this point, it is good to clarify the terminology that is used in cosmological literature. Focussing on scalar perturbations, there are several of them as discussed in Chap. 7 (in the minimal case, metric perturbations, $\delta\varphi$, $v \equiv \delta v$, and δT), but they are coupled to each other, via the Einstein equations and energy-momentum conservation. A special role is played by gauge-invariant linear combinations of scalar perturbations that are called curvature perturbations, or *adiabatic perturbations*. The latter term may sound confusing, since at the time of their generation the system is normally assumed cold (non-thermal), so that entropy should play no role. Indeed the nomenclature comes more from the side of the later universe, when the same perturbations re-enter the Hubble horizon (cf. Chap. 9), at a time when the universe is radiation-dominated. Then adiabatic perturbations are the ones which have a 'common origin', coming from a single curvature perturbation, which at late times splits into many separate components, no longer in equilibrium with each other (dark matter, neutrinos, baryonic matter, and photons).

Non-adiabatic perturbations, in contrast, are defined as differences of various types of curvature perturbations, and are also referred to as isocurvature perturbations, or simply *entropy perturbations*. As we discuss in this chapter, when the modes are outside of the Hubble horizon, isocurvature perturbations normally vanish. But before they exit, or after they re-enter, they are non-zero. We thus have to watch out for the period that is considered. If it is said that 'there is no isocurvature' or that 'perturbations are adiabatic', this normally refers to the time at which the perturbations are outside of the Hubble horizon. In other words, even if the primordial scalar perturbations were adiabatic, entropy perturbations still take a non-zero value at late times, when the modes re-enter inside the Hubble horizon. 'Adiabaticity' then refers to their *initial conditions*.

Before proceeding to the detailed discussion, we recall an intuitive argument about the nature of a background solution and the corresponding perturbations, and contrast it with the much stronger freeze-out statement that applies to curvature perturbations. Let Q be a field or an ensemble of fields, which satisfies the non-linear evolution equation

$$(\partial_t^2 - \nabla_{\mathbf{r}}^2)Q + \mathcal{A}(Q)\partial_t Q + \mathcal{B}(Q) = 0 \, . \tag{8.1}$$

We write $Q = \bar{Q} + \delta Q$, where $\nabla_{\mathbf{r}} \bar{Q} = 0$. The background and first-order equations read

$$\partial_t^2 \bar{Q} + \mathcal{A}(\bar{Q})\partial_t \bar{Q} + \mathcal{B}(\bar{Q}) = 0 \, , \tag{8.2}$$

$$\left[\partial_t^2 - \nabla_{\mathbf{r}}^2 + \mathcal{A}(\bar{Q})\partial_t + \mathcal{A}'(\bar{Q})\partial_t \bar{Q} + \mathcal{B}'(\bar{Q}) \right]\delta Q = 0 \, , \tag{8.3}$$

respectively. On the other hand, if we take a time derivative of Eq. (8.2), we find

$$\left[\partial_t^2 + \mathcal{A}(\bar{Q})\partial_t + \mathcal{A}'(\bar{Q})\partial_t \bar{Q} + \mathcal{B}'(\bar{Q}) \right]\partial_t \bar{Q} = 0 \, . \tag{8.4}$$

Comparing Eqs. (8.3) and (8.4), we see that the time derivative of the background solution, $\partial_t \bar{Q}$, already determines one of the possible perturbations, δQ, specifically a mode satisfying $\nabla_{\mathbf{r}}^2 \delta Q = 0$. In the cosmological context, this corresponds to comoving momenta that have exited the Hubble horizon, $k/a \ll H$.

Now, in the slow-roll regime, $\dot{\bar{\varphi}}$ is almost constant (cf. Sect. 6.2), in the sense that its relative change within a Hubble time ($\Delta t \equiv H^{-1}$) is small,

$$\frac{\Delta \dot{\bar{\varphi}}}{\dot{\bar{\varphi}}} \approx \frac{\Delta t \, \ddot{\bar{\varphi}}}{\dot{\bar{\varphi}}} \underset{(6.22)}{\overset{\Delta t \equiv H^{-1}}{=}} \approx \epsilon_V - \eta_V \ll 1 \,. \tag{8.5}$$

Then, if we insert φ in the role of Q in Eq. (8.3), and consider modes with $k/a \ll H$ in the slow-roll regime, we see that $\delta\varphi$ is almost constant.

However, what actually happens for curvature perturbations is much more remarkable: they stay constant even when $\bar{\varphi}$ starts to oscillate fast, as long as $k/a \ll H$. To see this, we have to inspect in detail their evolution equation, which in fact does *not* have exactly the same form as the equation satisfied by $\dot{\bar{\varphi}}$.

The rest of this chapter is organized as follows. In Sect. 8.2, we show that in single-field inflationary models, the curvature perturbation $\mathcal{R}_\varphi$ freezes out while it is outside of the Hubble horizon. In Sect. 8.3, we elaborate on how this implies that isocurvature perturbations vanish outside of the Hubble horizon. Despite the prediction of single-field models that isocurvature perturbations should vanish, they are broadly discussed in the literature. Therefore, in Sect. 8.4, we show how isocurvature perturbations can be non-vanishing in inflationary setups involving multiple scalar fields.

8.2 Freezing Out of Curvature Perturbations

We start this section on a general note. Given the very simple freeze-out result, it may be wondered if there is a simple intuitive explanation for this behaviour. Indeed physical arguments are presented in the literature, but unfortunately they depend on the gauge chosen. Therefore standard proofs rely on a mathematical determination of all independent solutions of the underlying differential equations (cf., e.g., Refs. [1, 2]). This is then also the strategy that we adopt in the following.

Let us consider the evolution equation for the gauge-invariant curvature perturbation $\mathcal{R}_\varphi$ from Eq. (7.81). We convert it into an equation for the corresponding *mode function*, $\mathcal{R}_k$. The mode functions are related, but not exactly equivalent, to Fourier transforms as defined in Eq. (9). An example of a mode expansion is given in Eq. (5.31), with the mode function denoted by $\widehat{\mathcal{Q}}_k(\tau)$. If we take a Fourier transform of this expression, according to Eq. (9), we get

$$\widehat{\mathcal{Q}}(\tau, \mathbf{k}) \underset{(9)}{\overset{(5.31)}{=}} \sqrt{(2\pi)^3} \left[w_{\mathbf{k}} \, \widehat{\mathcal{Q}}_k(\tau) + w^\dagger_{-\mathbf{k}} \, \widehat{\mathcal{Q}}^*_k(\tau) \right] \,. \tag{8.6}$$

The vacuum expectation value, similar to Eq. (5.51) but in momentum space, gives

$$\langle 0 | \, \widehat{\mathcal{Q}}_\varphi(\tau, \mathbf{k}) \widehat{\mathcal{Q}}_\varphi(\tau, \mathbf{q}) \, | 0 \rangle \underset{(5.33)}{\overset{(8.6)}{=}} (2\pi)^3 \delta^{(3)}(\mathbf{k} + \mathbf{q}) \, |\widehat{\mathcal{Q}}_k(\tau)|^2 \,. \tag{8.7}$$

This can be compared with the corresponding statistical average, from Eq. (2.49). We see that, if we use mode functions, the momentum conservation constraint is factored out, so that the mode functions give directly the power spectra,

$$P_{\widehat{\mathcal{Q}}}(\mathbf{k}) \overset{(2.49)}{\underset{(8.7)}{=}} |\widehat{\mathcal{Q}}_k(\tau)|^2 \, , \quad \mathcal{P}_{\widehat{\mathcal{Q}}}(k) \overset{(2.51)}{=} \frac{k^3}{2\pi^2} |\widehat{\mathcal{Q}}_k(\tau)|^2 \, . \tag{8.8}$$

We remark that the latter relation appears also in Eq. (5.52) and subsequently in Eq. (6.13), where we make use of it to define a rescaled mode function, $[\,\mathcal{R}_k\,]_{\text{rescaled}}$.

Returning now to Eq. (7.81), on the left-hand side it makes no difference whether we carry out a Fourier transform or go to mode functions, as in both cases we just need to replace $\nabla^2 \to -k^2$. However, we have to be careful with the inhomogeneous terms on the right-hand side (originating from ϱ and the noise part of Π). Fourier transforming the noise correlator from Eq. (1.63), we find

$$\langle\, \varrho(t_1, \mathbf{k})\, \varrho(t_2, \mathbf{q})\, \rangle \overset{(1.63)}{\underset{(9)}{=}} \int \mathrm{d}^3\mathbf{x} \int \mathrm{d}^3\mathbf{y}\ \overbrace{\langle\, \varrho(t_1, \mathbf{x})\, \varrho(t_2, \mathbf{y})\, \rangle}^{\Omega\,\delta(t_1-t_2)\delta^{(3)}(\mathbf{x}-\mathbf{y})/a^3(t_1)}\ e^{-i(\mathbf{k}\cdot\mathbf{x}+\mathbf{q}\cdot\mathbf{y})}$$

$$= \frac{\Omega\,\delta(t_1 - t_2)}{a^3(t_1)} \int \mathrm{d}^3\mathbf{x}\, e^{-i(\mathbf{k}+\mathbf{q})\cdot\mathbf{x}}$$

$$= (2\pi)^3\delta^{(3)}(\mathbf{k} + \mathbf{q})\, \underbrace{\frac{\Omega\,\delta(t_1 - t_2)}{a^3(t_1)}}_{\equiv\,\langle\, \varrho_k(t_1)\, \varrho_k(t_2)\, \rangle} \, , \tag{8.9}$$

where we inserted $\sqrt{-\bar{g}} = a^3$ for the determinant of the background metric in physical time. The definition of the *noise mode function*, ϱ_k, in Eq. (8.9), draws on an analogy with Eq. (8.7), i.e. on factoring out the momentum conservation constraint.

With these ingredients, we convert Eq. (7.81) into an equation for the curvature and noise mode functions, $\mathcal{R}_k$ and ϱ_k. In order to keep the discussion simple, we do this by first *assuming* that the isocurvature perturbations vanish, $\mathcal{S}_v, \mathcal{S}_T \to 0$. This assumption will be justified *a posteriori*, in Sect. 8.3. In the said limit, we get

$$\left\{ \partial_t^2 + \left(\Upsilon + 2\mathcal{F} + 3H \right)\partial_t + \frac{k^2}{a^2} \right\} \mathcal{R}_k - \frac{8\pi G}{3}k^2 \Pi \overset{(7.81)}{\underset{\mathcal{S}_v, \mathcal{S}_T \to 0}{\approx}} -\frac{\varrho_k H}{\dot{\varphi}} \, , \tag{8.10}$$

where $\mathcal{F}$ is defined in Eq. (6.12) and Π is the mode function corresponding to the anisotropic stress in Eq. (7.82). We put no subscript in Π as it is always preceded by k^2, so that the context should be clear. Our goal is to determine the solution of Eq. (8.10) after k/a goes outside of the Hubble horizon, $k/a \ll H$ (cf. Fig. 1.2 in Sect. 1.7 for an illustration). We recall from Eq. (3.134), and the discussion a few lines below Eq. (4.56) (or from Eq. (7.82)), that

$$\Pi \overset{(3.127)}{\supset} \Sigma \overset{(3.134)}{=} \frac{2\eta(v - \vartheta')}{a} \overset{(4.31)}{\underset{(4.61)}{=}} -\frac{2\eta(\psi + \mathcal{R}_v)}{a^2 H} \, , \tag{8.11}$$

so that the momentum appears as k^2/a^2 also in front of viscous corrections.

Given that Eq. (8.10) is a linear differential equation, its solution can be represented as a sum of the general homogeneous solution and a special solution of the inhomogeneous equation. Adopting the notation from Eq. (7.33), we denote this splitup as

$$\mathcal{R}_k = \mathcal{R}_k^{\text{vac}} + \mathcal{R}_k^{\text{cl}}, \tag{8.12}$$

where the 'vac' part refers to the homogeneous equation. Its integration constants are fixed through the initial conditions from Eqs. (6.15) and (6.17).

If we set $k/a \to 0$, the homogeneous equation takes the form

$$\left\{ \partial_t + \left(\Upsilon + 2\mathcal{F} + 3H \right) \right\} \dot{\mathcal{R}}_k^{\text{vac}} \underset{(8.10),\,(8.11)}{\overset{k \ll aH}{\approx}} 0. \tag{8.13}$$

There is a trivial solution,

$$\dot{\mathcal{R}}_k^{\text{vac}}(t) = 0, \tag{8.14}$$

and a non-trivial one,

$$\dot{\mathcal{R}}_k^{\text{vac}}(t) \approx \dot{\mathcal{R}}_k^{\text{vac}}(t_*) \exp\left\{ -\int_{t_*}^{t} dt' \left(\Upsilon + 2\mathcal{F} + 3H \right) \right\}, \tag{8.15}$$

where $t_* = t_*(k)$ is the moment at which a mode exits the horizon, i.e. $k \equiv a(t_*)H(t_*)$. To analyse Eq. (8.15), we specialize to the slow-roll regime, even if a full numerical solution of Eq. (8.10) shows that the result applies more generally. In the slow-roll regime, as is shown below Eq. (6.22), $\mathcal{F} \approx (2\epsilon_V - \eta_V)H \ll H$. Therefore

$$\Upsilon + 2\mathcal{F} + 3H \approx \Upsilon + 3H \overset{(7.28)}{\geq} 3H > 0. \tag{8.16}$$

This implies that Eq. (8.15) decreases exponentially, if H is approximately constant and $(t - t_*) H \gg 1$. Therefore, the *only* solution outside of the Hubble horizon reads

$$\mathcal{R}_k^{\text{vac}} \underset{(8.14),\,(8.15)}{\overset{k \ll aH}{\approx}} \text{constant}. \tag{8.17}$$

Let us then turn to the special solution of the inhomogeneous equation, i.e. the function $\mathcal{R}_k^{\text{cl}}$ from Eq. (8.12). As the average of the noise vanishes, the average of $\mathcal{R}_k^{\text{cl}}$ also vanishes. Therefore, the noise term cannot change the freeze-out dynamics in a qualitative way.

We can say a bit more by estimating the size of the mean-squared noise kicks. After the rescaling in Eq. (6.13), which leads to the initial conditions in Eqs. (6.15) and (6.17), the noise appearing in Eq. (8.10) has the *autocorrelator*

$$\frac{H(t_1)H(t_2)}{\dot{\varphi}(t_1)\dot{\varphi}(t_2)} \left\langle \left[\varrho_k(t_1)\, \varrho_k(t_2) \right]_{\text{rescaled}} \right\rangle \underset{(7.44),(8.9)}{\overset{(6.13)}{=}} \delta(t_1 - t_2) \left[2\,T\Upsilon\, \frac{H^2}{\dot{\varphi}^2}\, \frac{(k/a)^3}{2\pi^2} \right]_{t=t_1},$$

$$\tag{8.18}$$

where we inserted $\Omega = 2T\Upsilon$ from Eq. (7.44). The coefficient $H/\dot{\bar{\varphi}}$ is slowly varying, as long as we are in the slow-roll regime. If we are in the oscillatory regime, so that $\dot{\bar{\varphi}}$ has zeros, the autocorrelator in Eq. (8.18) looks singular, and needs to be regularized; however, after doing this with the help of complexified variables, the average value of the noise autocorrelator is small [3, Eq. (4.41)]. As discussed around Eq. (7.52), the plasma temperature, T, could also be slowly varying, to the extent that we can define it. Instead, the momentum dependence redshifts away rapidly in Eq. (8.18), as k^3/a^3. We therefore conclude that the thermal noise has a fast decreasing amplitude outside of the Hubble horizon, so that it adds only very small ripples on top of $\mathcal{R}_k^{\mathrm{vac}}$ (cf. Eq. (8.17)).

To summarize, Eq. (8.17) represents actually the general solution outside of the Hubble horizon, showing that curvature perturbations indeed freeze out.

8.3   Vanishing of Isocurvature Perturbations

For the argument in Sect. 8.2, we assumed the isocurvature perturbations, $\mathcal{S}_v$ and $\mathcal{S}_T$, to vanish. This assumption is justified rather succinctly in the literature, again from the solutions of gauge-fixed evolution equations, rather than from an intuitively clear physical phenomenon. In the following, even if we do not succeed in offering the latter, we provide a gauge-invariant demonstration for why the isocurvature perturbations vanish at the same time as $\mathcal{R}_\varphi$ freezes out. The implicit assumption we make is that a non-vanishing background temperature exists, so that $\mathcal{S}_v$ and $\mathcal{S}_T$, defined in Eqs. (7.61) and (7.62), are in principle non-trivial variables. However, the result is more general, and in Sect. 8.4 we demonstrate it for a system involving two scalar fields but no temperature at all.

Let us return back to Eq. (7.81), but keep now the isocurvature perturbations in place. We again go to mode functions, like in Eq. (8.10), and outside of the Hubble horizon, setting $k/a \ll H$. In order to avoid double subscripts, we add no index k to $\mathcal{S}_v$ and $\mathcal{S}_T$. Furthermore, we complement Eq. (7.81) with the two other evolution equations, from Eqs. (7.90) and (7.97), inspected under the same assumptions. We note that the anisotropic stress contribution is proportional to k^2/a^2, cf. Eqs. (8.10) and (8.11), so it drops out. Consequently the Bardeen potentials, ϕ and ψ, decouple from the dynamics, and we do not need to include them.

Now, we have demonstrated in Sect. 8.2 that, outside the Hubble horizon, when $k/a \ll H$, $\mathcal{R}_k$ is constant. Then suppose that we search for a *stationary solution* also for $\mathcal{S}_T$ and $\mathcal{S}_v$, setting all time derivatives to zero. Then the coupled set of our three equations becomes

$$\left\{ \frac{4\pi G}{H}\left(1 - \frac{\bar{p}_{,T}}{\bar{e}_{,T}}\right) + \frac{\Upsilon_{,T}\dot{\bar{\varphi}} + V_{,\varphi T}}{\bar{e}_{,T}\dot{\bar{\varphi}}} \right\} \mathcal{S}_T - \left\{ \frac{4\pi G(\Upsilon + 2\mathcal{F})}{H} \right\} \mathcal{S}_v \overset{(7.81)}{\underset{\partial_t,\frac{k}{a}\to 0}{=}} - \frac{\varrho_k H}{\dot{\bar{\varphi}}}, \quad (8.19)$$

$$\left\{ 3H + \frac{4\pi G\dot{\bar{\varphi}}^2}{H} \right\} \mathcal{S}_v - \left\{ \frac{\bar{p}_{,T}}{\bar{e}_{,T}} \right\} \mathcal{S}_T \overset{(7.90)}{\underset{\partial_t,\frac{k}{a}\to 0}{=}} 0, \quad (8.20)$$

$$\left\{ 3H\left(1 + \frac{\bar{p}_{,T}}{\bar{e}_{,T}}\right) + \frac{4\pi G(\bar{e} + \bar{p}) - \dot{H}}{H} - \frac{\dot{\bar{\varphi}}\,(\Upsilon_{,T}\dot{\bar{\varphi}} + V_{,\varphi T})}{\bar{e}_{,T}} \right\} \mathcal{S}_T$$

$$+ \frac{4\pi G}{H} \left\{ \Upsilon \dot{\bar{\varphi}}^2 + \frac{8\pi G (\bar{e} + \bar{p})\,\bar{e}}{H} \right\} \mathcal{S}_v \underset{\partial_t, \frac{k}{a} \to 0}{\overset{(7.97)}{=}} \varrho_k \dot{\bar{\varphi}} H \,. \tag{8.21}$$

This is a system of three linear equations for two unknowns, $\mathcal{S}_v$ and $\mathcal{S}_T$. The coefficients are, in general, non-degenerate. The average of the noise is zero, and its mean-squared amplitude is small outside of the Hubble horizon, as discussed below Eq. (8.18). Omitting the right-hand side, the only solution is zero, $\mathcal{S}_v = \mathcal{S}_T = 0$, because the system is overconstrained. This completes our argument that isocurvature perturbations vanish outside of the Hubble horizon. The vanishing of $\mathcal{S}_v$ and $\mathcal{S}_T$ implies that $\mathcal{R}_\varphi$, $\mathcal{R}_v$ and $\mathcal{R}_T$ freeze out to the same value (cf. Eqs. (7.61) and (7.62)).

8.4   Multi-field Inflation, Spectator Fields, and Dark Matter Isocurvature

The arguments presented in Sects. 8.2 and 8.3 are specific to *single-field inflation*. The case that multiple scalar fields participate in inflationary dynamics was considered early on (cf., e.g., Refs. [4–6]), and the situation was noticed to change. This type of models gained prominence particularly in the context of the *curvaton scenario* [7–9], in which the late-time adiabatic perturbations are generated not by the quantum fluctuations of the inflaton, but by the fluctuations of another scalar field, which is subdominant at the time of inflation. The latter field is conjectured to have a very small equilibration rate, $\Upsilon_{\text{curvaton}} \ll \Upsilon_{\text{inflaton}}$. Therefore it comes to dominate the energy density at a late time, when $\Upsilon_{\text{curvaton}} \ll H \ll \Upsilon_{\text{inflaton}}$ (cf. Fig. 7.1 in Sect. 7.1), and can generate cosmological perturbations then. In this setup, one of the scalar fields always acts as a *spectator*, but the roles change as time goes by.

From the point of view of model building, it may be particularly interesting if the curvaton field represents, or decays into, *dark matter*. In such a situation, it might also be natural to consider simultaneously multiple fields *and* fluids (cf., e.g., Ref. [10]). In the present section, we restrict ourselves to the simplest non-standard case, illustrating how the evolution of curvature and isocurvature perturbations changes if we introduce a second scalar field, denoted by χ.

We thus repeat the analysis from Sect. 5.1, but now looking for evolution equations for two (minimally coupled) scalar fields. The equations of motion, from Eq. (5.1), become

$$\varphi^{;\mu}{}_{;\mu} - V_{,\varphi} \overset{(5.1)}{=} 0 \,, \quad \chi^{;\mu}{}_{;\mu} - V_{,\chi} = 0 \,, \tag{8.22}$$

where $V = V(\varphi, \chi)$ yields an inflationary solution for φ or χ. Both fields are expanded around their background values, $\varphi = \bar{\varphi} + \delta\varphi$ and $\chi = \bar{\chi} + \delta\chi$. The background equations then take the form in Eq. (5.2),

$$\bar{\varphi}'' + 2\mathcal{H}\bar{\varphi}' + a^2\, V_{,\varphi} \overset{(5.2)}{=} 0 \,, \quad \bar{\chi}'' + 2\mathcal{H}\bar{\chi}' + a^2\, V_{,\chi} = 0 \,. \tag{8.23}$$

Rescaling the fields like before, $\delta\widehat{\varphi} \equiv a\delta\varphi$ and $\delta\widehat{\chi} \equiv a\delta\chi$, the first-order perturbations satisfy a generalization of Eq. (5.7),

$$\delta\widehat{\varphi}'' - \nabla^2\delta\widehat{\varphi} + \left(a^2 V_{,\varphi\varphi} - \frac{a''}{a}\right)\delta\widehat{\varphi} + a^2 V_{,\varphi\chi}\,\delta\widehat{\chi}$$
$$\stackrel{(5.7)}{=} \left(h_0' + 3h_{\mathrm{D}}' + \nabla^2 h\right)a\bar{\varphi}' - 2h_0\,a^3\,V_{,\varphi}\,. \tag{8.24}$$

The other equation is obtained with the exchange $\varphi \leftrightarrow \chi$.

On the side of the Einstein equations, the source terms for the metric and its perturbations change, because both scalar fields appear in the energy-momentum tensor. At the background level, Eq. (5.8) gets replaced with

$$\mathcal{H}^2 \stackrel{(5.8)}{=} \frac{4\pi G}{3}\left[(\bar{\varphi}')^2 + (\bar{\chi}')^2 + 2a^2 V\right]\,, \quad \mathcal{H}' \stackrel{(5.8)}{=} \frac{8\pi G}{3}\left[-(\bar{\varphi}')^2 - (\bar{\chi}')^2 + a^2 V\right]\,. \tag{8.25}$$

We also need linear combinations of these relations, as well as $\mathcal{H}''$ (cf. Eq. (5.9)), finding

$$\mathcal{H}^2 - \mathcal{H}' \stackrel{(8.25)}{=} 4\pi G\left[(\bar{\varphi}')^2 + (\bar{\chi}')^2\right]\,, \quad 2\mathcal{H}^2 + \mathcal{H}' \stackrel{(8.25)}{=} 8\pi Ga^2 V \tag{8.26}$$
$$\Rightarrow\ 2\mathcal{H}\mathcal{H}' - \mathcal{H}'' \stackrel{(8.26)}{=} 8\pi G\left(\bar{\varphi}'\bar{\varphi}'' + \bar{\chi}'\bar{\chi}''\right)\,. \tag{8.27}$$

At the first order in perturbations, we employ the shorthand notation $Y \equiv h_{\mathrm{D}} + \nabla^2\vartheta/3$ from Eq. (5.12). Then Eqs. (5.13) and (5.14) get replaced with

$$h_0 \stackrel{(5.13)}{=} -\frac{Y'}{\mathcal{H}} + \frac{4\pi G}{a\mathcal{H}}\left(\bar{\varphi}'\delta\widehat{\varphi} + \bar{\chi}'\delta\widehat{\chi}\right)\,, \tag{8.28}$$

$$h_0' + 3h_{\mathrm{D}}' + \nabla^2 h \stackrel{(5.14)}{=} -2\left(2\mathcal{H} + \frac{\mathcal{H}'}{\mathcal{H}}\right)h_0 - \frac{(\partial_\tau^2 + 2\mathcal{H}\partial_\tau - \nabla^2)Y}{\mathcal{H}}$$
$$- \frac{8\pi Ga}{\mathcal{H}}\left(V_{,\varphi}\,\delta\widehat{\varphi} + V_{,\chi}\,\delta\widehat{\chi}\right)\,. \tag{8.29}$$

The field perturbations are reparametrized like in Eq. (5.15),

$$\delta\widehat{\varphi} \stackrel{(5.15)}{=} \widehat{Q}_\varphi - \frac{a\bar{\varphi}'}{\mathcal{H}}Y\,, \quad \delta\widehat{\chi} = \widehat{Q}_\chi - \frac{a\bar{\chi}'}{\mathcal{H}}Y\,. \tag{8.30}$$

Inserting Eqs. (8.28)–(8.30) into (8.24), the first task is once again to verify the cancellation of Y. This goes like in Eqs. (5.18)–(5.20). The coefficient of Y' can be seen to cancel with the help of the background identity from Eq. (5.21), which remains unchanged,

$$\left(\frac{a\bar{\varphi}'}{\mathcal{H}}\right)' \stackrel{(8.23)}{\underset{(5.21)}{=}} a\bar{\varphi}'\left(-1 - \frac{\mathcal{H}'}{\mathcal{H}^2}\right) - \frac{a^3 V_{,\varphi}}{\mathcal{H}}\,. \tag{8.31}$$

To verify the cancellation of the coefficient of Y, we need to take a second derivative from Eq. (8.31). Proceeding like in Eqs. (5.22)–(5.25), we find

$$\left(\frac{a\bar{\varphi}'}{\mathcal{H}}\right)'' \stackrel{\substack{(8.31)\\(5.25)}}{=} \frac{a\bar{\varphi}'}{\mathcal{H}}\left\{ \mathcal{H}^2 + \mathcal{H}' - \left(2 + \frac{\mathcal{H}'}{\mathcal{H}^2}\right)8\pi G\left[(\bar{\varphi}')^2 + (\bar{\chi}')^2\right] \right.$$

$$\left. - \frac{16\pi Ga^2 V_{,\varphi}\bar{\varphi}'}{\mathcal{H}} - a^2 V_{,\varphi\varphi} \right\}$$

$$- \frac{a\bar{\chi}'}{\mathcal{H}}\left\{ \frac{8\pi Ga^2(V_{,\varphi}\bar{\chi}' + V_{,\chi}\bar{\varphi}')}{\mathcal{H}} + a^2 V_{,\varphi\chi} \right\}. \tag{8.32}$$

When this is inserted, the coefficient of Y cancels exactly.

Subsequently, we insert the physical (i.e. non-Y) terms from Eqs. (8.28)–(8.30) into Eq. (8.24), generalizing on Eqs. (5.26)–(5.28). The expressions can be simplified significantly by making use of Eq. (8.32), and the corresponding relation with $\varphi \leftrightarrow \chi$. In particular, we are led to define effective masses like in Eq. (5.29),

$$\widehat{m}_\varphi^2(\tau) \stackrel{(5.29)}{\equiv} -\frac{\mathcal{H}}{a\bar{\varphi}'}\left(\frac{a\bar{\varphi}'}{\mathcal{H}}\right)'', \quad \widehat{m}_\chi^2(\tau) \equiv -\frac{\mathcal{H}}{a\bar{\chi}'}\left(\frac{a\bar{\chi}'}{\mathcal{H}}\right)''. \tag{8.33}$$

Then Eq. (5.28) gets generalized into

$$0 = \left[\partial_\tau^2 - \nabla^2 + \widehat{m}_\varphi^2(\tau)\right]\widehat{\mathcal{Q}}_\varphi \tag{8.34}$$

$$+ \left[a^2 V_{,\varphi\chi} + \underbrace{\left(2 + \frac{\mathcal{H}'}{\mathcal{H}^2}\right)}_{\text{from (8.26): } \frac{8\pi Ga^2 V}{\mathcal{H}^2}}8\pi G\bar{\varphi}'\bar{\chi}' + \frac{8\pi Ga^2(V_{,\varphi}\bar{\chi}' + V_{,\chi}\bar{\varphi}')}{\mathcal{H}}\right]\left(\widehat{\mathcal{Q}}_\chi - \frac{\bar{\chi}'}{\bar{\varphi}'}\widehat{\mathcal{Q}}_\varphi\right).$$

There is a corresponding equation for $\widehat{\mathcal{Q}}_\chi$, obtained with $\varphi \leftrightarrow \chi$.

Finally we go over to curvature perturbations, like in Eqs. (5.16) and (5.17),

$$\widehat{\mathcal{Q}}_\varphi \stackrel{\substack{(5.16)\\(5.17)}}{=} -\frac{a\bar{\varphi}'}{\mathcal{H}}\mathcal{R}_\varphi, \quad \widehat{\mathcal{Q}}_\chi = -\frac{a\bar{\chi}'}{\mathcal{H}}\mathcal{R}_\chi. \tag{8.35}$$

Time derivatives transform in this substitution as indicated in Eq. (6.7). Subsequently, we multiply Eq. (8.34) with $-\mathcal{H}/(a\bar{\varphi}')$, yielding a generalization of Eq. (6.8). Finally we transform to physical time, according to Eqs. (6.9) and (6.10), and multiply the equation with $1/a^2$. Defining

$$\mathcal{F}_\varphi \stackrel{(6.12)}{\equiv} \frac{\ddot{\bar{\varphi}}}{\dot{\bar{\varphi}}} - \frac{\dot{H}}{H}, \quad \mathcal{F}_\chi \equiv \frac{\ddot{\bar{\chi}}}{\dot{\bar{\chi}}} - \frac{\dot{H}}{H}, \tag{8.36}$$

and introducing a mixed combination of the potential and its derivatives,

$$\mathcal{V}_{\varphi\chi} \equiv V_{,\varphi\chi} + \frac{8\pi G}{H}\left(V_{,\varphi}\dot{\bar{\chi}} + V_{,\chi}\dot{\bar{\varphi}}\right) + \left(\frac{8\pi G}{H}\right)^2 V\dot{\bar{\varphi}}\dot{\bar{\chi}}, \tag{8.37}$$

yields the pair of equations

$$\left[\partial_t^2 + \left(2\mathcal{F}_\varphi + 3H \right) \partial_t - \frac{\nabla^2}{a^2} \right] \mathcal{R}_\varphi + \mathcal{V}_{\varphi\chi} \frac{\dot{\chi}}{\dot{\varphi}} \left(\mathcal{R}_\chi - \mathcal{R}_\varphi \right) = 0 , \tag{8.38}$$

$$\left[\partial_t^2 + \left(2\mathcal{F}_\chi + 3H \right) \partial_t - \frac{\nabla^2}{a^2} \right] \mathcal{R}_\chi + \mathcal{V}_{\varphi\chi} \frac{\dot{\varphi}}{\dot{\chi}} \left(\mathcal{R}_\varphi - \mathcal{R}_\chi \right) = 0 . \tag{8.39}$$

Let us elaborate on the meaning of Eqs. (8.38) and (8.39). First of all, if we look at a stationary solution, with $\ddot{\mathcal{R}}_\varphi = \dot{\mathcal{R}}_\varphi = 0 = \ddot{\mathcal{R}}_\chi = \dot{\mathcal{R}}_\chi$, and consider modes outside of the Hubble horizon, with $k/a \ll H$, then the first terms of Eqs. (8.38) and (8.39) vanish. Then, if both scalar fields are displaced from their minima and are evolving, so that $\dot{\varphi} \neq 0$ and $\dot{\chi} \neq 0$, we find $\mathcal{V}_{\varphi\chi} \neq 0$. In this situation the last terms of Eqs. (8.38) and (8.39) guarantee that $\mathcal{R}_\varphi = \mathcal{R}_\chi$. So, we recover the result of Sect. 8.3, namely that isocurvature perturbations vanish outside of the Hubble horizon.

On the other hand, suppose that χ is at its minimum or oscillates around it, so that $\dot{\chi}^2$ carries little energy density. Then the last term in Eq. (8.38), weighted by $\dot{\chi}$, should have little influence, and we expect that $\mathcal{R}_\varphi$ evolves independently of $\mathcal{R}_\chi$. Conversely, if $\dot{\varphi}^2$ does not carry much energy density, Eq. (8.39) indicates that $\mathcal{R}_\chi$ evolves independently of $\mathcal{R}_\varphi$. This is how a curvaton field could play a role after the inflaton field has ceased to be active (i.e. $H \ll \Upsilon_{\text{inflaton}}$).

In model building, the coefficients of the last terms in Eqs. (8.38) and (8.39) might be non-zero during some period of time, but so small that they do not efficiently drive $\mathcal{R}_\varphi - \mathcal{R}_\chi$ to zero. Then some amount of an isocurvature perturbation is generated. Consequently, it requires a quantitative study to verify whether this amount remains within the observationally allowed window (cf. the discussion around Eq. (2.13)).

References

1. Weinberg, S.: Adiabatic modes in cosmology. Phys. Rev. D **67**, 123504 (2003). https://doi.org/10.1103/PhysRevD.67.123504, astro-ph/0302326
2. Lyth, D.H., Malik, K.A., Sasaki, M.: A general proof of the conservation of the curvature perturbation. JCAP **05**, 004 (2005). https://doi.org/10.1088/1475-7516/2005/05/004, astro-ph/0411220
3. Laine, M., Procacci, S., Rogelj, A.: Evolution of coupled scalar perturbations through smooth reheating. Part I. Dissipative regime. JCAP **10**, 040 (2024). https://doi.org/10.1088/1475-7516/2024/10/040, 2407.17074
4. Kofman, L.A., Linde, A.D.: Generation of density perturbations in the inflationary cosmology. Nucl. Phys. B **282**, 555 (1987). https://doi.org/10.1016/0550-3213(87)90698-5
5. Mollerach, S.: On the primordial origin of isocurvature perturbations. Phys. Lett. B **242**, 158 (1990). https://doi.org/10.1016/0370-2693(90)91453-I
6. Turner, M.S., Wilczek, F.: Inflationary axion cosmology. Phys. Rev. Lett. **66**, 5 (1991). https://doi.org/10.1103/PhysRevLett.66.5
7. Enqvist, K., Sloth, M.S.: Adiabatic CMB perturbations in pre-Big-Bang string cosmology. Nucl. Phys. B **626**, 395 (2002). https://doi.org/10.1016/S0550-3213(02)00043-3, hep-ph/0109214

8. Lyth, D.H., Wands, D.: Generating the curvature perturbation without an inflaton. Phys. Lett. B **524**, 5 (2002). https://doi.org/10.1016/S0370-2693(01)01366-1, hep-ph/0110002
9. Moroi, T., Takahashi, T.: Effects of cosmological moduli fields on cosmic microwave background. Phys. Lett. B **522**, 215 (2001); *ibid* **539**, 303 (2002) (erratum). https://doi.org/10.1016/S0370-2693(01)01295-3, hep-ph/0110096
10. Hwang, J.-c., Noh, H.: Cosmological perturbations with multiple fluids and fields. Class. Quant. Grav. **19**, 527 (2002). https://doi.org/10.1088/0264-9381/19/3/308, astro-ph/0103244

What Happens When Scalar Modes Re-enter the Hubble Horizon?

9

Abstract

Eventually, all the modes that have observational significance today, need to cross the horizon again, 're-entering' the domain governed by causal physics. This happens earlier for larger momenta, so that modes above the CMB scale (corresponding to shorter distances) re-enter before recombination. We derive the evolution equations satisfied by temperature and velocity perturbations after re-entry, and show how their solution leads to the emergence of acoustic oscillations. We discuss how the original curvature perturbation, $\mathcal{R}_\varphi$, decouples from the evolution, thereby exiting the stage. We indicate how non-relativistic perturbations can undergo gravitational collapse via Jeans instability, initiating the formation of the first large-scale structures in the universe. Finally, we provide references to frequently used softwares for a realistic study of CMB physics.

Keywords

Acoustic oscillations · Gravitational collapse · Jeans instability · Transition from radiation to matter domination · Dark matter · Scalar transfer function · Einstein-Boltzmann solver · Multicomponent universe · CMB spectrum · N-body simulations

9.1 Overview

The key result of Chap. 8 is that, when modes exit the Hubble horizon, they experience kind of a memory loss, at least in single-field inflationary models: many types of initial curvature perturbations merge into a single non-oscillating function of the comoving momentum, k. Furthermore, this function can be computed as the solution of a single differential equation (cf. Eq. (8.10)), or a set of equations (cf. Eqs. (7.81), (7.90), (7.97)). When the modes re-enter the Hubble horizon later on, the opposite dynamics takes place: the original curvature mode proliferates into many coupled perturbations, which undergo fast oscillations, with evolving phases and gradual

179

M. Laine and S. Procacci, *From Inflation to Hot Big Bang*, Lecture Notes in Physics 1047, https://doi.org/10.1007/978-3-032-09893-1_9

damping. At the end, these oscillations are observed as the consecutive peaks in the CMB spectrum (cf., e.g., Ref. [1]).

In the present chapter, we illustrate this late dynamics for a system in which three curvature perturbations are present, namely $\mathcal{R}_\varphi$, $\mathcal{R}_v$ and $\mathcal{R}_T$, as defined in Eqs. (4.60)–(4.62). In Sect. 9.2, we derive a coupled set of evolution equations for $\mathcal{R}_v$ and $\mathcal{R}_T$, under the assumption that $\bar{\varphi}$ gives a subdominant contribution to the overall energy density. In Sect. 9.3, we show how the coupled set leads to quasi-periodic solutions, known as acoustic oscillations. In Sect. 9.4, we return to $\mathcal{R}_\varphi$, and show that while still present, it decouples from the other evolutions. In Sect. 9.5, we enrich the setup by introducing a second fluid, and show how it can experience Jeans instability (cf., e.g., Refs. [2–4]). Finally, in Sect. 9.6, we briefly describe more complicated frameworks, normally based on Boltzmann equations, that are being used for describing the post-inflationary evolution of scalar perturbations in the real world, to be matched onto observed CMB spectra.

9.2   Evolution Equations for Temperature and Velocity Perturbations

Let us focus on a situation in which the modes considered re-enter inside the Hubble horizon once the inflaton field no longer dominates the energy density. Physically, this means that we exclude the very largest momenta (shortest distances) from the consideration, as they may re-enter already during an earlier epoch, for instance when inflaton oscillations dominate the energy density (cf. Sect. 1.7). Under this assumption, the scalar perturbations $v \equiv \delta v$ and δT are more important than $\delta\varphi$, and we can therefore consider a simplified set of evolution equations.

For the simplified setup, the starting point is given by Eqs. (3.119) and (3.121),

$$-\delta e' - 3\mathcal{H}(\delta e + \delta p) + (\bar{e} + \bar{p})(3h'_{\mathrm{D}} + \nabla^2 v) \stackrel{(3.119)}{=} 0 , \qquad (9.1)$$

$$\delta p + (\bar{e} + \bar{p})h_0 + (\partial_\tau + 4\mathcal{H})[(\bar{e} + \bar{p})(h - v)] + \frac{2}{3}\nabla^2\Pi \stackrel{(3.121)}{=} 0 . \qquad (9.2)$$

The anisotropic stress Π could be expressed in terms of dissipative coefficients as in Eqs. (3.134) and (3.141), but it is easier to display the expressions with Π, keeping in mind that it is a gauge-invariant variable.

Let us first consider Eq. (9.2). We insert here h_0 from Eq. (3.93), making use of the notation $Y \equiv h_{\mathrm{D}} + \nabla^2\vartheta/3$ from Eq. (5.12), and omit the background $\bar{\varphi}'$, whereby h_0 reads

$$h_0 \underset{(5.12)}{\overset{(3.93)}{\approx}} -\frac{Y'}{\mathcal{H}} + \frac{4\pi Ga^2(\bar{e} + \bar{p})(v - h)}{\mathcal{H}} . \qquad (9.3)$$

From Eqs. (4.61) and (4.62), the other perturbations can be expressed as

$$h - v \stackrel{(4.61)}{=} \frac{\mathcal{R}_v + Y}{\mathcal{H}} \, , \tag{9.4}$$

$$\delta T \stackrel{(4.62)}{=} -\frac{T'(\mathcal{R}_T + Y)}{\mathcal{H}} \, . \tag{9.5}$$

We express δp in terms of δT via $\delta p \approx \bar{p}_{,T}\, \delta T$, and collect together the terms acting on Y on the left-hand side (L) of Eq. (9.2),

$$(9.2)_{\mathrm{L}} \supset \left\{ -\overbrace{\frac{\bar{p}_{,T}\,T'}{\mathcal{H}}}^{\text{from }\delta p} \underbrace{-\frac{(\bar{e} + \bar{p})\partial_\tau}{\mathcal{H}} - \frac{4\pi G a^2(\bar{e} + \bar{p})^2}{\mathcal{H}^2}}_{\text{from }h_0 \text{ via (9.3)}} + \overbrace{\frac{\bar{e}' + \bar{p}'}{\mathcal{H}} - \frac{(\bar{e} + \bar{p})\mathcal{H}'}{\mathcal{H}^2} + \frac{(\bar{e} + \bar{p})\partial_\tau}{\mathcal{H}}}^{\text{from }h-v \text{ via (9.4)}} \right.$$

$$\left. + 4(\bar{e} + \bar{p}) \right\} Y \, . \tag{9.6}$$

As background identities, we can make use of Eq. (1.70), with $\kappa \to 0$ and $(\bar{\varphi}')^2 \ll a^2(\bar{e} + \bar{p})$, and of Eq. (1.72),

$$4\pi G a^2(\bar{e} + \bar{p}) \stackrel{(1.70)}{\approx} \mathcal{H}^2 - \mathcal{H}' \, , \quad \bar{e}' + 3\mathcal{H}(\bar{e} + \bar{p}) \stackrel{(1.72)}{\approx} 0 \, . \tag{9.7}$$

Inserting these in Eq. (9.6), all terms cancel. This demonstrates that the evolution equation is gauge invariant, and offers for a valuable crosscheck of the computation.

For the remaining terms, multiplying the whole equation with $\mathcal{H}$, we get

$$\mathcal{H} \times (9.2)_{\mathrm{L}} \stackrel{Y \to 0}{=} -\overbrace{\bar{p}'\,\mathcal{R}_T}^{\text{from }\delta p} \overbrace{-(\bar{e} + \bar{p})\left(\mathcal{H} - \frac{\mathcal{H}'}{\mathcal{H}}\right)\mathcal{R}_v}^{\text{from }h_0 \text{ via (9.3),(9.7)}} + \overbrace{(\bar{e}' + \bar{p}')\mathcal{R}_v + (\bar{e} + \bar{p})\mathcal{R}_v'}^{\text{from }h-v \text{ via (9.4)}}$$

$$+ \overbrace{(\bar{e} + \bar{p})\left(4\mathcal{H} - \frac{\mathcal{H}'}{\mathcal{H}}\right)\mathcal{R}_v}^{\text{from }h-v \text{ via (9.4)}} + \frac{2}{3}\mathcal{H}\,\nabla^2\Pi$$

$$\stackrel{(9.7)}{=} \bar{p}'\left(\mathcal{R}_v - \mathcal{R}_T\right) + (\bar{e} + \bar{p})\mathcal{R}_v' + \frac{2}{3}\mathcal{H}\,\nabla^2\Pi = 0 \, . \tag{9.8}$$

Going over to physical time ($\bar{p}' = a\dot{\bar{p}}$, $\mathcal{H} = aH$) and to momentum space ($\nabla^2\Pi \to -k^2\Pi$), Eq. (9.8) turns into the evolution equation that is shown in Eq. (9.13).

Turning then to Eq. (9.1), we rewrite the last term in a longer but more convenient form,

$$3h_{\mathrm{D}}' + \nabla^2 v = \nabla^2(v - h) + \underbrace{\nabla^2 h + 3h_{\mathrm{D}}' + h_0'}_{\text{from (3.96)}} - \underbrace{h_0'}_{\text{from (9.3)}} \, . \tag{9.9}$$

Therefore, we need to evaluate

$$\nabla^2 h + 3h'_{\mathrm{D}} + h'_0 \overset{(3.96)}{\approx} -2\left(2\mathcal{H} + \frac{\mathcal{H}'}{\mathcal{H}}\right)h_0 - \frac{1}{\mathcal{H}}(\partial_\tau^2 + 2\mathcal{H}\partial_\tau - \nabla^2)\,Y$$

$$+ \frac{4\pi G a^2}{\mathcal{H}}\left[\left(\bar{p}_{,\tau} - \bar{e}_{,\tau}\right)\delta T + \frac{2}{3}\nabla^2\Pi\right], \qquad (9.10)$$

$$h'_0 \overset{(9.3)}{\underset{(9.7)}{\approx}} \frac{\mathcal{H}' - \mathcal{H}\partial_\tau}{\mathcal{H}^2}\,Y' + \left(1 - \frac{\mathcal{H}'}{\mathcal{H}^2}\right)[\mathcal{H}(v - h)]'$$

$$+ \left(\frac{2\mathcal{H}'^2}{\mathcal{H}^3} - \frac{\mathcal{H}''}{\mathcal{H}^2}\right)[\mathcal{H}(v - h)]. \qquad (9.11)$$

Given that $\mathcal{H}''$ appears, we also need an additional background identity, given in Eq. (7.75).

When we insert Eqs. (9.10) and (9.11) into (9.9), and then the whole into Eq. (9.1), the expression becomes quite complicated. In this situation, it is all the more important to have the powerful principle of gauge invariance as a crosscheck that we are not making mistakes along the way. Without displaying the details, we note that it is remarkable how all different structures in the differential operator acting on Y cancel, after we make use of the background identities in Eqs. (9.7) and (7.75).

Let us then proceed to the physical terms. Inserting Eqs. (9.3)–(9.5) and (9.10), (9.11) into (9.1), we find

$$(9.1)_{\mathrm{L}} \overset{Y \to 0}{=} \overbrace{\left(\frac{\bar{e}'\mathcal{R}_T}{\mathcal{H}}\right)'}^{\text{from } \delta e'} + \overbrace{3(\bar{e}' + \bar{p}')\mathcal{R}_T}^{\text{from } -3\mathcal{H}(\delta e + \delta p)} - \overbrace{\frac{(\bar{e} + \bar{p})\nabla^2\mathcal{R}_v}{\mathcal{H}}}^{\text{from } \nabla^2(v - h) \text{ via } (9.4)}$$

$$+ (\bar{e} + \bar{p})\Bigg\{ \overbrace{2\left(2\mathcal{H} + \frac{\mathcal{H}'}{\mathcal{H}}\right)\left(1 - \frac{\mathcal{H}'}{\mathcal{H}^2}\right)\mathcal{R}_v}^{\text{from } h_0 \text{ via } (9.10),(9.3)} + \overbrace{\frac{4\pi G a^2(\bar{e}' - \bar{p}')}{\mathcal{H}^2}\mathcal{R}_T}^{\text{from } \delta T \text{ via } (9.10)}$$

$$+ \overbrace{\left(1 - \frac{\mathcal{H}'}{\mathcal{H}^2}\right)\mathcal{R}'_v + \left(\frac{2\mathcal{H}'^2}{\mathcal{H}^3} - \frac{\mathcal{H}''}{\mathcal{H}^2}\right)\mathcal{R}_v}^{\text{from } h'_0 \text{ via } (9.11),(9.4)} \Bigg\} + \frac{2}{3}\frac{4\pi G a^2(\bar{e} + \bar{p})}{\mathcal{H}}\nabla^2\Pi$$

$$\overset{(9.7)}{=} -3[(\bar{e} + \bar{p})\mathcal{R}_T]' + 3(\bar{e}' + \bar{p}')\mathcal{R}_T + (\bar{e} + \bar{p})\Bigg\{ -\frac{\nabla^2\mathcal{R}_v}{\mathcal{H}} + \left(1 - \frac{\mathcal{H}'}{\mathcal{H}^2}\right)\mathcal{R}'_v$$

$$+ \left(4\mathcal{H} - \frac{2\mathcal{H}'}{\mathcal{H}} - \frac{2\mathcal{H}'^2}{\mathcal{H}^3} + \frac{2\mathcal{H}'^2}{\mathcal{H}^3} - \frac{\mathcal{H}''}{\mathcal{H}^2}\right)\mathcal{R}_v + \frac{4\pi G a^2(\bar{e}' - \bar{p}')}{\mathcal{H}^2}\mathcal{R}_T\Bigg\}$$

$$+ \frac{8\pi G a^2(\bar{e} + \bar{p})}{3\mathcal{H}}\nabla^2\Pi$$

$$\overset{(9.7)}{\underset{(7.75)}{=}} (\bar{e} + \bar{p})\left\{ -3\mathcal{R}_T' - \frac{\nabla^2 \mathcal{R}_v}{\mathcal{H}} + \frac{4\pi G a^2 (\bar{e} + \bar{p})}{\mathcal{H}^2}\mathcal{R}_v' + \frac{4\pi G a^2 (\bar{e}' - \bar{p}')}{\mathcal{H}^2}(\mathcal{R}_T - \mathcal{R}_v)\right\}$$

$$+ \frac{8\pi G a^2 (\bar{e} + \bar{p})}{3\mathcal{H}}\nabla^2 \Pi$$

$$\overset{(9.8)}{=} (\bar{e} + \bar{p})\left\{ -3\mathcal{R}_T' - \frac{\nabla^2 \mathcal{R}_v}{\mathcal{H}} + \frac{4\pi G a^2 \bar{e}'}{\mathcal{H}^2}(\mathcal{R}_T - \mathcal{R}_v)\right\} = 0 . \tag{9.12}$$

Going subsequently to physical time ($\tau \to t$) and to comoving momentum space ($\mathbf{x} \to \mathbf{k}$), and making use of the background identity in Eq. (9.7) once more, we obtain Eq. (9.14).

To summarize, the evolution equations for curvature perturbations in a plasma-dominated universe read

$$\dot{\mathcal{R}}_v \overset{(9.8)}{\underset{(9.33)}{\approx}} \underbrace{\frac{\dot{\bar{p}}}{\bar{e} + \bar{p}}}_{\approx \dot{T}/T} \left(\mathcal{R}_T - \mathcal{R}_v \right) + \frac{2H\,k^2 \Pi}{3(\bar{e} + \bar{p})} , \tag{9.13}$$

$$\dot{\mathcal{R}}_T \overset{(9.12)}{\underset{(9.34)}{\approx}} \frac{\mathcal{R}_v}{3H}\frac{k^2}{a^2} - \underbrace{\frac{4\pi G(\bar{e} + \bar{p})}{H}}_{\approx +\dot{H}/H}\left(\mathcal{R}_T - \mathcal{R}_v \right) . \tag{9.14}$$

Here we also anticipated that an alternative derivation can be found in Eqs. (9.33) and (9.34). For the terms proportional to $\mathcal{R}_T - \mathcal{R}_v$, we indicated simpler interpretations for the coefficients, by making use of thermodynamic identities from Eq. (7.51) as well as a background identity from Eq. (1.71), once again simplified by taking $\kappa \to 0$ and $\dot{\bar{\varphi}}^2 \ll \bar{e} + \bar{p}$.

Equations (9.13) and (9.14) manifest the property discussed in Sect. 8.3. Namely, if we look at momenta outside the Hubble horizon, $k/a \ll H$, then viscous corrections (cf. Eq. (8.11)) as well as the first term on the right-hand side of Eq. (9.14) drop out. Then the system has a stationary solution, i.e. $\dot{\mathcal{R}}_v \approx 0 \approx \dot{\mathcal{R}}_T$, with vanishing isocurvature perturbations, i.e. $\mathcal{R}_v \approx \mathcal{R}_T$. According to Chap. 8, the initial value for $\mathcal{R}_v$ and $\mathcal{R}_T$ can be equated with $\mathcal{R}_\varphi$, which in turn can be determined from the inflationary dynamics.

We end this section by noting that the anisotropic stress, Π, could become important in Eq. (9.13) once the momenta re-enter the Hubble horizon, i.e. $k/a \gg H$. As argued around Eq. (3.143), the most important part of the anisotropic stress is probably a term proportional to the shear viscosity, given in Eq. (8.11). That term contains the Bardeen potential ψ, so we need an equation for ψ.

One of the equations for the Bardeen potentials is given in Eq. (4.56) (see also Eq. (9.19)), but it contains both ϕ and ψ, and does not suffice on its own. To derive another equation, we can take Eq. (9.3) as a starting point (we could also proceed from Eq. (7.102), by considering the limit $\dot{\bar{\varphi}}^2 \ll \bar{e} + \bar{p}$). In order to simplify the equations, we make use of the notation in Eq. (5.12). On the left-hand side of Eq. (9.3), we can then insert

$$h_0 \overset{(4.30)}{\underset{(5.12)}{=}} \phi - X' - \mathcal{H}X \, , \tag{9.15}$$

and on the right-hand side of Eq. (9.3), we use the two representations

$$Y \overset{(4.31)}{\underset{(5.12)}{=}} \psi + \mathcal{H}X \, , \tag{9.16}$$

$$v - h \overset{(9.4)}{=} -\frac{1}{\mathcal{H}}\left(Y + \mathcal{R}_v\right) \, . \tag{9.17}$$

In addition we need the background identity from Eq. (9.7). All in all, Eq. (9.3) multiplied by $\mathcal{H}$ yields

$$\overbrace{\mathcal{H}\phi - \mathcal{H}X' - \mathcal{H}^2X}^{\text{from } \mathcal{H}h_0 \text{ via (9.15)}} \overset{\mathcal{H}\times(9.3)}{=} \overbrace{-\psi' - \mathcal{H}X - \mathcal{H}X'}^{\text{from } -Y' \text{ via (9.16)}}$$

$$- \overbrace{(\mathcal{H}^2 - \mathcal{H}')}^{4\pi G a^2(\bar{e}+\bar{p})} \times \frac{\overbrace{\psi + \mathcal{H}X + \mathcal{R}_v}^{\text{from } v-h \text{ via (9.16),(9.17)}}}{\mathcal{H}} \, . \tag{9.18}$$

Going over to physical time, with $\psi' = a\dot{\psi}$, $\mathcal{H} = aH$, and $\mathcal{H}' - \mathcal{H}^2 = a^2\dot{H}$, we get

$$\phi - \psi + 8\pi G a^2 \Pi \overset{(4.56)}{=} 0 \, , \tag{9.19}$$

$$\dot{\psi} + H\phi - \frac{\dot{H}}{H}\left(\psi + \mathcal{R}_v\right) \overset{(9.18)}{=} 0 \, . \tag{9.20}$$

Combining Eqs. (9.19) and (9.20) with Eqs. (9.13) and (9.14), we now have a complete set of equations also for the situation that anisotropic stress plays a role. It is interesting that in the constraint relation in Eq. (9.19), Π does *not* come with a coefficient k^2, unlike in the time evolution equations, and a^2 rather cancels against the $1/a^2$ in Eq. (8.11).

9.3   The Origin of Acoustic Oscillations

The next task is to solve Eqs. (9.13) and (9.14). To simplify our life a bit, we omit the viscous corrections appearing in the anisotropic stress, so that Eqs. (9.19) and (9.20) do not need to be included. The physics that anisotropic stress describes is nevertheless interesting: it leads to the damping of the acoustic oscillations, with the damping rate growing as k^2 with the momentum. This is related to the *Silk damping* discussed in Sect. 2.2.

We now focus on times $t \geq t_{\text{out}} \gg t_e$, where t_e denotes the end of inflation, and t_{out} a time at which the mode considered is well outside of the Hubble horizon. Following Eqs. (1.85) and (1.86), we adopt a simplified equation of state to describe the background evolution. Moreover, it is convenient to define the parameter

$$
\alpha \equiv \frac{2}{3(1+w)} = \begin{cases} \frac{1}{2}, & w = \frac{1}{3} \ \text{(radiation domination)} \\ \frac{2}{3}, & w = 0 \ \text{(matter domination)} \end{cases}.
\tag{9.21}
$$

The scale factor and the Hubble rate then behave as

$$
a \overset{(1.85)}{\approx} a_{\text{out}}\left(\frac{t}{t_{\text{out}}}\right)^{\alpha}, \qquad H \overset{(1.86)}{\approx} \frac{\alpha}{t}, \qquad t \geq t_{\text{out}} \gg t_e .
\tag{9.22}
$$

Furthermore, from Eq. (2.34), we know that $T \propto 1/a$. Therefore the ratios appearing in Eqs. (9.13) and (9.14) scale as

$$
\frac{\dot{T}}{T} \overset{(2.34)}{\underset{(9.22)}{\approx}} -\frac{\alpha}{t}, \qquad \frac{\dot{H}}{H} \overset{(9.22)}{\approx} -\frac{1}{t} .
\tag{9.23}
$$

For the momentum appearing in Eq. (9.14), we write

$$
\frac{k^2}{a^2 H} = \left(\frac{k}{aH}\right)_{\text{out}}^{2}\left(\frac{a_{\text{out}}}{a}\right)^{2}\frac{H_{\text{out}}^2}{H} \overset{(9.22)}{\approx} \left(\frac{k}{aH}\right)_{\text{out}}^{2}\left(\frac{t_{\text{out}}}{t}\right)^{2\alpha}\frac{\alpha t}{t_{\text{out}}^2} .
\tag{9.24}
$$

It is also helpful to take

$$
x \equiv \ln\left(\frac{t}{t_{\text{out}}}\right) \geq 0, \qquad \mathrm{d}x = \frac{\mathrm{d}t}{t}, \qquad \partial_t = \frac{1}{t}\partial_x ,
\tag{9.25}
$$

as an integration variable. Then Eqs. (9.13) and (9.14) become

$$
\partial_x \mathcal{R}_v \overset{(9.13)}{\underset{(9.23),\,(9.25)}{\approx}} \alpha\left(\mathcal{R}_v - \mathcal{R}_T\right) ,
\tag{9.26}
$$

$$
\partial_x \mathcal{R}_T \overset{(9.14)}{\underset{(9.24),\,(9.25)}{\approx}} \mathcal{R}_v - \mathcal{R}_T + \frac{\alpha}{3}\underbrace{\left(\frac{t}{t_{\text{out}}}\right)^{2(1-\alpha)}}_{e^{2(1-\alpha)x}}\left(\frac{k}{aH}\right)_{\text{out}}^{2}\mathcal{R}_v .
\tag{9.27}
$$

Equations (9.26) and (9.27) are transparent enough that the qualitative features of their solution can be discussed. The last term in Eq. (9.27) is small at $t \to t_{\text{out}}$, if $k \ll (aH)_{\text{out}}$. Then we find a stationary solution $\mathcal{R}_v = \mathcal{R}_T$, which serves as an initial condition. When $x > 0$, the importance of the last term in Eq. (9.27) grows exponentially. In spite of this large contribution, the function $\mathcal{R}_T$ cannot grow much, because of the damping term $-\mathcal{R}_T$. Therefore the magnitude of $\mathcal{R}_v$, multiplying the growing coefficient, has to decrease exponentially. These features are confirmed by a numerical solution of Eqs. (9.26) and (9.27), which is shown for $\alpha = \frac{1}{2}$ in Fig. 9.1

(left). With some more effort, incorporating a numerical background solution like in Fig. 7.2 (left) in Sect. 7.7, it can be verified that Fig. 9.1 (left) represents a good approximation also to the solution of the un-approximated Eqs. (9.13) and (9.14) [5].

Let us finally discuss the physical meaning of $\mathcal{R}_T$. Knowing the scaling of H and T from Eqs. (9.22) and (9.23), respectively, we rewrite Eq. (4.62) as

$$\mathcal{R}_T \overset{(4.62)}{=} -\left(h_{\mathrm{D}} + \frac{\nabla^2 \vartheta}{3}\right) - \mathcal{H} \frac{\delta T}{T'} = -\left(h_{\mathrm{D}} + \frac{\nabla^2 \vartheta}{3}\right) - \underbrace{\frac{HT}{\dot{T}}}_{\overset{(9.22)}{\underset{(9.23)}{\approx}} 1} \frac{\delta T}{T} . \tag{9.28}$$

We see that $\mathcal{R}_T$ represents a gauge-invariant version of a relative temperature fluctuation. In addition, given that $\delta e / \bar{e}' \approx \delta T / T'$,

$$\mathcal{R}_T \overset{(4.63)}{\approx} \mathcal{R}_e \overset{(4.65)}{=} \zeta , \tag{9.29}$$

so it also captures energy density perturbations. Furthermore we note that another measure of energy density perturbations, from Eq. (4.64), becomes

$$\Delta \overset{(4.64)}{=} \frac{\delta e}{\bar{e}} + \frac{\bar{e}'}{\bar{e}}(h - v) \approx \frac{\bar{e}_{,T}\delta T}{\bar{e}} + \frac{\bar{e}_{,T}T'}{\bar{e}}(h - v) = \frac{\bar{e}_{,T}T'}{\bar{e}}\left(h - v + \frac{\delta T}{T'}\right)$$

$$= \frac{\dot{\bar{e}}}{\bar{e}H}\left(\mathcal{R}_v - \mathcal{R}_T\right) \overset{(1.44)}{\underset{(9.21)}{\approx}} \frac{2}{\alpha}\left(\mathcal{R}_T - \mathcal{R}_v\right) \overset{x \gg 1}{\underset{(9.27)}{\approx}} \frac{2}{\alpha}\mathcal{R}_T . \tag{9.30}$$

So, $\mathcal{R}_T$ determines this function as well, up to a coefficient of order unity.

In non-relativistic physics, propagating fluctuations in the temperature and longitudinal flow velocity are sound waves. Given this analogy, the oscillations in $\mathcal{R}_T$ (and in $\mathcal{R}_v$) that we see in Fig. 9.1 (left) can be called *acoustic oscillations*.

9.4   What Happens to $\mathcal{R}_\varphi$?

The key variable in the determination of the curvature perturbations in Chap. 8 was $\mathcal{R}_\varphi$, defined in Eq. (4.60). Recalling the definition of the Hubble time as $\Delta t \equiv H^{-1}$, it contains

$$\mathcal{R}_\varphi \overset{(4.60)}{\supset} -H\frac{\delta\varphi}{\dot{\bar{\varphi}}} = -\frac{\delta\varphi}{\Delta t\,\dot{\bar{\varphi}}} \approx -\frac{\delta\varphi}{\Delta\bar{\varphi}} . \tag{9.31}$$

As we enter the regime $\Upsilon \gg H$, the background value $\bar{\varphi}$ decreases, but $\delta\varphi$ also decreases, so it is not clear how the ratio in Eq. (9.31) behaves. Physically, for small $\bar{\varphi}$, the distinction between $\bar{\varphi}$ and $\delta\varphi$ loses its significance, and in the end second-order perturbations determine the contribution of φ to the plasma energy density and pressure (cf. Sect. 7.6). However, for the consistency of our framework, it is essential that this transition is reflected by the equations that we are solving,

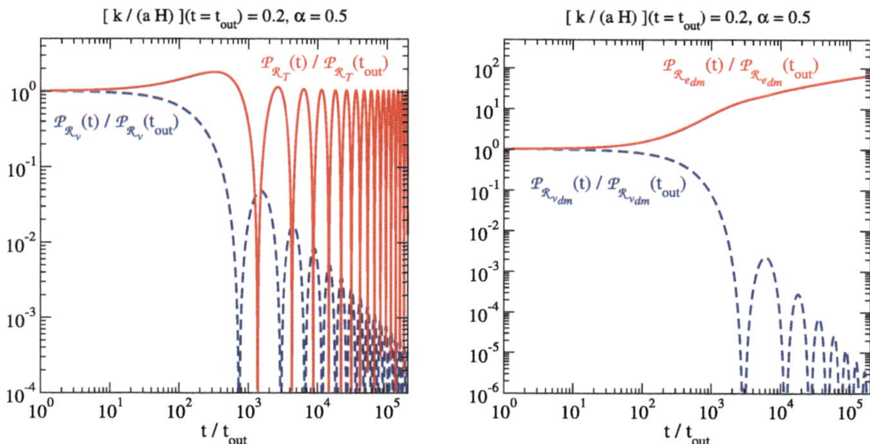

Fig. 9.1 Left: An example of a numerical solution of Eqs. (9.26) and (9.27), for $\alpha = \frac{1}{2}$, corresponding to radiation domination (cf. Eq. (9.21)) (see Sect. 9.7 for more details). The solid curve, displaying the power spectrum corresponding to $\mathcal{R}_T$, can be interpreted as a determination of a transfer function from Eq. (9.60). We remark that this is a good approximation to the solution of the full Eqs. (9.13) and (9.14). Right: Prototypical dark matter perturbations, from Eqs. (9.58) and (9.59). The power spectrum $\mathcal{P}_{\mathcal{R}_{e_{\rm dm}}}$ grows as $\sim \ln^2(t/t_{\rm out})$ in physical time, once the mode is inside the Hubble horizon, however the approximate forms in Eqs. (9.58) and (9.59) only apply when dark matter is subdominant compared to radiation (otherwise Eqs. (9.52) and (9.53) need to be solved)

rather than having to be put in by hand. The purpose of this section is to demonstrate how this happens.

Let us anticipate the outcome. If we solve Eq. (7.81), the solution $\mathcal{R}_\varphi$ remains non-trivial and large until late times. However, when $\dot{\bar{\varphi}}^2 \ll \bar{e}$, $\mathcal{R}_\varphi$ decouples from the evolution of $\mathcal{R}_\nu$ and $\mathcal{R}_T$, so that as far as the latter two are considered, it is sufficient to solve Eqs. (9.13) and (9.14). Therefore, we are free to stop solving for $\mathcal{R}_\varphi$, and simply include its second-order contribution in our thermodynamic functions if it equilibrates.

In order to demonstrate the decoupling explicitly, we take the full Eqs. (7.90) and (7.97) for $\mathcal{S}_\nu$ and $\mathcal{S}_T$ as a starting point. We then proceed by noting that the background identities from Eqs. (1.69), (1.71) and (1.73) can be simplified when $\dot{\bar{\varphi}}^2$ carries little energy density,

$$3H^2 \overset{(1.69)}{\approx} 8\pi G\bar{e}\,, \quad \dot{H} \overset{(1.71)}{\approx} -4\pi G(\bar{e}+\bar{p})\,, \quad \dot{\bar{e}} \overset{(1.73)}{\approx} -3H(\bar{e}+\bar{p})\,. \tag{9.32}$$

We also note that in this situation there is only one important variable, i.e. T plays a role for the dynamics but φ not, so we can write $\bar{p}_{,T}/\bar{e}_{,T} \approx \dot{\bar{p}}/\dot{\bar{e}}$. Furthermore we can approximate $4\pi G\dot{\bar{\varphi}}^2/H \ll 4\pi G\bar{e}/H \sim H$. Therefore Eq. (7.90) becomes

$$0 \overset{(7.90)}{\approx} (\partial_t + 3H)\,\overbrace{[(\bar{e} + \bar{p})(\mathcal{R}_v - \mathcal{R}_\varphi)]}^{(7.61):\ \text{was}\ \mathcal{S}_v} - \frac{\dot{\bar{p}}}{\dot{\bar{e}}}\,\overbrace{[\dot{\bar{e}}(\mathcal{R}_T - \mathcal{R}_\varphi)]}^{(7.62):\ \text{was}\ \mathcal{S}_T} + (\bar{e} + \bar{p})\dot{\mathcal{R}}_\varphi + \frac{2\,H\nabla^2\Pi}{3}$$

$$\overset{(9.32)}{=} (\bar{e} + \bar{p})\dot{\mathcal{R}}_v + [\dot{\bar{e}} + \dot{\bar{p}} + 3H(\bar{e} + \bar{p})](\mathcal{R}_v - \mathcal{R}_\varphi) + \dot{\bar{p}}\,(\mathcal{R}_\varphi - \mathcal{R}_T) + \frac{2H\nabla^2\Pi}{3}$$

$$= (\bar{e} + \bar{p})\dot{\mathcal{R}}_v + \dot{\bar{p}}\,(\mathcal{R}_v - \mathcal{R}_T) + \frac{2\,H\,\nabla^2\Pi}{3}\,. \tag{9.33}$$

We see that $\mathcal{R}_\varphi$ has dropped out, and Eq. (9.13) is recovered.

Turning to Eq. (7.97), we can similarly drop explicit appearances of $\dot{\bar{\varphi}}$ from the coefficients, given that $\dot{\bar{\varphi}}^2 \ll \bar{e}$. Then we find

$$0 \overset{(7.97)}{\approx} \left\{ \partial_t + \overbrace{\frac{4\pi G(\bar{e} + \bar{p}) - \dot{H}}{H}}^{(9.32):\ \text{becomes}\ -2\dot{H}/H} \right\} \overbrace{[-3\,H(\bar{e} + \bar{p})(\mathcal{R}_T - \mathcal{R}_\varphi)]}^{(7.62),\ (9.32):\ \text{was}\ \mathcal{S}_T}$$

$$+ \left\{ 3\,H\left(1 + \frac{\dot{\bar{p}}}{\dot{\bar{e}}}\right) \right\} \overbrace{[\dot{\bar{e}}(\mathcal{R}_T - \mathcal{R}_\varphi)]}^{(7.62):\ \text{was}\ \mathcal{S}_T} - \frac{\nabla^2}{a^2} \left\{ \overbrace{(\bar{e} + \bar{p})\mathcal{R}_\varphi + \mathcal{S}_v}^{(7.61):\ \text{becomes}\ (\bar{e}+\bar{p})\mathcal{R}_v} \right\}$$

$$- \left\{ \overbrace{\frac{8\pi G(\bar{e} + \bar{p})\,\bar{e}}{H}}^{(9.32):\ \text{becomes}\ 3\,H(\bar{e}+\bar{p})} \right\} \left\{ \dot{\mathcal{R}}_\varphi - \frac{4\pi G}{H} \overbrace{[(\bar{e} + \bar{p})(\mathcal{R}_v - \mathcal{R}_\varphi)]}^{(7.61):\ \text{was}\ \mathcal{S}_v} \right\}$$

$$= \{3 - 3\} H(\bar{e} + \bar{p})\dot{\mathcal{R}}_\varphi - \{3\,H(\bar{e} + \bar{p})\}\dot{\mathcal{R}}_T$$

$$+ \left\{ (3 - 6)\dot{H}(\bar{e} + \bar{p}) + (3 - 3)H(\dot{\bar{e}} + \dot{\bar{p}}) - 12\pi G(\bar{e} + \bar{p})^2 \right\} \mathcal{R}_\varphi$$

$$+ \left\{ (6 - 3)\dot{H}(\bar{e} + \bar{p}) + (3 - 3)H(\dot{\bar{e}} + \dot{\bar{p}}) \right\} \mathcal{R}_T$$

$$+ \left\{ 12\pi G(\bar{e} + \bar{p})^2 - (\bar{e} + \bar{p})\frac{\nabla^2}{a^2} \right\} \mathcal{R}_v\,. \tag{9.34}$$

Again, $\dot{\mathcal{R}}_\varphi$ and $\mathcal{R}_\varphi$ drop out. Dividing the remainder by $3H(\bar{e} + \bar{p})$ and using Eq. (9.32) once more, Eq. (9.14) is reproduced. Therefore, we have shown that $\mathcal{R}_\varphi$ has no influence on the evolution of $\mathcal{R}_v$ and $\mathcal{R}_T$, if $\dot{\bar{\varphi}}^2 \ll \bar{e}$.

To summarize, the quantity that we identify as the scalar curvature perturbation, when determining observable quantities like A_s or n_s (cf. Sect. 2.2), is the curvature perturbation $\mathcal{R}_T$ (cf. Eqs. (9.28), (9.29)). The curvature perturbation $\mathcal{R}_\varphi$ is valuable as it determines the constant value of $\mathcal{R}_T$ outside of the Hubble horizon, from the initial inflationary dynamics. However, strictly speaking $\mathcal{R}_\varphi$ is not the observable that we measure in the end.

9.5   The Origin of Jeans Instability

A physical reason for the acoustic oscillations in the CMB spectrum that we discussed in Sect. 9.3 is sometimes explained as follows. If we have a region of overdensity, gravity pulls it together, so that it tends to collapse. But as it does so, its pressure increases. This prohibits a collapse, and rather leads to oscillations.

To obtain a realistic picture, giving rise to proper structure formation, requires the addition of more ingredients. The first is that there are at least two matter components which behave differently: *dark matter*, which does not feel pressure ($p_{\mathrm{dm}} \ll e_{\mathrm{dm}}$), and Standard Model matter, which does ($p_r \approx e_r/3$). The second is that in the physical universe, the nature of the background evolution changes as the collapse proceeds, from radiation-dominated to matter-dominated. The energy density of radiation in a late universe, e_r, is carried by photons and neutrinos, whereas that of matter, e_m, resides in dark matter, e_{dm}, and in baryons, e_b. At temperatures well below the electron mass, $T \ll m_e$, we have

$$
e_r \underset{(2.22),(7.55)}{\overset{T \ll m_e}{\approx}} \frac{\pi^2 T^4}{30} \left[\overbrace{2}^{\text{from } e_\gamma} + \frac{7}{4}\left(\frac{4}{11}\right)^{4/3} \overbrace{N_{\mathrm{eff}}}^{\text{from } e_\nu} \right], \tag{9.35}
$$

$$
e_m \overset{T \ll m_e}{\approx} e_{\mathrm{dm}} + \underbrace{m_p n_\gamma \left(\frac{n_b}{n_\gamma}\right)}_{e_b\ (p_b > 0)}. \tag{9.36}
$$

Here m_p is the proton mass, $n_\gamma \equiv 2 \int_{\mathbf{p}} n_{\mathrm{B}}(p) = 2\zeta(3) T^3/\pi^2$ is the photon number density, and $n_b/n_\gamma \approx 6.1 \times 10^{-10}$, referred to as *baryon asymmetry*, is among the most important and mysterious parameters of post-inflationary cosmology. Inserting a realistic number for dark matter, $e_{\mathrm{dm}} \simeq 5.4\, e_b$, *matter-radiation equality* is reached at $T \sim$ eV, whereas CMB decoupling and electron-proton recombination take place a bit later, at $T \sim 0.3$ eV.

The 'common wisdom' for the early stages of gravitational collapse can now be formulated as follows. Given that dark matter does not feel radiation pressure, perturbations in dark matter can start to grow earlier than in visible matter [4]. Initially, baryons and photons oscillate independently of dark matter. At some point, their oscillations start to be influenced by dark matter-induced potential wells. After recombination, photons decouple and start streaming freely, constituting the CMB. Baryons stop feeling radiation pressure, and can therefore now collapse, eventually catching up with the dark matter. The *gravitational collapse* is sometimes referred to as a *Jeans instability* (cf., e.g., Refs. [2, 3]).

To illustrate some features of this dynamics, we consider an early epoch in which the universe is still radiation-dominated, and we follow a perturbation in the subdominant dark matter component (this assumption is introduced in Eq. (9.54), in order to decouple the evolution of radiation perturbations from those of dark matter, so that some qualitative features of the solution can be understood analytically). We stress that this is just an 'anticipatory' phase. Later on, dark matter is expected to collapse to form early galaxies during matter domination, a process commonly referred to as *halo formation*.

To proceed, let us generalize Eqs. (9.13) and (9.14) to a situation in which two different fluids are present (cf., e.g., Ref. [6]). Before focussing on dark matter, we keep the equations general, so that they may apply to other problems as well; the fluids are labelled with the indices $i = 1, 2$. For simplicity, we assume that the fluids interact only via gravity, so that their energy-momentum tensors are conserved separately,

$$T_{\mu\nu}^{(i);\mu} = 0 , \quad i = 1, 2 . \tag{9.37}$$

At the background level, this implies

$$\bar{e}_i' + 3\mathcal{H}(\bar{e}_i + \bar{p}_i) = 0 , \quad i = 1, 2 . \tag{9.38}$$

At first order in perturbations, Eqs. (9.1) and (9.2) become

$$-\delta e_1' - 3\mathcal{H}(\delta e_1 + \delta p_1) + (\bar{e}_1 + \bar{p}_1)(3h_{\mathrm{D}}' + \nabla^2 v_1) \overset{(9.1)}{=} 0 , \tag{9.39}$$

$$\delta p_1 + (\bar{e}_1 + \bar{p}_1)h_0 + (\partial_\tau + 4\mathcal{H})[(\bar{e}_1 + \bar{p}_1)(h - v_1)] + \frac{2}{3}\nabla^2 \Pi_1 \overset{(9.2)}{=} 0 , \tag{9.40}$$

and similarly for δe_2, δp_2 (for dark matter, $\delta e_2 \to \delta e_{\mathrm{dm}}$, $\delta p_2 \to \delta p_{\mathrm{dm}} \ll \delta e_{\mathrm{dm}}$).

In order to align our discussion with the one in Sect. 9.2, we assume that both fluids can be parametrized by the respective temperatures, T_1 and T_2. However, this does not need to be the case literally, as long as there is a unique equation of state in the dark matter fluid, which allows us to replace δT_2 through the more physical δe_2 in the end. We return to a further elaboration of this point around Eq. (9.55).

We now express the matter and metric perturbations in terms of curvature perturbations, defined as in Eqs. (4.61) and (4.62), as well as gauge degrees of freedom, abbreviated as in Eq. (5.12). If we write $\delta p_1 = \bar{p}_{1,T_1}\delta T_1$, and then express δT_1 like in Eq. (9.5), we can factor out the combination $\bar{p}_{1,T_1} T_1' = \bar{p}_1'$. Therefore,

$$h - v_1 \overset{(9.4)}{=} \frac{\mathcal{R}_{v_1} + Y}{\mathcal{H}} , \quad \delta p_1 \overset{(9.5)}{=} -\frac{\bar{p}_1'}{\mathcal{H}}\left(\mathcal{R}_{T_1} + Y\right) , \tag{9.41}$$

$$\delta e_1 \overset{(9.5)}{=} -\frac{\bar{e}_1'}{\mathcal{H}}\left(\mathcal{R}_{T_1} + Y\right) \overset{(9.38)}{=} 3(\bar{e}_1 + \bar{p}_1)\left(\mathcal{R}_{T_1} + Y\right) . \tag{9.42}$$

On the side of the Einstein equations, the right-hand side is a sum of the two energy-momentum tensors. Therefore, the background identity from Eq. (9.7) becomes

$$\mathcal{H}^2 - \mathcal{H}' \overset{(9.7)}{=} 4\pi Ga^2(\bar{e}_1 + \bar{e}_2 + \bar{p}_1 + \bar{p}_2) , \tag{9.43}$$

whereas $\mathcal{H}''$ can now be expressed as

$$\mathcal{H}'' + 2\mathcal{H}\mathcal{H}' - 4\mathcal{H}^3 \overset{(7.55)}{=} 4\pi Ga^2\left(\bar{e}_1' + \bar{e}_2' - \bar{p}_1' - \bar{p}_2'\right) . \tag{9.44}$$

Turning to metric perturbations, h_0 appearing in Eq. (9.40) reads

$$h_0 \overset{(9.3)}{\underset{(5.12)}{=}} -\frac{Y'}{\mathcal{H}} + \frac{4\pi G a^2}{\mathcal{H}}\left[(\bar{e}_1 + \bar{p}_1)(v_1 - h) + (\bar{e}_2 + \bar{p}_2)(v_2 - h)\right]$$

$$\overset{(9.41)}{\underset{(9.43)}{=}} -\frac{Y'}{\mathcal{H}} + \left(\frac{\mathcal{H}'}{\mathcal{H}^2} - 1\right)Y - \frac{4\pi G a^2}{\mathcal{H}^2}\left[(\bar{e}_1 + \bar{p}_1)\mathcal{R}_{v_1} + (\bar{e}_2 + \bar{p}_2)\mathcal{R}_{v_2}\right]. \qquad (9.45)$$

For the metric perturbations appearing in Eq. (9.39), we make use of Eq. (9.9), and write

$$\nabla^2 h + 3h'_\mathrm{D} + h'_0 \overset{(9.10)}{=} -2\left(2\mathcal{H} + \frac{\mathcal{H}'}{\mathcal{H}}\right)h_0 - \frac{Y'' + 2\mathcal{H}Y' - \nabla^2 Y}{\mathcal{H}}$$

$$+ \frac{4\pi G a^2}{\mathcal{H}}\left[\delta p_1 + \delta p_2 - \delta e_1 - \delta e_2 + \frac{2}{3}\nabla^2(\Pi_1 + \Pi_2)\right], \qquad (9.46)$$

as well as

$$h'_0 \overset{(9.45)}{=} -\frac{Y''}{\mathcal{H}} + \left(\frac{2\mathcal{H}'}{\mathcal{H}^2} - 1\right)Y' + \left(\frac{\mathcal{H}''}{\mathcal{H}^2} - \frac{2\mathcal{H}'^2}{\mathcal{H}^3}\right)Y$$

$$+ 2\left(\frac{\mathcal{H}'}{\mathcal{H}} - \mathcal{H}\right)\frac{4\pi G a^2}{\mathcal{H}^2}\left[(\bar{e}_1 + \bar{p}_1)\mathcal{R}_{v_1} + (\bar{e}_2 + \bar{p}_2)\mathcal{R}_{v_2}\right] \qquad (9.47)$$

$$- \frac{4\pi G a^2}{\mathcal{H}^2}\left[(\bar{e}'_1 + \bar{p}'_1)\mathcal{R}_{v_1} + (\bar{e}_1 + \bar{p}_1)\mathcal{R}'_{v_1} + (\bar{e}'_2 + \bar{p}'_2)\mathcal{R}_{v_2} + (\bar{e}_2 + \bar{p}_2)\mathcal{R}'_{v_2}\right].$$

The task now is to substitute Eqs. (9.41)–(9.47) into (9.39) and (9.40). We start with the latter, which is somewhat simpler. Collecting first the terms containing Y on the left-hand side (L), we find

$$(9.40)_\mathrm{L} \supset \overbrace{-\frac{\bar{p}'_1 Y}{\mathcal{H}}}^{\text{from } \delta p_1 \text{ via (9.41)}} + (\bar{e}_1 + \bar{p}_1)\overbrace{\left[-\frac{Y'}{\mathcal{H}} + \left(\frac{\mathcal{H}'}{\mathcal{H}^2} - 1\right)Y\right]}^{\text{from } h_0 \text{ via (9.45)}} + (\partial_\tau + 4\mathcal{H})\overbrace{\left[\frac{(\bar{e}_1 + \bar{p}_1)Y}{\mathcal{H}}\right]}^{\text{from } (\bar{e}_1 + \bar{p}_1)(h - v_1) \text{ via (9.41)}}$$

$$= \left[-\frac{\bar{p}'_1}{\mathcal{H}} + (\bar{e}_1 + \bar{p}_1)\left(\frac{\mathcal{H}'}{\mathcal{H}^2} - 1 - \frac{\mathcal{H}'}{\mathcal{H}^2} + 4\right) + \frac{\bar{e}'_1 + \bar{p}'_1}{\mathcal{H}}\right]Y \overset{(9.38)}{=} 0. \qquad (9.48)$$

Proceeding to the physical terms, the substitutions yield

$$
\mathcal{H} \times (9.40)_{\mathrm{L}} \stackrel{Y \to 0}{=} \overbrace{-\bar{p}_1' \mathcal{R}_{T_1}}^{\text{from } \delta p_1 \text{ via } (9.41)} -(\bar{e}_1 + \bar{p}_1) \frac{4\pi G a^2}{\mathcal{H}} \overbrace{\big[(\bar{e}_1 + \bar{p}_1)\mathcal{R}_{v_1} + (\bar{e}_2 + \bar{p}_2)\mathcal{R}_{v_2} \big]}^{\text{from } h_0 \text{ via } (9.45)}
$$

$$
+ \mathcal{H}\,(\partial_\tau + 4\mathcal{H}) \left[\underbrace{\frac{(\bar{e}_1 + \bar{p}_1)\mathcal{R}_{v_1}}{\mathcal{H}}}_{\text{from } (\bar{e}_1 + \bar{p}_1)(h - v_1) \text{ via } (9.41)} \right] + \frac{2\mathcal{H}}{3} \nabla^2 \Pi_1
$$

$$
\stackrel{\mathcal{R}_{v_2} = \mathcal{R}_{v_1} + \mathcal{R}_{v_2} - \mathcal{R}_{v_1}}{=} \; +\bar{p}_1'\big(\mathcal{R}_{v_1} - \mathcal{R}_{T_1} \big) - \frac{4\pi G a^2 (\bar{e}_1 + \bar{p}_1)(\bar{e}_2 + \bar{p}_2)}{\mathcal{H}} \big(\mathcal{R}_{v_2} - \mathcal{R}_{v_1} \big)
$$

$$
+ (\bar{e}_1 + \bar{p}_1)\mathcal{R}_{v_1}' + \frac{2\mathcal{H}}{3} \nabla^2 \Pi_1 \tag{9.49}
$$

$$
+ \left[\bar{e}_1' + (\bar{e}_1 + \bar{p}_1)\bigg(-\mathcal{H} + \overbrace{\frac{\mathcal{H}'}{\mathcal{H}} - \frac{\mathcal{H}'}{\mathcal{H}}}^{\text{via } (9.43)} + 4\mathcal{H} \bigg) \right] \mathcal{R}_{v_1} = 0 \, .
$$

$$
\underbrace{\phantom{+ \left[\bar{e}_1' + (\bar{e}_1 + \bar{p}_1)\bigg(-\mathcal{H} + \frac{\mathcal{H}'}{\mathcal{H}} - \frac{\mathcal{H}'}{\mathcal{H}} + 4\mathcal{H} \bigg) \right]}}_{\text{vanishes due to } (9.38)}
$$

Finally, we transform to physical time, with $\mathcal{R}_{v_1}' = a\dot{\mathcal{R}}_{v_1}$ and $\mathcal{H} = aH$, and to comoving momentum space. Dividing the whole by $a(\bar{e}_1 + \bar{p}_1)$, yields the result given in Eq. (9.52).

Let us then proceed to Eq. (9.39). For the terms containing Y, we find

$$
(9.39)_{\mathrm{L}} \supset \overbrace{-3(\bar{e}_1' + \bar{p}_1') Y}^{-\delta e_1' \text{ via } (9.42)} - 3(\bar{e}_1 + \bar{p}_1) Y' + \overbrace{3(\bar{e}_1' + \bar{p}_1') Y}^{\delta e_1 + \delta p_1 \text{ via } (9.41),(9.42)}
$$

$$
+ (\bar{e}_1 + \bar{p}_1) \Bigg\{ \overbrace{-\frac{\nabla^2 Y}{\mathcal{H}}}^{\nabla^2 (v_1 - h) \text{ via } (9.41)} + \overbrace{2\bigg(2\mathcal{H} + \frac{\mathcal{H}'}{\mathcal{H}} \bigg)\bigg[\frac{Y'}{\mathcal{H}} + \bigg(1 - \frac{\mathcal{H}'}{\mathcal{H}^2} \bigg) Y \bigg]}^{\text{from } h_0 \text{ in } (9.46) \text{ via } (9.45)}
$$

$$
- \overbrace{\frac{Y'' + 2\mathcal{H} Y' - \nabla^2 Y}{\mathcal{H}}}^{\text{from } (9.46)} + \overbrace{\frac{4\pi G a^2 (\bar{e}_1' + \bar{e}_2' - \bar{p}_1' - \bar{p}_2')}{\mathcal{H}^2} Y}^{\text{from } \delta p_i,\, \delta e_i \text{ in } (9.46)}
$$

$$
\underbrace{+ \frac{Y''}{\mathcal{H}} + \bigg(1 - \frac{2\mathcal{H}'}{\mathcal{H}^2} \bigg) Y' + \bigg(\frac{2\mathcal{H}'^2}{\mathcal{H}^3} - \frac{\mathcal{H}''}{\mathcal{H}^2} \bigg) Y \Bigg\}}_{\text{from } -h_0' \text{ via } (9.47)}
$$

$$
\stackrel{(9.44)}{=} (\bar{e}_1 + \bar{p}_1) \left\{ \bigg(-3 + 4 + \frac{2\mathcal{H}'}{\mathcal{H}^2} - 2 + 1 - \frac{2\mathcal{H}'}{\mathcal{H}^2} \bigg) Y' \right. \tag{9.50}
$$

$$+\left(4\mathcal{H}-\frac{2\mathcal{H}'}{\mathcal{H}}-\frac{2\mathcal{H}'^2}{\mathcal{H}^3}+\frac{\mathcal{H}''+2\mathcal{H}\mathcal{H}'-4\mathcal{H}^3}{\mathcal{H}^2}+\frac{2\mathcal{H}'^2}{\mathcal{H}^3}-\frac{\mathcal{H}''}{\mathcal{H}^2}\right)Y\right\}=0\,.$$

So all appearances of Y cancel, and therefore gauge independence has been verified. For the physical terms, we find

$$(9.39)_{\mathrm{L}}\ \overset{Y\,\rightarrow\,0}{\cong}\ \overbrace{-3(\bar{e}_1'+\bar{p}_1')\mathcal{R}_{T_1}}^{-\delta e_1'\ \text{via (9.42)}}-3(\bar{e}_1+\bar{p}_1)\mathcal{R}_{T_1}'+\overbrace{3(\bar{e}_1'+\bar{p}_1')\mathcal{R}_{T_1}}^{\delta e_1+\delta p_1\ \text{via (9.41),(9.42)}}$$

$$+(\bar{e}_1+\bar{p}_1)\left\{-\overbrace{\frac{\nabla^2\mathcal{R}_{v_1}}{\mathcal{H}}}^{\nabla^2(v_1-h)\ \text{via (9.41)}}\right.$$

$$+\overbrace{2\left(2\mathcal{H}+\frac{\mathcal{H}'}{\mathcal{H}}\right)\frac{4\pi Ga^2}{\mathcal{H}^2}\left[(\bar{e}_1+\bar{p}_1)\mathcal{R}_{v_1}+(\bar{e}_2+\bar{p}_2)\mathcal{R}_{v_2}\right]}^{\text{from }h_0\ \text{in (9.46) via (9.45)}}$$

$$+\frac{4\pi Ga^2}{\mathcal{H}^2}\left[\overbrace{(\bar{e}_1'-\bar{p}_1')\mathcal{R}_{T_1}+(\bar{e}_2'-\bar{p}_2')\mathcal{R}_{T_2}}^{\text{from }\delta p_i,\,\delta e_i\ \text{in (9.46)}}+\overbrace{\frac{2\mathcal{H}}{3}\nabla^2(\Pi_1+\Pi_2)}^{\text{from (9.46)}}\right]$$

$$+\underbrace{\left(2\mathcal{H}-\frac{2\mathcal{H}'}{\mathcal{H}}\right)\frac{4\pi Ga^2}{\mathcal{H}^2}\left[(\bar{e}_1+\bar{p}_1)\mathcal{R}_{v_1}+(\bar{e}_2+\bar{p}_2)\mathcal{R}_{v_2}\right]}_{\text{from }-h_0'\ \text{via (9.47)}}$$

$$+\underbrace{\frac{4\pi Ga^2}{\mathcal{H}^2}\left[(\bar{e}_1'+\bar{p}_1')\mathcal{R}_{v_1}+(\bar{e}_1+\bar{p}_1)\mathcal{R}_{v_1}'+(\bar{e}_2'+\bar{p}_2')\mathcal{R}_{v_2}+(\bar{e}_2+\bar{p}_2)\mathcal{R}_{v_2}'\right]}_{\text{from }-h_0'\ \text{via (9.47)}}\right\}$$

$$=(\bar{e}_1+\bar{p}_1)\left\{-3\mathcal{R}_{T_1}'-\frac{\nabla^2\mathcal{R}_{v_1}}{\mathcal{H}}\right.$$

$$+\frac{4\pi Ga^2}{\mathcal{H}^2}\left[6\mathcal{H}(\bar{e}_1+\bar{p}_1)\mathcal{R}_{v_1}+6\mathcal{H}(\bar{e}_2+\bar{p}_2)\mathcal{R}_{v_2}\right.$$

$$+(\bar{e}_1'-\bar{p}_1')\mathcal{R}_{T_1}+(\bar{e}_2'-\bar{p}_2')\mathcal{R}_{T_2}+\frac{2\mathcal{H}}{3}\nabla^2(\Pi_1+\Pi_2)$$

$$\left.\left.+(\bar{e}_1'+\bar{p}_1')\mathcal{R}_{v_1}+(\bar{e}_1+\bar{p}_1)\mathcal{R}_{v_1}'+(\bar{e}_2'+\bar{p}_2')\mathcal{R}_{v_2}+(\bar{e}_2+\bar{p}_2)\mathcal{R}_{v_2}'\right]\right\}$$

$$\overset{\substack{(9.49)\ \text{for }\mathcal{R}_{v_1}'\\ \text{and for }\mathcal{R}_{v_2}'}}{=}(\bar{e}_1+\bar{p}_1)\left\{-3\mathcal{R}_{T_1}'-\frac{\nabla^2\mathcal{R}_{v_1}}{\mathcal{H}}\right.$$

$$+\frac{4\pi Ga^2}{\mathcal{H}^2}\left[6\mathcal{H}(\bar{e}_1+\bar{p}_1)\mathcal{R}_{v_1}+6\mathcal{H}(\bar{e}_2+\bar{p}_2)\mathcal{R}_{v_2}\right.$$

$$+ \bar{e}'_1 (\mathcal{R}_{T_1} + \mathcal{R}_{v_1}) + \bar{e}'_2 (\mathcal{R}_{T_2} + \mathcal{R}_{v_2}) \Big] \Big\}$$

$$\overset{(9.38)}{=} (\bar{e}_1 + \bar{p}_1) \Big\{ -3\mathcal{R}'_{T_1} - \frac{\nabla^2 \mathcal{R}_{v_1}}{\mathcal{H}} \tag{9.51}$$

$$+ \frac{12\pi G a^2}{\mathcal{H}} \Big[(\bar{e}_1 + \bar{p}_1)(\mathcal{R}_{v_1} - \mathcal{R}_{T_1}) + (\bar{e}_2 + \bar{p}_2)(\mathcal{R}_{v_2} - \mathcal{R}_{T_2}) \Big] \Big\} = 0 \,.$$

Going subsequently to physical time and to comoving momentum space, and dividing by $3a$, we obtain Eq. (9.53).

To summarize, the evolution equations for two fluids read

$$\dot{\mathcal{R}}_{v_1} \overset{(9.49)}{=} \frac{\dot{\bar{p}}_1}{\bar{e}_1 + \bar{p}_1} \left(\mathcal{R}_{T_1} - \mathcal{R}_{v_1} \right) + \frac{2\,H\,k^2 \Pi_1}{3(\bar{e}_1 + \bar{p}_1)} + \frac{4\pi G(\bar{e}_2 + \bar{p}_2)}{H} \left(\mathcal{R}_{v_2} - \mathcal{R}_{v_1} \right) , \tag{9.52}$$

$$\dot{\mathcal{R}}_{T_1} \overset{(9.51)}{=} \frac{\mathcal{R}_{v_1}}{3\,H} \frac{k^2}{a^2} + \frac{4\pi G}{H} \Big[(\bar{e}_1 + \bar{p}_1) \left(\mathcal{R}_{v_1} - \mathcal{R}_{T_1} \right) + (\bar{e}_2 + \bar{p}_2) \left(\mathcal{R}_{v_2} - \mathcal{R}_{T_2} \right) \Big] . \tag{9.53}$$

The evolution equations for $\mathcal{R}_{v_2}$ and $\mathcal{R}_{T_2}$ are obtained through the interchange $1 \leftrightarrow 2$.

We note that if we look for a stationary solution ($\dot{\mathcal{R}}_{v_i} = \dot{\mathcal{R}}_{T_i} = 0$) and modes outside of the Hubble horizon ($k \ll aH$), then the non-vanishing terms in Eqs. (9.52) and (9.53) are proportional to various isocurvature perturbations. There are three independent ones, $\mathcal{R}_{v_1} - \mathcal{R}_{T_1}$, $\mathcal{R}_{v_2} - \mathcal{R}_{T_2}$, and $\mathcal{R}_{v_2} - \mathcal{R}_{v_1}$. The homogeneous system that relates them only has a trivial solution. In accordance with Sect. 8.3, this implies that the isocurvature perturbations vanish outside of the Hubble horizon.

We now return to the physical setting described around Eqs. (9.35) and (9.36). We consider a time before matter-radiation equality, so that $e_r \gg e_m$. Fluid 1 is identified as radiation and fluid 2 as dark matter. Baryonic matter interacts with radiation and should be viewed as part of fluid 1, however in this epoch $e_b \ll e_r$ and $p_b \ll p_r$, so it plays no role in practice. The dark matter is assumed cold, with $p_{\mathrm{dm}} \approx 0$. It follows that

$$\frac{4\pi G(e_{\mathrm{dm}} + p_{\mathrm{dm}})}{H} \approx \frac{4\pi G e_{\mathrm{dm}}}{H} \ll \frac{4\pi G(e_r + p_r)}{H} \overset{(9.32)}{\approx} -\frac{\dot{H}}{H} \,. \tag{9.54}$$

Since $\dot{\bar{p}}_1/(\bar{e}_1 + \bar{p}_1) \approx \dot{T}_1/T_1$, and since $\dot{T}_1/T_1 \sim \dot{H}/H$ according to Eq. (9.23), the last terms of Eqs. (9.52) and (9.53) represent small corrections as far as the evolutions of $\mathcal{R}_{v_1}$ and $\mathcal{R}_{T_1}$ are concerned. Therefore, the radiation fluid undergoes acoustic oscillations, as described in Sect. 9.3 and illustrated numerically in Fig. 9.1 (left).

Let us then consider perturbations in the dark matter sector. The equations are as in Eqs. (9.52) and (9.53), after interchanging $1 \leftrightarrow 2$. So far we have used T_2 to parametrize the energy density e_2, but we can now undo this parametrization,

$$\frac{\delta T_2}{T_2'} \approx \frac{\bar{e}_{2,T}\,\delta T_2}{\bar{e}_{2,T}\,T_2'} = \frac{\delta e_2}{\bar{e}_2'}\,. \tag{9.55}$$

Therefore, in accordance with Eqs. (4.62) and (4.63), we can replace $\mathcal{R}_{T_2} \to \mathcal{R}_{e_2} \equiv \mathcal{R}_{e_{\mathrm{dm}}}$. This representation is more useful than that in terms of temperature, because in many models, dark matter particles become non-relativistic at low temperatures, and fall out of chemical equilibrium. Then their energy density fluctuations originate dominantly from those in their number density, rather than in kinetic energy. We remark that if both energy types are important simultaneously, we need to introduce both temperature and chemical potential fluctuations, and fixing the dynamics requires a further equation, the conservation law for particle number, $J_\mu^{(2);\mu} = 0$.

Working under the assumption of Eq. (9.55); omitting small terms in accordance with Eq. (9.54); and denoting $\mathcal{R}_{v_2} \to \mathcal{R}_{v_{\mathrm{dm}}}$, $\mathcal{R}_{v_1} \to \mathcal{R}_v$, and $\mathcal{R}_{T_1} \to \mathcal{R}_T$, the evaluation of Eqs. (9.52) and (9.53) after the interchange $1 \leftrightarrow 2$ yields

$$\dot{\mathcal{R}}_{v_{\mathrm{dm}}} \underset{1\leftrightarrow 2}{\overset{(9.52),\,(9.54)}{\approx}} \frac{2H\,k^2\Pi_{\mathrm{dm}}}{3\bar{e}_{\mathrm{dm}}} + \overbrace{\frac{4\pi G(\bar{e}_r + \bar{p}_r)}{H}}^{\text{from (9.32):}\ \approx\,-\dot{H}/H} \left(\mathcal{R}_v - \mathcal{R}_{v_{\mathrm{dm}}}\right), \tag{9.56}$$

$$\dot{\mathcal{R}}_{e_{\mathrm{dm}}} \underset{1\leftrightarrow 2}{\overset{(9.53),\,(9.54)}{\approx}} \frac{\mathcal{R}_{v_{\mathrm{dm}}}}{3H}\frac{k^2}{a^2} + \underbrace{\frac{4\pi G(\bar{e}_r + \bar{p}_r)}{H}}_{\text{from (9.32):}\ \approx\,-\dot{H}/H} \left(\mathcal{R}_v - \mathcal{R}_T\right). \tag{9.57}$$

We focus on the dynamics at fairly early times, when $k/(aH)$ is not yet huge. Then we may ignore the anisotropic stress as well. Going over to the variables in Eq. (9.25), Eqs. (9.56) and (9.57) can be converted into

$$\partial_x \mathcal{R}_{v_{\mathrm{dm}}} \underset{(9.23),(9.25)}{\overset{(9.56)}{\approx}} \mathcal{R}_v - \mathcal{R}_{v_{\mathrm{dm}}}\,, \tag{9.58}$$

$$\partial_x \mathcal{R}_{e_{\mathrm{dm}}} \underset{(9.24),(9.25)}{\overset{(9.57)}{\approx}} \mathcal{R}_v - \mathcal{R}_T + \frac{\alpha\,e^{2(1-\alpha)x}}{3}\left(\frac{k}{aH}\right)^2_{\mathrm{out}} \mathcal{R}_{v_{\mathrm{dm}}}\,. \tag{9.59}$$

How does the solution of Eqs. (9.58) and (9.59) differ from that of Eqs. (9.26) and (9.27)? From Eq. (9.58), we see that $\mathcal{R}_{v_{\mathrm{dm}}}$ grows if $\mathcal{R}_{v_{\mathrm{dm}}} < \mathcal{R}_v$, and decreases if $\mathcal{R}_{v_{\mathrm{dm}}} > \mathcal{R}_v$. Therefore, $\mathcal{R}_{v_{\mathrm{dm}}}$ traces the oscillations of $\mathcal{R}_v$ (one form of matter 'drags' the other through gravity), however the amplitude and frequency of its oscillations are smaller.

In contrast, the evolutions of $\mathcal{R}_{e_{\mathrm{dm}}}$ and $\mathcal{R}_T$ differ considerably from each other. Whereas in Eq. (9.27) there is a term counteracting the growth of $\mathcal{R}_T$, there is no such term for $\mathcal{R}_{e_{\mathrm{dm}}}$ in Eq. (9.59). All the individual terms on the right-hand side of Eq. (9.59) oscillate. Though it is not easy to see this without a numerical solution, their sum does not average to zero, but rather oscillates around a constant positive value. Therefore, $\mathcal{R}_{e_{\mathrm{dm}}}$ grows linearly in x. This implies logarithmic growth in t

(cf. Eq. (9.25)), and $\sim \ln^2(t/t_{\mathrm{out}})$ for the corresponding power spectrum, $\mathcal{P}_{\mathcal{R}_{e_{\mathrm{dm}}}}$ (the growth is illustrated numerically in Fig. 9.1 (right)). Such a growth means that we soon leave the domain of validity of linear perturbation theory, and a non-linear analysis becomes necessary.

Let us stress that in structure formation, the dark matter energy density becomes the dominant component after matter-radiation equality (cf. Eqs. (9.35) and (9.36)), and then the approximations in Eq. (9.54) break down. However, Eqs. (9.52) and (9.53) and their interchange $1 \leftrightarrow 2$ (with $\mathcal{R}_{T_2} \to \mathcal{R}_{e_2}$) remain valid, as long as we stay in the linear regime.

9.6   Numerical Softwares for the Physical Multicomponent Universe

In Sect. 9.3, we have solved for gauge-invariant temperature and flow-velocity perturbations for a system consisting of just one plasma, and in Sect. 9.5, we have added a pressureless fluid to mimic the presence of dark matter. However, this is not yet enough to capture the real world. As discussed in Sects. 2.2 and 2.4, the modes that are seen in the CMB re-enter inside the Hubble horizon only rather late in the history of the universe, around or later than the BBN epoch ($T \sim 0.1\,\mathrm{MeV}$). By this time, neutrinos have decoupled from the Standard Model plasma (this happens at $T \sim 2\,\mathrm{MeV}$), and should be represented as free-streaming particles. Moreover, dark matter has probably decoupled from the Standard Model plasma much earlier. The precise moment of its decoupling is model-dependent, and in principle it is possible that it never entered equilibrium in the first place. In any case, simulations of large-scale structure formation suggest that gravitational collapse in the dark sector should start earlier than in the baryonic one, and therefore these degrees of freedom need to be handled separately [4].

To treat this complex system quantitatively, requires that we solve a coupled system of equations, in which the various degrees of freedom evolve in a gravitational background, and also source new metric perturbations through the Einstein equations. The equations to be solved depend on the degree of freedom in question. For particle-like degrees of freedom out of equilibrium, like neutrinos and some models of dark matter, it is natural to solve a Boltzmann equation for a phase-space distribution. For very light bosonic degrees of freedom, for instance some variants of axions, we may assume that they can more efficiently be described by a field (or a 'condensate'), and solve the corresponding field equation. Finally, for species that do equilibrate, like baryons or photons, we could consider either hydrodynamic equations as we have done above (incorporating viscosities to handle shorter wavelengths), or Boltzmann equations. It is perhaps appropriate to mention that hydrodynamic equations are valid in a narrower kinematic range (only for the smallest momenta and frequencies), but within that range they offer for a powerful tool, able to incorporate radiative corrections from particle interactions.

In any case, this late dynamics can be disentangled from the early one that yields the solution outside of the Hubble horizon. Within the linear regime, the outcome of the late dynamics can be represented by a *transfer function*, because the initial condition depends on one single function. For instance, for $\mathcal{R}_T$, in analogy with Eq. (2.5), we could write

$$\mathcal{R}_T(t, k) \; \equiv \; X_{\mathcal{R}_T}(t, t_{\text{out}}, k)\, \mathcal{R}_T(t_{\text{out}}, k) \,, \tag{9.60}$$

with the initial conditions

$$X_{\mathcal{R}_T}(t, t_{\text{out}}, k)\big|_{t=t_{\text{out}}} \; = \; 1 \,, \quad \partial_t X_{\mathcal{R}_T}(t, t_{\text{out}}, k)\big|_{t=t_{\text{out}}} \; = \; 0 \,. \tag{9.61}$$

Then the power spectrum at photon decoupling ($t = t_{\text{dec}}$) is given by

$$\mathcal{P}_{\mathcal{R}_T}(t_{\text{dec}}, k) \; = \; \left[X_{\mathcal{R}_T}(t_{\text{dec}}, t_{\text{out}}, k) \right]^2 \mathcal{P}_{\mathcal{R}_T}(t_{\text{out}}, k) \,. \tag{9.62}$$

In practice, however, there is no need to stop at photon decoupling: we rather want to connect the initial curvature perturbations to today's CMB observations (cf. Eq. (2.5)). Then we need to address also the physics of the last Thomson scattering, which induces polarization and spectral distortions (cf. Sect. 2.2). The appropriate variable for this is not the hydrodynamic temperature, but rather the microscopic photon phase-space distribution function, $f_\gamma(t, \mathbf{x}, \mathbf{p})$ [1]. The photon temperature may be viewed as a derived quantity, obtained like in the *Stefan-Boltzmann law*, from an integral over the phase-space distribution,

$$\frac{2 \times \pi^2 T_\gamma^4(t, \mathbf{x})}{30} \; \overset{\substack{(7.55) \\ (7.110)}}{\equiv} \; 2 \int \frac{d^3\mathbf{p}}{(2\pi)^3} \, |\mathbf{p}| \, f_\gamma(t, \mathbf{x}, \mathbf{p}) \,. \tag{9.63}$$

Here the factor 2 counts the photon polarization states, which could also be treated as separate degrees of freedom. We remark that the choice of a microscopic Boltzmann approach, in terms of $f_\gamma(t, \mathbf{x}, \mathbf{p})$, comes with a price, namely that the framework is guaranteed to be correct only up to leading order in the electromagnetic fine-structure constant, α_{em}, which controls the scattering cross section (specifically, plasma-induced modifications to dispersion relations are not easy to accommodate systematically).

The desired physical observable is the deviation of $T_\gamma(\mathbf{x})$ from its average value, $\bar{T}_\gamma \equiv T_0$ (cf. Sect. 2.2), where we have dropped the time argument $t = t_0$ for simplicity. In theoretical considerations, this is represented in Fourier space, as

$$\delta T_\gamma(\mathbf{k}) \; \overset{(9)}{=} \; \int_{\mathbf{x}} \delta T_\gamma(\mathbf{x}) \, e^{-i\mathbf{k}\cdot\mathbf{x}} \,. \tag{9.64}$$

Observations of δT_γ are made as a function of a photon propagation direction, $\mathbf{n}$. 2-point correlators are evaluated by combining observations in two different directions, $\mathbf{n}_i$, $i = 1, 2$. In a first step, we can represent the directional dependence of δT_γ as a function of the relative angle between $\mathbf{n}_i$ and $\mathbf{k}$, defining $\cos\theta_i \equiv \mathbf{n}_i \cdot \mathbf{k}/k$. The angular dependence is represented as a series in spherical harmonics, $Y_{\ell_i m_i}(\theta_i, \phi_i)$.

The coefficients appearing in Eq. (2.3) describe differences of two observation directions, $\mathbf{n}_1 \cdot \mathbf{n}_2 \equiv \cos\theta$, where the θ-dependence is parametrized through the overall multipole, ℓ, obtained from ℓ_1 and ℓ_2. Therefore we are forced to do angular momentum additions in the domain $\ell \sim 10^3$, which is computationally quite expensive. Even if this may not be obvious, large comoving momenta k contribute to rapid angular variations in θ, and therefore to large angular momenta ℓ.

The equation governing the time evolution of $f_\gamma(t, \mathbf{x}, \mathbf{p})$ is a Boltzmann equation, with *Thomson scatterings* appearing as a collision term [1]. The Boltzmann equation is solved in an expanding background, in the presence of metric perturbations (which lead to the *Sachs-Wolfe effect* mentioned in Sect. 2.2). The softwares that have been written for this purpose are generally referred to as *Einstein-Boltzmann solvers*, and include as additional variables the phase-space distribution of neutrinos, the density of cold dark matter, and a tightly coupled plasma of electrons and baryons. Perhaps a first popular solver was CMBFAST [7], somewhat more recent frameworks are CAMB [8] and CLASS [9].

The physics that the Einstein-Boltzmann solvers incorporate; the approximations that they adopt; as well as the historical developments that led to the current state of the art, are reviewed in one of the publications that lays out the foundations for CAMB [10]. This reference employs a gauge independent approach, recalling the confusions that specific gauge choices may cause. That said, CMBFAST and CLASS rely on gauge-fixed evolution equations, originally derived in Ref. [11]. As a technical remark, we note that the solvers actually operate in conformal time τ, measured in units of Mpc.

For us, perhaps the most important issue about the solvers is to appreciate their choice of initial conditions. To this aim, a radiation-dominated era is considered (in terms of Eqs. (9.35) and (9.36), $T \equiv T_\gamma \gg$ eV). The initial moment is chosen early enough that the modes of interest are deep inside the Hubble horizon ($|k\tau| \ll 1$, cf. Fig. 1.1 in Sect. 1.4). In this domain, the evolution equations can be solved analytically, as a power series in $k^2\tau^2$. As we have seen in Sect. 8.3, the solution is described by a single function, the power spectrum related to the curvature perturbation (though, traditionally, a Bardeen potential is employed). Having fixed the initial condition, the remaining evolution is uniquely determined. For the photon phase-space distribution at values of k relevant for the CMB, a solution obtained at linear order in perturbations represents a good approximation all the way till today.

Finally, as alluded to at the end of Sect. 9.5, if we consider cold dark matter, density perturbations start to grow rapidly once they are inside the Hubble horizon. For baryonic matter, the growth starts after recombination, when baryons no longer feel radiation pressure. To treat this system, relevant for large-scale structure formation, we need to go beyond the linear order in perturbations. Determining the corresponding dynamics represents a field of its own, traditionally based on so-called N-body simulations, requiring high-performance computing resources. In recent years, alternative (partly more analytic) approaches have been developed as well.

At this point, it is appropriate to briefly return to the assumption that we made at the very beginning, in Sect. 1.2, namely that the background solution is homogeneous and isotropic. In the end, as is the case in empirical science, this assumption cannot be justified *a priori*, but needs rather to be verified from a comparison with data. Even though tensions always arise and cause excitement, at the time of writing the framework of a homogeneous and isotropic initial background, on top of which perturbations are solved at linear order as outlined above, or beyond it when larger momenta are considered, provides for a viable and accurate description of the observed universe.

9.7   Numerics for Acoustic Oscillations and Jeans Instability

In this section, we show how Eqs. (9.26) and (9.27) can be solved numerically. This amounts to a determination of the early stage of a transfer function for scalar perturbations, describing the effect of acoustic oscillations. Simultaneously, we solve Eqs. (9.58) and (9.59), as an example of a dark matter component, illustrating the origin of a Jeans instability (unattenuated growth). The results are illustrated in Fig. 9.1. A `python` script producing these solutions is given below.

```python
# acoustic oscillations and Jeans instability after re-entry [numerics_acoustic_jeans.py]
#
# import basic tools and integration routines
import numpy as np
from scipy.integrate import solve_ivp

# parameters
koaH = 0.2    # initial momentum over Hubble rate
alpha = 0.5   # rate of background expansion (radiation dominated)

# sources for [Rv, RT, Rvdm, Redm]
# derivatives are taken with respect to x = ln(t/t_out)
def derivatives(x,y):
    Rv, RT, Rvdm, Redm = y
    dRv_dx = alpha*(Rv - RT)
    dRT_dx = Rv - RT + alpha/3*np.exp(2*(1-alpha)*x)*koaH*koaH*Rv
    dRvdm_dx = Rv - Rvdm
    dRedm_dx = Rv - RT + alpha/3*np.exp(2*(1-alpha)*x)*koaH*koaH*Rvdm
    dy_dx = [dRv_dx, dRT_dx, dRvdm_dx, dRedm_dx]
    return dy_dx

# initial conditions
Rv_0 = 1.0; RT_0 = 1.0; Rvdm_0 = 1.0; Redm_0 = 1.0

# integrate up to chosen point
sol = solve_ivp(derivatives,
                [0, 15/2/(1-alpha)],
                [Rv_0, RT_0, Rvdm_0, Redm_0],method='DOP853',
                t_eval=np.linspace(0, 15/2/(1-alpha),3000))
x_grid = sol.t
Rv_sol = sol.y[0]; RT_sol = sol.y[1]; Rvdm_sol = sol.y[2]; Redm_sol = sol.y[3]

# print to file
np.savetxt('numerics_acoustic_jeans.dat',
           np.c_[np.exp(x_grid),Rv_sol**2, RT_sol**2, Rvdm_sol**2, Redm_sol**2],
           fmt='%.6e',newline='\n',
           header='columns: t/t_out, Rv^2, RT^2, Rvdm^2, Redm^2')
```

References

1. Peebles, P.J.E., Yu, J.T.: Primeval Adiabatic Perturbation in an Expanding Universe. Astrophys. J. **162**, 815 (1970). https://doi.org/10.1086/150713
2. Peebles, P.J.E., Dicke, R.H.: Origin of the Globular Star Clusters. Astrophys. J. **154**, 891 (1968). https://doi.org/10.1086/149811
3. Zeldovich, Ya.B.: Gravitational Instability: An Approximate Theory for Large Density Perturbations. Astron. Astrophys. **5**, 84 (1970)
4. Davis, M., Efstathiou, G., Frenk, C.S., White, S.D.M.: The evolution of large-scale structure in a universe dominated by cold dark matter. Astrophys. J. **292**, 371 (1985). https://doi.org/10.1086/163168
5. Laine, M., Procacci, S., Rogelj, A.: Evolution of coupled scalar perturbations through smooth reheating. Part I. Dissipative regime. JCAP **10**, 040 (2024). https://doi.org/10.1088/1475-7516/2024/10/040, 2407.17074
6. Malik, K.A., Wands, D., Ungarelli, C.: Large-scale curvature and entropy perturbations for multiple interacting fluids. Phys. Rev. D **67**, 063516 (2003). https://doi.org/10.1103/PhysRevD.67.063516, astro-ph/0211602
7. Seljak, U., Zaldarriaga, M.: A Line-of-Sight Integration Approach to Cosmic Microwave Background Anisotropies. Astrophys. J. **469**, 437 (1996). https://doi.org/10.1086/177793, astro-ph/9603033
8. Lewis, A., Challinor, A., Lasenby, A.: Efficient Computation of Cosmic Microwave Background Anisotropies in Closed Friedmann-Robertson-Walker Models. Astrophys. J. **538**, 473 (2000). https://doi.org/10.1086/309179, astro-ph/9911177
9. Blas, D., Lesgourgues, J., Tram, T.: The Cosmic Linear Anisotropy Solving System (CLASS). Part II: Approximation schemes. JCAP **07**, 034 (2011). https://doi.org/10.1088/1475-7516/2011/07/034, 1104.2933
10. Challinor, A., Lasenby, A.: Cosmic Microwave Background Anisotropies in the CDM Model: A Covariant and Gauge-invariant Approach. Astrophys. J. **513**, 1 (1999). https://doi.org/10.1086/306841, astro-ph/9804301
11. Ma, C.-P., Bertschinger, E.: Cosmological Perturbation Theory in the Synchronous and Conformal Newtonian Gauges. Astrophys. J. **455**, 7 (1995). https://doi.org/10.1086/176550, astro-ph/9506072

Gravitational-Wave Probes of the Inflationary and Reheating Epochs   10

Abstract

While the motivation for inflationary cosmology comes from scalar perturbations (the source of anisotropies in the CMB and of structure formation, cf. Sect. 2.2), a future probe of this period might be offered by tensor perturbations, manifesting themselves as gravitational waves (cf. Eq. (3.106)). A key property of gravitational waves is that they propagate almost freely until present time, i.e. that their transfer function is simple (cf. Eq. (2.18)). Apart from the inflationary epoch, we discuss how gravitational waves could originate at or after reheating, from hydrodynamic fluctuations, second-order scalar perturbations, or elementary particle decays and scatterings. Many other sources have been proposed in the literature, such as preheating, topological defects, or phase transitions, but these are strongly model-dependent, and not discussed here. As gravitational-wave science will grow in importance in the next decades, we end the book with a summary of the various frequency domains that can hopefully be empirically investigated one day.

Keywords

Gravitational energy density · Tensor perturbations · Tensor-to-scalar ratio · Anisotropic stress · Scalar-induced gravitational waves · Gravitational waves from particle scatterings and decays · Transfer function in the tensor channel · Gravitational strain

10.1   What Is Gravitational-Wave Energy Density?

A physical manifestation of gravitational waves is that they carry energy density. When we discuss various sources of gravitational waves from the early universe, we normalize the *gravitational-wave energy density*, e_{gw}, to the current critical energy density, e_{crit} (cf. Eq. (2.18)). The critical energy density is the one which would yield

© The Author(s) 2026   201

M. Laine and S. Procacci, *From Inflation to Hot Big Bang*, Lecture Notes in Physics 1047, https://doi.org/10.1007/978-3-032-09893-1_10

the *current Hubble rate*, H_0, if $\kappa = 0$,

$$H_0^2 \overset{(1.43)}{\underset{(1)}{\equiv}} \frac{8\pi}{3} \frac{e_{\text{crit}}}{m_{\text{pl}}^2} \,. \tag{10.1}$$

Measuring H_0 is not easy, and therefore the result is often parametrized as

$$H_0 = h \times \bar{H}_0 \,, \quad \bar{H}_0 \equiv 100 \, \frac{\text{km}}{\text{s Mpc}} \,, \tag{10.2}$$

where the observed value now appears as the *reduced Hubble rate*, h. The fractional gravitational energy density is expressed as

$$\Omega_{\text{gw},0} \equiv \frac{e_{\text{gw},0}}{e_{\text{crit}}} \overset{(10.1)}{=} \frac{8\pi e_{\text{gw},0}}{3m_{\text{pl}}^2 H_0^2} \,. \tag{10.3}$$

Given Eq. (10.2), the quantity normally plotted is $h^2 \Omega_{\text{gw},0}$.

Obtaining a formal expression for e_{gw} is non-trivial. We first determine the gravitational contribution to the Einstein-Hilbert action, in Eq. (1.45). Then, from the action, we identify the energy density. In order to arrive at this result, we need to expand the Ricci scalar, R, and the determinant of the metric, $\sqrt{-g}$, to *second order in perturbations*. As a simplification, we may adopt the role of an observer rather than a source. Then we may restrict ourselves to modes well within the horizon, and consider a locally Minkowskian space-time, so that the scale factor a can be taken to be constant (i.e. its time derivative is omitted). In the notation of Eq. (3.34), we therefore consider

$$g_{\mu\nu} \equiv \underbrace{a^2}_{\text{constant}} (\eta_{\mu\nu} + h_{\mu\nu}) \,, \quad \eta_{\mu\nu} \equiv \text{diag}(-+++) \,. \tag{10.4}$$

The inverse metric is given by Eq. (3.35), *viz.*

$$g^{\mu\nu} \overset{(3.35)}{=} a^{-2} \left(\eta^{\mu\nu} - h^{\mu\nu} \right) + \mathcal{O}(h_{\mu\nu}^2) \,. \tag{10.5}$$

Next, we compute the determinant of the metric. It can be represented with the help of the antisymmetric *Levi-Civita symbol*, $\epsilon^{\mu\nu\rho\sigma}$, as

$$g = \det g_{\mu\nu} = \frac{1}{24} \epsilon^{\mu\nu\rho\sigma} \epsilon^{\alpha\beta\kappa\lambda} g_{\mu\alpha} g_{\nu\beta} g_{\rho\kappa} g_{\sigma\lambda} \,. \tag{10.6}$$

As discussed around Eq. (10.15), for constant a the Ricci scalar starts at first order in $h_{\mu\nu}$, and the Einstein-Hilbert action contains $\sqrt{-g}\, R$, so we only need g up to first order in $h_{\mu\nu}$. Then at least three of the metric tensors are represented by $\eta_{\mu\nu}$, and we encounter

$$\epsilon^{\mu\nu\rho\sigma} \epsilon^{\alpha\beta\kappa\lambda} \eta_{\mu\alpha} \eta_{\nu\beta} \eta_{\rho\kappa} = -6 \eta^{\sigma\lambda} \,, \tag{10.7}$$

$$\epsilon^{\mu\nu\rho\sigma} \epsilon^{\alpha\beta\kappa\lambda} \eta_{\mu\alpha} \eta_{\nu\beta} \eta_{\rho\kappa} \eta_{\sigma\lambda} = -24 \,. \tag{10.8}$$

It follows that

$$
g \overset{(10.4)}{\underset{(10.6)}{=}} \frac{a^8}{24} \epsilon^{\mu\nu\rho\sigma} \epsilon^{\alpha\beta\kappa\lambda} \left(\eta_{\mu\alpha} + h_{\mu\alpha} \right) \left(\eta_{\nu\beta} + h_{\nu\beta} \right) \left(\eta_{\rho\kappa} + h_{\rho\kappa} \right) \left(\eta_{\sigma\lambda} + h_{\sigma\lambda} \right)
$$

$$
\overset{(10.7)}{\underset{(10.8)}{=}} \frac{a^8}{24} \left(-24 - 4 \times 6\, \eta^{\sigma\lambda} h_{\sigma\lambda} \right) + \mathcal{O}(h_{\mu\nu}^2) \;=\; -a^8 \left(1 + h_\alpha^\alpha \right) + \mathcal{O}(h_{\mu\nu}^2) \, . \tag{10.9}
$$

The square root reads

$$
\sqrt{-g} \overset{(10.9)}{=} a^4 \left(1 + \tfrac{1}{2} h_\alpha^\alpha \right) + \mathcal{O}(h_{\mu\nu}^2) \, . \tag{10.10}
$$

Next, we compute the *Christoffel symbols* from Eq. (1.9),

$$
\Gamma^\rho_{\mu\nu} \overset{(1.9)}{=} \frac{1}{2} g^{\rho\sigma} \left(g_{\sigma\mu,\nu} + g_{\sigma\nu,\mu} - g_{\mu\nu,\sigma} \right)
$$

$$
\overset{(10.4)}{\underset{(10.5)}{=}} \frac{1}{2} \left(\eta^{\rho\sigma} - h^{\rho\sigma} \right) \left(h_{\sigma\mu,\nu} + h_{\sigma\nu,\mu} - h_{\mu\nu,\sigma} \right) + \mathcal{O}(h_{\mu\nu}^3) \, . \tag{10.11}
$$

With their help, the *Ricci scalar* follows from Eqs. (1.20) and (1.28),

$$
R \overset{(1.20)}{\underset{(1.28)}{=}} g^{\mu\nu} \left(\Gamma^\alpha_{\mu\nu,\alpha} - \Gamma^\alpha_{\mu\alpha,\nu} + \Gamma^\beta_{\mu\nu} \Gamma^\alpha_{\beta\alpha} - \Gamma^\beta_{\mu\alpha} \Gamma^\alpha_{\beta\nu} \right)
$$

$$
\overset{(10.5)}{=} a^{-2} \left(\eta^{\mu\nu} - h^{\mu\nu} \right) \left(\Gamma^\alpha_{\mu\nu,\alpha} - \Gamma^\alpha_{\mu\alpha,\nu} + \Gamma^\beta_{\mu\nu} \Gamma^\alpha_{\beta\alpha} - \Gamma^\beta_{\mu\alpha} \Gamma^\alpha_{\beta\nu} \right) + \mathcal{O}(h_{\mu\nu}^3) \, . \tag{10.12}
$$

Let us start at $\mathcal{O}(h_{\mu\nu})$. The Christoffel symbols become

$$
\Gamma^\rho_{\mu\nu} \overset{(10.11)}{=} \frac{1}{2} \left(h^\rho_{\mu,\nu} + h^\rho_{\nu,\mu} - h_{\mu\nu}^{;\rho} \right) + \mathcal{O}(h_{\mu\nu}^2) \, , \tag{10.13}
$$

and then

$$
a^2 R \overset{(10.12)}{=} \eta^{\mu\nu} \left(\Gamma^\alpha_{\mu\nu,\alpha} - \Gamma^\alpha_{\mu\alpha,\nu} \right) + \mathcal{O}(h_{\mu\nu}^2) \tag{10.14}
$$

$$
\overset{(10.13)}{=} \frac{1}{2} \eta^{\mu\nu} \left(\cancel{h^\alpha_{\mu,\nu\alpha}} + h^\alpha_{\nu,\mu\alpha} - h^\alpha_{\mu\nu,\alpha} - \cancel{h^\alpha_{\mu,\alpha\nu}} - h^\alpha_{\alpha,\mu\nu} + h^\alpha_{\mu\alpha,\nu} \right) + \mathcal{O}(h_{\mu\nu}^2)
$$

$$
= \frac{1}{2} \left(h^{\alpha,\nu}_{\nu,\alpha} - h^{\mu,\alpha}_{\mu,\alpha} - h^{\alpha,\mu}_{\alpha,\mu} + h^{\nu,\alpha}_{\alpha,\nu} \right) + \mathcal{O}(h_{\mu\nu}^2)
$$

$$
= h^{\alpha,\nu}_{\nu,\alpha} - h^{\mu,\alpha}_{\mu,\alpha} + \mathcal{O}(h_{\mu\nu}^2) \, . \tag{10.15}
$$

When we multiply this with the square root from Eq. (10.10), we find

$$
\sqrt{-g}\, R \overset{(10.10)}{\underset{(10.15)}{=}} a^2 \left(h^{\alpha,\nu}_{\nu,\alpha} - h^{\mu,\alpha}_{\mu,\alpha} \right) + \mathcal{O}(h_{\mu\nu}^2) \, . \tag{10.16}
$$

However, for constant a, this is a total derivative, and plays no role in Eq. (1.45).

We then proceed to $\mathcal{O}(h_{\mu\nu}^2)$. The first non-trivial term originates when we multiply Eq. (10.15) with the first-order contribution from Eq. (10.10), yielding

$$\sqrt{-g}\,R \quad \underset{(10.15)}{\overset{(10.10)}{\supset}} \quad \frac{1}{2}a^2\,h_\beta^\beta\left(h_{\nu,\alpha}^{\alpha,\nu} - h_{\mu,\alpha}^{\mu,\alpha}\right)$$

$$\overset{\text{IBP}}{=} \frac{1}{2}a^2\left(h_{\beta,\alpha}^\beta h_\mu^{\mu,\alpha} - h_\beta^{\beta,\nu} h_{\nu,\alpha}^\alpha\right). \tag{10.17}$$

Here IBP stands for *integration by parts* (i.e. the omission of total derivatives, leading to boundary terms in Eq. (1.45)), which we employ in order to set the expression in a form in which only single derivatives appear. The second term comes by replacing $\eta^{\mu\nu} \to -h^{\mu\nu}$ in Eq. (10.14), producing

$$\sqrt{-g}\,R \quad \underset{(10.14)}{\overset{(10.5)}{\supset}} \quad -\frac{1}{2}a^2 h^{\mu\nu}\left(h_{\nu,\mu\alpha}^\alpha - h_{\mu\nu,\alpha}^{;\alpha} - h_{\alpha,\mu\nu}^\alpha + h_{\mu\alpha,\nu}^{;\alpha}\right)$$

$$= -\frac{1}{2}a^2 h_\mu^\nu\left(h_{\nu,\alpha}^{\alpha,\mu} - h_{\nu,\alpha}^{\mu,\alpha} - h_{\alpha,\nu}^{\alpha,\mu} + h_{\alpha,\nu}^{\mu,\alpha}\right)$$

$$\overset{\text{IBP}}{=} \frac{1}{2}a^2\left(2\,h_\mu^{\nu,\mu} h_{\nu,\alpha}^\alpha - h_\mu^{\nu,\alpha} h_{\nu,\alpha}^\mu - h_\mu^{\nu,\mu} h_{\alpha,\nu}^\alpha\right), \tag{10.18}$$

where in the last step we have renamed indices, in order to combine identical structures. The third term originates from the contributions quadratic in Christoffel symbols in Eq. (10.12), which after the insertion of Eq. (10.13) give

$$\sqrt{-g}\,R \quad \underset{(10.13)}{\overset{(10.12)}{\supset}} \quad \frac{1}{4}a^2\eta^{\mu\nu}\left[\left(h_{\mu,\nu}^\beta + h_{\nu,\mu}^\beta - h_{\mu\nu}^{;\beta}\right)\left(\cancel{h_{\beta,\alpha}^\alpha} + h_{\alpha,\beta}^\alpha - \cancel{h_{\beta\alpha}^{;\alpha}}\right)\right.$$

$$\left. -\left(h_{\mu,\alpha}^\beta + h_{\alpha,\mu}^\beta - h_{\mu\alpha}^{;\beta}\right)\left(h_{\beta,\nu}^\alpha + h_{\nu,\beta}^\alpha - h_{\beta\nu}^{;\alpha}\right)\right]$$

$$= \frac{1}{4}a^2\left[\left(2h_\mu^{\beta,\mu} - h_\mu^{\mu,\beta}\right)h_{\alpha,\beta}^\alpha - \left(h_{\mu,\alpha}^\beta + h_{\alpha,\mu}^\beta - h_{\mu\alpha}^{;\beta}\right)\left(h_\beta^{\alpha,\mu} + h_{,\beta}^{\alpha\mu} - h_\beta^{\mu,\alpha}\right)\right]$$

$$\overset{\text{IBP}}{=} \frac{1}{4}a^2\left[h_\mu^{\beta,\mu} h_{\alpha,\beta}^\alpha(+2) + h_\mu^{\mu,\beta} h_{\alpha,\beta}^\alpha(-1) + h_\mu^{\beta,\mu} h_{\beta,\alpha}^\alpha(-1 - \cancel{1} - \cancel{1} + \cancel{1} + \cancel{1} - 1)\right.$$

$$\left. + h_{\mu,\alpha}^\beta h_\beta^{\mu,\alpha}(+\cancel{1} - \cancel{1}) + h_{\mu\alpha}^{;\beta} h_{,\beta}^{\mu\alpha}(+1)\right]. \tag{10.19}$$

In principle there is also a fourth term, originating by inserting the $\mathcal{O}(h_{\mu\nu}^2)$ expressions of $\Gamma_{\mu\nu}^\alpha$ and $\Gamma_{\mu\alpha}^\alpha$ in Eq. (10.14), however it is a total derivative, and can therefore be omitted.

Now we can sum together Eqs. (10.17)–(10.19). Relabelling indices, and lowering and raising them where it makes the appearance more elegant (which also permits for us to see that the last two structures in Eq. (10.19) are equivalent), but retaining the order of appearances from Eq. (10.19), we get [1,2]

$$\sqrt{-g}\, R \overset{(10.17)-(10.19)}{=} \frac{1}{4}a^2\Big[\, h^{\mu,\beta}_{\beta}\, h^{\alpha}_{\alpha,\mu}\, \overset{-2}{\overbrace{(-2-2+2)}} + h^{\beta,\mu}_{\beta}\, h^{\alpha}_{\alpha,\mu}\, \overset{+1}{\overbrace{(+2+0-1)}} \tag{10.20}$$

$$+\, h^{\mu,\beta}_{\beta}\, h^{\alpha}_{\mu,\alpha}\, \underbrace{(+0+4-2)}_{+2} + h^{\beta}_{\mu,\alpha}\, h^{\mu,\alpha}_{\beta}\, \underbrace{(+0-2+1)}_{-1}\,\Big] + \mathcal{O}(h^3_{\mu\nu})\;.$$

Here the contributions originate from Eqs. (10.17), (10.18), and (10.19), respectively.

Next we would like to extract the contribution of tensor modes to Eq. (10.20). We now denote the *tensor perturbations* by h^{t}_{ij}, whereby it is good to keep in mind the relation to our previous notation from Eq. (3.40),

$$h^{\mathrm{t}}_{ij} \overset{(3.40)}{\underset{(10.4)}{\equiv}} 2\vartheta^{\mathrm{t}}_{ij}\;. \tag{10.21}$$

Recalling from Eq. (3.45) that $\delta^{ij} h^{\mathrm{t}}_{ij} = 0 = \partial^i h^{\mathrm{t}}_{ij}$, trace parts h^{α}_{α} or divergences $h^{\alpha}_{\mu,\alpha}$ contain no tensor part. Consequently, only the last term of Eq. (10.20) has a tensor part. The *Einstein-Hilbert action* from Eq. (1.45) therefore contains

$$S \overset{(1.45)}{\underset{(10.20)}{\supset}} \frac{1}{32\pi G}\int_{\mathcal{X}} a^2 \left(-\frac{1}{2}\partial_{\alpha}h^{\mathrm{t}}_{ij}\, \partial^{\alpha}h^{\mathrm{t}}_{ij}\right)$$

$$\overset{(10.5)}{\underset{(10.23)}{=}} \int_{\mathcal{X}} a^4 \left(-\frac{g^{\mu\nu}}{2}\partial_{\mu}\mathrm{h}^{\mathrm{t}}_{ij}\, \partial_{\nu}\mathrm{h}^{\mathrm{t}}_{ij}\right) + \mathcal{O}(h^3_{\mu\nu})\;. \tag{10.22}$$

In the second step we have rescaled

$$h^{\mathrm{t}}_{ij} \equiv \sqrt{32\pi G}\,\mathrm{h}^{\mathrm{t}}_{ij}\;, \tag{10.23}$$

where $\mathrm{h}^{\mathrm{t}}_{ij}$ has the canonical dimension of a bosonic field, for us measured in GeV.

We observe that $\mathrm{h}^{\mathrm{t}}_{ij}$ appears in exactly the same way in Eq. (10.22) as φ appears in Eq. (1.45), provided that the potential $V(\varphi)$ is omitted. Therefore, we can take over results from a massless minimally coupled scalar field, and apply them to $\mathrm{h}^{\mathrm{t}}_{ij}$, being only careful with the counting of the independent degrees of freedom. In particular, the scalar field energy density is given by T_{00} in Eq. (7.108), so that

$$\boxed{\; e_{\mathrm{gw}} \overset{(10.22)}{\underset{(7.108)}{\approx}} \sum_{ij} \frac{1}{2a^2}\big[(\mathrm{h}^{\mathrm{t}\prime}_{ij})^2 + |\nabla\mathrm{h}^{\mathrm{t}}_{ij}|^2\big] \overset{(10.23)}{\underset{\substack{\text{oscillation}\\\text{average}}}{\simeq}} \sum_{ij} \frac{\langle(h^{\mathrm{t}\prime}_{ij})^2\rangle}{32\pi G a^2}\;. \;} \tag{10.24}$$

It is customary to take the second step, but it requires additional assumptions, namely that h^{t}_{ij} has the form of a plane wave, where spatial gradients and time derivatives contribute equally on average; and that we do take that average. Instead, the first form requires no average, given that oscillatory terms cancel for waves: $[\partial_{\tau}\sin(k\tau)]^2 + k^2\sin^2(k\tau) = k^2$.

10.2　Gravitational Waves from Vacuum Fluctuations

We have seen in Sect. 10.1 that in local Minkowskian coordinates, tensor perturbations have, up to overall normalization, the same action as a free (minimally coupled) massless scalar field (cf. Eq. (10.22)). Moreover, the Einstein equation satisfied by tensor perturbations, in Eq. (3.142), has the same form as that for scalar perturbations, in Eq. (3.114), provided that we remove the potential V and the metric perturbations h_0, h_{D} and h from the latter. This analogy implies that we can take over parts of the discussion from Chap. 5, and derive the spectrum of tensor perturbations produced by vacuum fluctuations.

Before we proceed, there is however one difference to scalar perturbations that is worth anticipating. While for the scalar perturbations that affect the CMB and large-scale structure, we are interested in very small momenta or frequencies, which did exit the Hubble horizon at some point, and re-entered only relatively late in the history of the universe (cf. Sect. 2.4), for tensor perturbations, future experiments give access also to much larger momenta and frequencies (cf. Sect. 10.8). Such modes may have never crossed outside of the Hubble horizon. In this case, the way we treat them differs from that for scalar perturbations (cf. Sect. 10.7). In the remainder of the present section, we assume that the modes did exit the Hubble horizon.

Let us next clarify the role of polarization (or *helicity*) in tensor perturbations. The *projector to tensor perturbations* has been given in Eq. (3.24), and we now rewrite it in momentum space,

$$
\mathbb{T}^{mn}_{ij}(\mathbf{k}) \overset{(3.24)}{\underset{\mathbf{x}\,\to\,\mathbf{k}}{\equiv}} \frac{1}{2}\left(\mathbb{K}^m_i \mathbb{K}^n_j + \mathbb{K}^n_i \mathbb{K}^m_j - \mathbb{K}_{ij}\mathbb{K}^{mn} \right), \quad \mathbb{K}_{ij} \equiv \delta_{ij} - \frac{k_i k_j}{\mathbf{k}^2}.
$$
(10.25)

If we consider the vector space of symmetric 3×3 matrices, its projection by Eq. (10.25) defines a two-dimensional invariant subspace. We can imagine finding an orthonormal basis in this subspace, spanned by the *polarization vectors* $\epsilon^\lambda_{ij}(\mathbf{k})$, $\lambda \in \{+, \times\}$. It is not necessary for us to give an explicit representation of these vectors, but we need their orthonormality and completeness relations. The basis vectors can be complex, and it is convenient to reverse the index positions for complex conjugates, so that

$$
\sum_{ij} \epsilon^\lambda_{ij}(\mathbf{k})\,[\epsilon^{ij}_{\lambda'}(\mathbf{k})]^* = \delta^\lambda_{\lambda'}, \quad \sum_\lambda \epsilon^\lambda_{ij}(\mathbf{k})\,[\epsilon^{mn}_\lambda(\mathbf{k})]^* = \mathbb{T}^{mn}_{ij}(\mathbf{k}).
$$
(10.26)

This implies that a general tensor perturbation can be written as

$$
h^{\mathrm{t}}_{ij} = \sum_{mn} \mathbb{T}^{mn}_{ij} h^{\mathrm{t}}_{mn} \overset{(10.26)}{=} \sum_\lambda \epsilon^\lambda_{ij} \sum_{mn} (\epsilon^{mn}_\lambda)^* h^{\mathrm{t}}_{mn} \equiv \sum_\lambda \epsilon^\lambda_{ij}\, h^{\mathrm{t}}_\lambda.
$$
(10.27)

A sum over i, j like in Eqs. (10.22) and (10.24) is, after going to momentum space, equivalent to a sum over λ,

$$\sum_{ij} h_{ij}^{\mathrm{t}} h_{\mathrm{t}}^{ij*} \overset{(10.26)}{\underset{(10.27)}{=}} \sum_{\lambda} h_{\lambda}^{\mathrm{t}} h_{\mathrm{t}}^{\lambda*} \, . \tag{10.28}$$

We have included complex conjugates because relations such as Eq. (2.51) imply that, in the presence of translational invariance or volume average, a local product of real quantities turns into a product of a Fourier transform and its complex conjugate.

The *power spectrum of tensor perturbations* is now defined as

$$\mathcal{P}_{\mathrm{t}}(k) \equiv \sum_{\lambda} \mathcal{P}_{h_{\lambda}^{\mathrm{t}}}(k) \, . \tag{10.29}$$

Given that h_{λ}^{t} satisfies the same equation, the power spectrum $\mathcal{P}_{h_{\lambda}^{\mathrm{t}}}$ is related to the one of a massless scalar field, $\mathcal{P}_{\mathcal{Q}_{\varphi}}$, which can be obtained from Eq. (6.71),

$$\mathcal{P}_{\mathcal{Q}_{\varphi}}(k) \overset{(6.71)}{\approx} \left(\frac{H_*}{2\pi}\right)^2 \bigg|_{H_*=k/a_*} \, . \tag{10.30}$$

The subscript $(\ldots)_*$ denotes the time at which the horizon-crossing condition $H_* = k/a_*$ is satisfied. We also recall that a massless scalar field corresponds to the rescaled $\mathrm{h}_{ij}^{\mathrm{t}}$, related to h_{ij}^{t} via Eq. (10.23). Putting all these ingredients together yields

$$\mathcal{P}_{\mathrm{t}}(k) \approx \underbrace{2}_{\sum_{\lambda}} \times \underbrace{32\pi G}_{(10.23)} \times \underbrace{\left(\frac{H_*}{2\pi}\right)^2 \bigg|_{H_*=k/a_*}}_{(10.30)} = \frac{16}{\pi}\left(\frac{H_*}{m_{\mathrm{pl}}}\right)^2 \bigg|_{H_*=k/a_*} \, . \tag{10.31}$$

This applies to times during which the modes are outside of the Hubble horizon ($t_{\mathrm{out}} > t_*$). Remarkably, the amplitude of the tensor power spectrum directly probes the magnitude of the Hubble rate in Planck units [3–6].

At small frequencies, the tensor power spectrum can be tested by the effect that it has on CMB polarization (cf. Sect. 2.3). No definite sign of primordial B-mode polarization has been observed, which sets an upper bound on the magnitude of the tensor power spectrum, as has been discussed around Eq. (2.17). A theoretical prediction for the *tensor-to-scalar ratio*, r, can be obtained by combining Eqs. (5.54) and (10.31),

$$r_{0.002} \overset{(2.17)}{\underset{\mathcal{R}_T \approx \mathcal{R}_\varphi}{=}} \frac{\mathcal{P}_{\mathrm{t}}(0.04\,k_*)}{\mathcal{P}_{\mathcal{R}_\varphi}(0.04\,k_*)} \overset{(10.31)}{\underset{(5.54)}{\approx}} \pi G\left(\frac{8\dot{\varphi}_*}{H_*}\right)^2 \overset{(2.17)}{<} 0.038 \, . \tag{10.32}$$

In terms of the slow-roll parameters, this becomes

$$r_{0.002} \approx \frac{64\pi G \dot{\varphi}_*^2}{H_*^2} \overset{(6.18)}{\approx} \frac{64\pi G V_{,\varphi}^2}{9H_*^4} \overset{(6.19)}{\approx} 16\left(\frac{8\pi G V}{3H_*^2}\right)^2 \epsilon_V \overset{(6.20)}{\approx} 16\epsilon_V \, . \tag{10.33}$$

A comparison with the numerical value from Eq. (10.32) then indicates that the slow-roll parameter ϵ_V needs to be small indeed, $\epsilon_V \lesssim 0.003$. For the *tensor tilt*, defined according to Eq. (2.16), a computation like in Eqs. (5.56)–(5.58) yields

$$n_{\rm t} \underset{(5.58),\,(10.31)}{\overset{(2.16)}{\approx}} \frac{2H_*}{H_*^2 + \dot{H}_*}\frac{\dot{H}_*}{H_*} \overset{(6.21)}{\approx} -2\epsilon_V \,. \tag{10.34}$$

The small value of $r_{0.002}$ thus guarantees the smallness of the tensor tilt.

10.3 Gravitational Waves from a Matter Energy-Momentum Tensor

Apart from vacuum fluctuations, gravitational waves can also be generated later on, for instance from hydrodynamic fluctuations according to the evolution equation (3.142). Re-phrasing the latter in terms of $h_{ij}^{\rm t}$, via Eq. (10.21), we are faced with

$$\left[\partial_\tau^2 + \left(2\mathcal{H} + 16\pi G a\eta\right)\partial_\tau - \nabla^2\right]h_{ij}^{\rm t} \underset{(10.21)}{\overset{(3.142)}{=}} 16\pi G a^2 S_{ij}^{\rm t} \,, \tag{10.35}$$

where η is the shear viscosity. This is an inhomogeneous linear partial differential equation. Even if the physical interpretation is different, formally the problem is similar to that considered for scalar perturbations in Sect. 7.3, with the role of the thermal noise, ϱ, now taken over by hydrodynamic fluctuations, $S_{ij}^{\rm t}$ (cf. Eq. (3.141)). The method of solution, with a Green's function, is also analogous.

The friction term in Eq. (10.35), $16\pi G a\eta$, was estimated in Eq. (3.143), and argued to be usually small compared with the Hubble rate. In order to permit for analytic manipulations, the friction will be omitted in the following. Nevertheless, it turns out that there is a particular epoch, at low temperatures in a late universe, where the matter-induced damping of gravitational waves does play a significant role, and we return to this below Eq. (10.143).

With the reservations mentioned, we reinstate a general energy-momentum tensor on the right-hand side of the gravitational-wave equation, considering then

$$(\partial_\tau^2 + 2\mathcal{H}\partial_\tau - \nabla^2)h_{ij}^{\rm t} \underset{(10.21)}{\overset{(3.106)}{=}} 16\pi G a^2 \Pi_{ij}^{\rm t} \,. \tag{10.36}$$

We stress that the right-hand side can depend on the unknown function $h_{ij}^{\rm t}$: the movement of the matter that generates anisotropic stress is influenced by the gravitational background in which it propagates. When this is ignored, we are working at zeroth order in the sense of linear-response theory (cf. Sect. 7.2).

In order to solve for a power spectrum, we go to momentum space in spatial directions, and introduce a *retarded Green's function* (cf. Eqs. (2.56) and (7.16)).

The Green's function satisfies

$$(\partial_\tau^2 + 2\mathcal{H}\partial_\tau + k^2)G_{\text{R}}(\tau, \tau_1, k) \equiv \delta(\tau - \tau_1) \,, \tag{10.37}$$

$$G_{\text{R}}(\tau, \tau_1, k) \equiv 0 \quad \text{for} \quad \tau < \tau_1 \,, \tag{10.38}$$

$$\lim_{\tau \to \tau_1^+} \partial_\tau G_{\text{R}}(\tau, \tau_1, k) = 1 \,, \tag{10.39}$$

where Eqs. (10.37) and (10.38) serve as definitions, and Eq. (10.39) follows from them, by integrating over the Dirac-δ. The formal solution of Eq. (10.36) then reads

$$h_{ij}^{\text{t}}(\tau, k) = h_{ij}^{\text{t}}(\tau, k)|_{\text{vac}} + 16\pi G \int_{-\infty}^{\tau} \mathrm{d}\tau_1 \, G_{\text{R}}(\tau, \tau_1, k) \, a^2(\tau_1) \Pi_{ij}^{\text{t}}(\tau_1, k) \,, \tag{10.40}$$

where $h_{ij}^{\text{t}}(\tau, k)|_{\text{vac}}$ is a general solution of the homogeneous equation. We consider the integration constants to be fixed like for quantum-mechanical mode functions, so that the power spectrum from the vacuum part corresponds to that in Eq. (10.31).

To make contact with the literature, we recall that, by rescaling variables, the basic equations can be set in different forms. For instance, if we define $\widehat{h}_{ij}^{\text{t}} \equiv a h_{ij}^{\text{t}}$, then we can read off from Eqs. (5.5) to (5.7) that Eq. (10.36) turns into

$$\left(\partial_\tau^2 - \nabla^2 - \frac{a''}{a}\right)\widehat{h}_{ij}^{\text{t}} \underset{(5.5)-(5.7)}{\overset{(10.36)}{=}} 16\pi G a^3 \Pi_{ij}^{\text{t}} \,. \tag{10.41}$$

As explained around Eq. (6.42), this form has the benefit that the Green's function can be expressed in a simple way in terms of normalized solutions of the homogeneous equation. Moreover, as will be discussed around Eqs. (10.95) and (10.96), (10.41) gets simplified in a radiation-dominated universe, given that a''/a vanishes in the limit that the plasma can be approximated as a collection of non-interacting massless particles. On the other hand, the un-rescaled form in Eq. (10.37) has its own advantages, notably that the small-k asymptotics of its solution is transparent. This will be important for us in the following, so we stick to Eqs. (10.37) and (10.40).

With Eq. (10.40), we can write down an expression for the tensor power spectrum, generalizing on Eq. (10.31). The vacuum part remains intact, as long as the modes are outside of the Hubble horizon in a certain time window, from which we choose a representative, denoted by τ_{out}. Manipulating the Green's function part like in Eq. (2.59) (and omitting mixed terms since they average to zero), the late-time tensor power spectrum becomes

$$\mathcal{P}_{\text{t}}(\tau, k) \underset{(10.31),(10.139)}{\overset{(2.59),(10.40)}{\approx}} X_{\text{t}}^2(\tau, \tau_{\text{out}}, k) \frac{16}{\pi}\left(\frac{H_*}{m_{\text{pl}}}\right)^2\Bigg|_{H_*=k/a_*}$$

$$+ \frac{k^3}{2\pi^2}(16\pi G)^2 \int_{-\infty}^{\tau} \mathrm{d}\tau_1 \int_{-\infty}^{\tau} \mathrm{d}\tau_2 \, G_{\text{R}}(\tau, \tau_1, k) \, G_{\text{R}}(\tau, \tau_2, k)$$

$$\times a^2(\tau_1)a^2(\tau_2) \, P_\Pi(\tau_1, \tau_2, \mathbf{k}) \,, \tag{10.42}$$

where $X_t^2(\tau, \tau_{out}, k)$ from Eq. (10.139) is a transfer function in the tensor channel (this is discussed at some length in Sect. 10.7). The energy-momentum correlator reads

$$P_\Pi(\tau_1, \tau_2, \mathbf{k}) \stackrel{(2.49)}{\equiv} \int d^3\mathbf{x}\, e^{-i\mathbf{k}\cdot\mathbf{x}} \langle \Pi_{ij}^t(\tau_1, \mathbf{x}) \Pi_{ij}^t(\tau_2, \mathbf{0}) \rangle$$

$$\stackrel{(10.25)}{=} \mathbb{T}_{ij}^{mn}(\mathbf{k}) \int d^3\mathbf{x}\, e^{-i\mathbf{k}\cdot\mathbf{x}} \langle \Pi_{ij}(\tau_1, \mathbf{x}) \Pi_{mn}(\tau_2, \mathbf{0}) \rangle \,. \quad (10.43)$$

Usually this correlator has some symmetry properties. For instance, assuming parity and translation invariance of the ensemble average, we can write

$$\langle \Pi_{ij}^t(\tau_1, \mathbf{x}) \Pi_{ij}^t(\tau_2, \mathbf{0}) \rangle \stackrel{\text{parity}}{=} \langle \Pi_{ij}^t(\tau_1, -\mathbf{x}) \Pi_{ij}^t(\tau_2, \mathbf{0}) \rangle$$

$$\stackrel{\text{translation}}{=} \langle \Pi_{ij}^t(\tau_1, \mathbf{y} - \mathbf{x}) \Pi_{ij}^t(\tau_2, \mathbf{y}) \rangle$$

$$\stackrel{\mathbf{y} \to \mathbf{x}}{=} \langle \Pi_{ij}^t(\tau_1, \mathbf{0}) \Pi_{ij}^t(\tau_2, \mathbf{x}) \rangle \,. \quad (10.44)$$

In the classical approximation, operators commute, implying that

$$P_\Pi(\tau_2, \tau_1, \mathbf{k}) \underset{\text{classical limit}}{\overset{(10.43),\,(10.44)}{\approx}} P_\Pi(\tau_1, \tau_2, \mathbf{k}) \,. \quad (10.45)$$

Let us now ask how Eq. (10.42) behaves as a function of k. In particular, we can investigate the small-k asymptotics (the large-k limit will be discussed in Sect. 10.6). As far as P_Π from Eq. (10.43) goes, if we consider very large distances, we are in a spacelike domain, and then causality dictates that correlations decay fast (in the tensor channel, there is no disconnected term like Eq. (2.52)). Therefore we can assume that the Fourier transform remains finite for $\mathbf{k} \to \mathbf{0}$.

The key is then to determine what happens with the Green's function. If we set $k \to 0$ in Eqs. (10.37)–(10.39), an explicit solution can be found,

$$G_R(\tau, \tau_i, k) \xrightarrow[k \to 0]{\tau \geq \tau_i} a^2(\tau_i) \int_{\tau_i}^{\tau} \frac{d\tau'}{a^2(\tau')} \,. \quad (10.46)$$

This demonstrates that the small-k limit of the Green's function is finite.

Inspecting now Eq. (10.42), we can draw an interesting conclusion. The first part, the vacuum contribution, is to a good approximation independent of k. Instead, the second part grows as k^3 at small k according to the discussion above, because we can set $k \to 0$ within the integrand (in G_R and P_Π) without meeting a singularity. So, if there is any anisotropic stress in the universe, this produces a *growing spectrum of gravitational waves*. Of course, the growth cannot continue forever, and we return to this in Sect. 10.6.

Finally, let us make Eq. (10.42) more concrete, by assuming that after it has heated up, the universe contains a thermalized plasma (cf. Chap. 7). In such a plasma, *hydrodynamic fluctuations* take place, as has been discussed in Sect. 3.7. In particular, the contribution to P_Π originates from the fluctuations in Eq. (3.141), as shown by Eq. (10.35). We also need to recall the relation between momentum space correlators from Eq. (2.49), which eliminates $(2\pi)^3 \delta^{(3)}(\mathbf{k} + \mathbf{q})$ from Eq. (3.141), yielding then

$$P_\Pi(\tau_1, \tau_2, \mathbf{k}) \underset{(3.141),(2.49)}{\overset{(10.35),\,(10.43)}{\approx}} \mathbb{T}_{ij}^{mn}(\mathbf{k}) \frac{2\,T\,\delta(\tau_1 - \tau_2)}{a^2(\tau_1)a^2(\tau_2)} \left[\eta\left(\delta_{im}\delta_{jn} + \delta_{in}\delta_{jm}\right) + \left(\zeta - \frac{2\eta}{3}\right)\delta_{ij}\delta_{mn} \right]$$

$$\overset{(10.25)}{=} \frac{8\,T\eta\,\delta(\tau_1 - \tau_2)}{a^2(\tau_1)\,a^2(\tau_2)}\,, \tag{10.47}$$

where η is the shear viscosity. We have shown time arguments for a (suppressing them for T and η), to make it clear that, when the expression is inserted in Eq. (10.42), the factor $a^2(\tau_1)\,a^2(\tau_2)$ cancels, and we get

$$\boxed{\begin{aligned}
\mathcal{P}_{\mathrm{t}}(\tau, k) \underset{(10.47)}{\overset{(10.42)}{\supset}} &\; X_{\mathrm{t}}^2(\tau, \tau_{\mathrm{out}}, k) \frac{16}{\pi}\left(\frac{H_*}{m_{\mathrm{pl}}}\right)^2 \Bigg|_{H_*=k/a_*} \\
&+ \frac{32^2 k^3}{m_{\mathrm{pl}}^4} \int_{-\infty}^{\tau} \mathrm{d}\tau_1\, G_{\mathrm{R}}^2(\tau, \tau_1, k)\, T(\tau_1)\eta(\tau_1)\,.
\end{aligned}} \tag{10.48}$$

We note that $H_*^2 \approx 8\pi e_*/(3m_{\mathrm{pl}}^2)$, cf. Eq. (1.43), so both the vacuum contribution on the first line of Eq. (10.48), and the hydrodynamic contribution on the second line, are formally of $\mathcal{O}(1/m_{\mathrm{pl}}^4)$, even if they display very different dependences on k. We return to Eq. (10.48) around Eq. (10.164), re-phrasing the result in terms of measurable quantities.

10.4   Scalar-Induced Gravitational Waves

In Eq. (10.48), we have given a prediction for the gravitational wave power spectrum originating from vacuum and hydrodynamic fluctuations. However, the most significant contributions might not originate from such expected 'equilibrium' sources, but rather from more peculiar phenomena, active perhaps only for a short period of time, but undergoing violent 'non-equilibrium' dynamics during that instance (cf., e.g. Ref. [7]). These phenomena are in general model-dependent, and thus speculative. In the present section, we discuss one generic source, which originates from the curvature perturbations that we have discussed in the previous chapters, and is therefore quite likely to be present, even if the amplitude of curvature perturbations at momenta larger than probed by CMB and structure formation, is not strongly constrained at the time of writing.

Let us consider Eq. (1.64) complemented by the first-order-in-gradients viscous corrections from Eq. (3.124),

$$T_{\mu\nu} \overset{(1.64)}{\underset{(3.124)}{\equiv}} \varphi_{,\mu}\varphi_{,\nu} - \frac{g_{\mu\nu}\,\varphi_{,\alpha}\varphi^{,\alpha}}{2} + (e + p)\,u_\mu u_\nu + p\,g_{\mu\nu} + \overbrace{a^2 \Pi_{\mu\nu}}^{\text{involves gradients}}. \tag{10.49}$$

At *linear order in perturbations*, the spatial part T_{ij} has two tensor components. One originates from $g_{ij} \supset 2a^2 \vartheta_{ij}^{\text{t}}$ (cf. Eq. (3.40)), which multiplies the second and fourth terms, but this cancels after the use of background identities, cf. the discussion leading up to Eq. (3.106). The other originates from Π_{ij} (cf. Eqs. (3.134) and (3.141)). The latter led to Eq. (10.48). However, we can talk about viscous hydrodynamics only under special conditions (cf. Sect. 7.1). This prompts us to ask how else anisotropic stress could arise?

One definite source for anisotropic stress is found by going up to *second order in perturbations*. Restricting to scalar perturbations on the matter side (although second-order effects from vector and tensor perturbations can also be considered), let us insert

$$\varphi \overset{(3.80)}{=} \bar{\varphi} + \delta\varphi \, , \quad u_\mu \overset{(3.72)}{\underset{(3.76)}{=}} a(-1 - h_0, h_{,i} - v_{,i}) + \mathcal{O}(\delta^2) \tag{10.50}$$

into Eq. (10.49). Recalling the metric from Eqs. (3.40) and (3.41), and omitting the parts proportional to δ_{ij}, which have no tensor component, this yields

$$\varphi_{,i}\,\varphi_{,j} \overset{(10.50)}{=} \delta\varphi_{,i}\,\delta\varphi_{,j} \, , \tag{10.51}$$

$$-\frac{g_{ij}\,\varphi_{,\alpha}\varphi^{,\alpha}}{2} \overset{(3.40)}{\underset{(3.85)}{\supset}} \vartheta_{ij}\Big[(1 - 2h_0)(\bar{\varphi}')^2 + 2\bar{\varphi}'\delta\varphi' \Big] + \mathcal{O}(\delta^3) \, , \tag{10.52}$$

$$(e + p)\,u_i u_j \overset{(10.50)}{=} a^2(\bar{e} + \bar{p})(h - v)_{,i}(h - v)_{,j} + \mathcal{O}(\delta^3) \, , \tag{10.53}$$

$$p\,g_{ij} \overset{(3.40)}{\underset{(3.65)}{\supset}} 2a^2 \vartheta_{ij}\big(\bar{p} + \delta p\big) \, . \tag{10.54}$$

We note that, within our definitions of ϑ_{ij}, δp and $\delta\varphi$ in Eqs. (3.40), (3.65), and (3.80), no perturbative expansion has been made in Eqs. (10.51) and (10.54).

Now, an important point is that second-order perturbations, notably those in Eqs. (10.51) and (10.53), by themselves *generate* anisotropic stress, as we will see in the remainder of this section. So, it can be dangerous to include simultaneously the first-order $a^2 \Pi_{ij}$ from Eq. (10.49), as this could lead to double counting. This issue will manifest itself concretely below, however for the moment we keep both contributions in the equations.

To proceed, we slightly deviate from our previous philosophy of working in a general gauge, and instead do a *partial* gauge fixing. This simplifies the derivation of the Einstein tensor, where we have to go up to second order in perturbations. A complete derivation, without any gauge fixing and with multiple fluids, can be found in Ref. [8].

In the language of Eqs. (4.49)–(4.54), we choose a partially Newtonian gauge,

$$\xi = -\vartheta \overset{(4.52)}{\Longrightarrow} \tilde{\vartheta} = 0 \,. \tag{10.55}$$

The gauge parameter ξ^0 is left free, permitting for a check of gauge dependence. The gauge-fixed quantities are denoted by $\tilde{h}_0$, $\tilde{h}_{\mathrm{D}}$, $\tilde{h}$, $\tilde{v}$, and $\delta\tilde{\varphi}$. With this choice, the metric from Eq. (3.40) becomes

$$g_{\mu\nu} \overset{(3.40)}{\underset{(10.55)}{=}} a^2 \begin{pmatrix} -1 - 2\tilde{h}_0 & \tilde{h}_{,j} \\ \tilde{h}_{,i} & (1 - 2\tilde{h}_{\mathrm{D}})\,\delta_{ij} + 2\vartheta^{\mathrm{t}}_{ij} \end{pmatrix} \,. \tag{10.56}$$

Here vector perturbations have been omitted.

The first task is to invert the metric. For a matrix that is of the form

$$g_{\mu\nu} = a^2 \left(\eta + A \right) \,, \quad \eta \equiv \begin{pmatrix} -1 & 0 \\ 0 & \mathbb{1} \end{pmatrix} \,, \tag{10.57}$$

the inverse reads

$$g^{\mu\nu} = \frac{1}{a^2} \left(\eta - \eta A \eta + \eta A \eta A \eta + \dots \right) \,. \tag{10.58}$$

This yields

$$g^{\mu\nu} \overset{(10.56)}{\underset{(10.58)}{=}} \frac{1}{a^2} \begin{pmatrix} -1 + 2\tilde{h}_0 - 4\tilde{h}_0^2 + \tilde{h}_{,k}\tilde{h}_{,k} & \tilde{h}_{,j}\,[1 + 2(\tilde{h}_{\mathrm{D}} - \tilde{h}_0)] \\ \tilde{h}_{,i}\,[1 + 2(\tilde{h}_{\mathrm{D}} - \tilde{h}_0)] & (1 + 2\tilde{h}_{\mathrm{D}} + 4\tilde{h}_{\mathrm{D}}^2)\,\delta_{ij} - \tilde{h}_{,i}\tilde{h}_{,j} - 2\vartheta^{\mathrm{t}}_{ij} \end{pmatrix} + \mathcal{O}(\delta^3) \,, \tag{10.59}$$

where $\vartheta^{\mathrm{t}}_{ij}$ has been counted as being of $\mathcal{O}(\delta^2)$, so that only its first-order appearance (without $\tilde{h}_0$, $\tilde{h}_{\mathrm{D}}$, $\tilde{h}$) has been retained.

Subsequently, we insert Eqs. (10.56) and (10.59) in (1.9), in order to determine the Christoffel symbols $\Gamma^{\rho}_{\mu\nu}$; then in Eq. (1.20), for the Ricci tensor $R_{\mu\nu}$; in Eq. (1.28), for the Ricci scalar R; and in Eq. (1.29), for the Einstein tensor. Given that the equations become lengthy, we implement them with a symbolic script, as illustrated in Sect. 10.9. For the spatial components $i \neq j$, contributing to the tensor channel, we find

$$G_{ij} \overset{i \neq j}{\supset} \overbrace{(\partial_\tau^2 + 2\mathcal{H}\partial_\tau - \nabla^2)\vartheta^{\mathrm{t}}_{ij} - 2(2\mathcal{H}' + \mathcal{H}^2)\vartheta^{\mathrm{t}}_{ij}}^{\text{from (3.105)}} - \overbrace{\left(\tilde{h}_0 - \tilde{h}_{\mathrm{D}} + \tilde{h}' + 2\mathcal{H}\tilde{h} \right)_{,ij}}^{\text{from the } i \neq j \text{ part of (3.91)}}$$

$$+ \tilde{h}_{0,i}\tilde{h}_{0,j} + 3\tilde{h}_{\mathrm{D},i}\tilde{h}_{\mathrm{D},j} + 2(\tilde{h}_0\tilde{h}_{0,ij} + \tilde{h}_{\mathrm{D}}\tilde{h}_{\mathrm{D},ij}) - (\tilde{h}_{0,i}\tilde{h}_{\mathrm{D},j} + \tilde{h}_{0,j}\tilde{h}_{\mathrm{D},i})$$

$$+ (\tilde{h}_0' + \tilde{h}_{\mathrm{D}}' + 4\mathcal{H}\tilde{h}_0)\tilde{h}_{,ij} + 2\tilde{h}_0\tilde{h}_{,ij}'$$

$$- (\partial_\tau + 2\mathcal{H})(\tilde{h}_{\mathrm{D},i}\tilde{h}_{,j} + \tilde{h}_{\mathrm{D},j}\tilde{h}_{,i}) + \tilde{h}_{,kk}\tilde{h}_{,ij} - \tilde{h}_{,ki}\tilde{h}_{,kj} + \mathcal{O}(\delta^3) \,. \tag{10.60}$$

Next, we go over to gauge-invariant variables. The Bardeen potentials from Eqs. (4.30) and (4.31) become

$$\phi \overset{(4.30)}{\underset{(10.55)}{=}} \tilde{h}_0 + \tilde{h}' + \mathcal{H}\tilde{h} \,, \tag{10.61}$$

$$\psi \overset{(4.31)}{\underset{(10.55)}{=}} \tilde{h}_{\mathrm{D}} - \mathcal{H}\tilde{h} \,. \tag{10.62}$$

Therefore we can substitute

$$\tilde{h}_0 \overset{(10.61)}{=} \phi - \tilde{h}' - \mathcal{H}\tilde{h} \qquad \Rightarrow \qquad \tilde{h}_0' = \phi' - \tilde{h}'' - \mathcal{H}'\tilde{h} - \mathcal{H}\tilde{h}' \,, \tag{10.63}$$

$$\tilde{h}_{\mathrm{D}} \overset{(10.62)}{=} \psi + \mathcal{H}\tilde{h} \qquad \Rightarrow \qquad \tilde{h}_{\mathrm{D}}' = \psi' + \mathcal{H}'\tilde{h} + \mathcal{H}\tilde{h}' \,. \tag{10.64}$$

Inserting these into Eq. (10.60), we get

$$G_{ij} \overset{(10.60)}{\underset{(10.63),(10.64)}{\supset}} (\partial_\tau^2 + 2\mathcal{H}\partial_\tau - \nabla^2)\vartheta_{ij}^{\mathrm{t}} - 2(2\mathcal{H}' + \mathcal{H}^2)\vartheta_{ij}^{\mathrm{t}} + (\psi - \phi)_{,ij}$$

$$+ \phi_{,i}\phi_{,j} + 3\psi_{,i}\psi_{,j} + 2(\phi\,\phi_{,ij} + \psi\,\psi_{,ij}) - (\phi_{,i}\,\psi_{,j} + \phi_{,j}\,\psi_{,i})$$

$$- (\phi_{,i}\tilde{h}' + \psi'\tilde{h}_{,i})_{,j} - (\phi_{,j}\tilde{h}' + \psi'\tilde{h}_{,j})_{,i} + (\phi' + 3\psi')\,\tilde{h}_{,ij}$$

$$+ 2\mathcal{H}[\,2\phi\,\tilde{h}_{,ij} + (\psi\,\tilde{h} - \phi\,\tilde{h})_{,ij}\,] + 2(\mathcal{H}^2 - \mathcal{H}')\tilde{h}_{,i}\tilde{h}_{,j}$$

$$+ (\tilde{h}_{,kk} - 2\mathcal{H}\tilde{h}' - \tilde{h}'')\,\tilde{h}_{,ij} + \tilde{h}_{,i}'\tilde{h}_{,j}' - \tilde{h}_{,ki}\tilde{h}_{,kj} + \mathcal{O}(\delta^3) \,. \tag{10.65}$$

For gravitational waves, we need to determine the tensor part of Eq. (10.65), denoting the result by G_{ij}^{t}. We recall that the tensor projector from Eq. (3.24) is transverse with respect to spatial derivatives (or, in momentum space, spatial momenta),

$$\mathbb{T}_{ij}^{mn}(\ldots)_{,mn} = \mathbb{T}_{ij}^{mn}(\ldots)_{,m} = 0 \,. \tag{10.66}$$

This implies that we can effectively carry out *integrations by parts* (IBP) in the tensor channel,

$$\mathbb{T}_{ij}^{mn}\phi\,\phi_{,mn} = \mathbb{T}_{ij}^{mn}[\,(\phi\,\phi_{,n})_{,m} - \phi_{,m}\,\phi_{,n}\,] \overset{(10.66)}{=} -\mathbb{T}_{ij}^{mn}\phi_{,m}\,\phi_{,n} \,. \tag{10.67}$$

Furthermore, recalling that we are considering waves, we can envisage taking a time average over an oscillation period, or a spatial average over a wavelength. Given that the background is time-dependent, this has to be implemented in a covariant way, as

$$\langle\, f'\, g'\,\rangle \;\equiv\; \frac{1}{\mathcal{N}}\int_{\mathcal{X}} \overbrace{\sqrt{-\bar{g}}}^{a^4}\; \overbrace{\bar{g}^{00}}^{-1/a^2}\,(-f'\,g') \;=\; \frac{1}{\mathcal{N}}\int_{\mathcal{X}} a^2\,\partial_\tau f\,\partial_\tau g$$

$$=\; \frac{1}{\mathcal{N}}\int_{\mathcal{X}} \Big[\, \overbrace{\partial_\tau(a^2 f\,\partial_\tau g)}^{\text{boundary terms}} -2aa'\, f\,\partial_\tau g - a^2\, f\,\partial_\tau^2 g\,\Big]$$

$$=\; \frac{1}{\mathcal{N}}\int_{\mathcal{X}} \sqrt{-\bar{g}}\,\bar{g}^{00}\, f\left(2\mathcal{H}g' + g''\right) \;+\; \text{(boundary terms)}$$

$$=\; \big\langle\, f\left(-2\mathcal{H}g' - g''\right)\big\rangle \;+\; \text{(boundary terms)}\,, \tag{10.68}$$

where $\mathcal{N}$ is a normalization factor. For spatial derivatives, the same logic gives

$$\langle\, f_{,k}\,g_{,k}\,\rangle \;\equiv\; \frac{1}{\mathcal{N}}\int_{\mathcal{X}} \overbrace{\sqrt{-\bar{g}}}^{a^4}\; \overbrace{\bar{g}^{kl}}^{\delta_{kl}/a^2}\,(f_{,k}\,g_{,l}) \;=\; \frac{1}{\mathcal{N}}\int_{\mathcal{X}} a^2\,\partial_k f\,\partial_k g$$

$$=\; \big\langle\, f\left(-g_{,kk}\right)\big\rangle \;+\; \text{(boundary terms)}\,. \tag{10.69}$$

Omitting the boundary terms, as would be viable for periodic motion, this implies that in the first term on the last line of Eq. (10.65), we can write

$$\big\langle\, (-2\mathcal{H}\tilde{h}' - \tilde{h}'')\,\tilde{h}_{,ij}\,\big\rangle \underset{g\to\tilde{h},\,f\to\tilde{h}_{,ij}}{\overset{(10.68)}{=}} \big\langle\, \tilde{h}'\,\tilde{h}'_{,ij}\,\big\rangle \overset{(10.67)}{\underset{\mathbb{T}^{mn}_{ij}}{=}} \big\langle\, -\tilde{h}'_{,i}\tilde{h}'_{,j}\,\big\rangle\,, \tag{10.70}$$

$$\big\langle\, \tilde{h}_{,kk}\,\tilde{h}_{,ij}\,\big\rangle \underset{g\to\tilde{h},\,f\to\tilde{h}_{,ij}}{\overset{(10.69)}{=}} \big\langle\, -\tilde{h}_{,k}\,\tilde{h}_{,kij}\,\big\rangle \overset{(10.67)}{\underset{\mathbb{T}^{mn}_{ij}}{=}} \big\langle\, \tilde{h}_{,ki}\,\tilde{h}_{,kj}\,\big\rangle\,. \tag{10.71}$$

With these, the last line of Eq. (10.65) drops out. In the following, we denote

$$\{(\dots)_{ij}\}^{\mathrm{t}} \;\equiv\; \mathbb{T}^{mn}_{ij}\langle\dots\rangle_{mn}\,. \tag{10.72}$$

Thereby Eq. (10.65) can be converted into

$$G^{\mathrm{t}}_{ij} \underset{(10.70)-(10.72)}{\overset{(10.65)}{=}} (\partial_\tau^2 + 2\mathcal{H}\partial_\tau - \nabla^2)\vartheta^{\mathrm{t}}_{ij} - 2(2\mathcal{H}' + \mathcal{H}^2)\vartheta^{\mathrm{t}}_{ij}$$

$$+\big\{\, -\phi_{,i}\,\phi_{,j} + \psi_{,i}\,\psi_{,j} - (\phi_{,i}\,\psi_{,j} + \phi_{,j}\,\psi_{,i})$$

$$+ (\phi' + 3\psi' + 4\mathcal{H}\phi)\,\tilde{h}_{,ij} + 2(\mathcal{H}^2 - \mathcal{H}')\tilde{h}_{,i}\tilde{h}_{,j}\,\big\}^{\mathrm{t}} + \mathcal{O}(\delta^3)\,. \tag{10.73}$$

Turning to the right-hand side of the Einstein equations, we express it in terms of curvature perturbations, defined in Eqs. (4.60) and (4.61),

$$\mathcal{R}_\varphi \overset{(4.60)}{\underset{(10.55)}{=}} -\tilde{h}_{\mathrm{D}} - \mathcal{H}\frac{\delta\tilde{\varphi}}{\bar{\varphi}'}\,, \tag{10.74}$$

$$\mathcal{R}_v \overset{(4.61)}{\underset{(10.55)}{=}} -\tilde{h}_{\mathrm{D}} + \mathcal{H}(\tilde{h} - \tilde{v})\,. \tag{10.75}$$

In Eqs. (10.51) and (10.53), we substitute

$$\delta\tilde{\varphi}_{,i}\,\delta\tilde{\varphi}_{,j} \overset{(10.74)}{=} \frac{(\bar{\varphi}')^2}{\mathcal{H}^2}(\mathcal{R}_\varphi + \tilde{h}_{\mathrm{D}})_{,i}(\mathcal{R}_\varphi + \tilde{h}_{\mathrm{D}})_{,j}\,, \tag{10.76}$$

$$(\tilde{h} - \tilde{v})_{,i}(\tilde{h} - \tilde{v})_{,j} \overset{(10.75)}{=} \frac{1}{\mathcal{H}^2}(\mathcal{R}_v + \tilde{h}_{\mathrm{D}})_{,i}(\mathcal{R}_v + \tilde{h}_{\mathrm{D}})_{,j}\,. \tag{10.77}$$

Then the right-hand side of the Einstein equations becomes

$$8\pi G\,T^{\mathrm{t}}_{ij} \overset{(10.49)-(10.54)}{\underset{(10.76),\,(10.77)}{=}} 8\pi G\left\{\left[(\bar{\varphi}')^2 + 2a^2\bar{p}\right]\vartheta^{\mathrm{t}}_{ij} + a^2\Pi^{\mathrm{t}}_{ij}\right\}$$

$$+ \frac{8\pi G}{\mathcal{H}^2}\left\{(\bar{\varphi}')^2\mathcal{R}_{\varphi,i}\mathcal{R}_{\varphi,j} + a^2(\bar{e} + \bar{p})\mathcal{R}_{v,i}\mathcal{R}_{v,j}\right.$$

$$+ [(\bar{\varphi}')^2\mathcal{R}_\varphi + a^2(\bar{e} + \bar{p})\mathcal{R}_v]_{,i}\,\tilde{h}_{\mathrm{D},j} + [(\bar{\varphi}')^2\mathcal{R}_\varphi + a^2(\bar{e} + \bar{p})\mathcal{R}_v]_{,j}\,\tilde{h}_{\mathrm{D},i}$$

$$\left. + [(\bar{\varphi}')^2 + a^2(\bar{e} + \bar{p})]\,\tilde{h}_{\mathrm{D},i}\,\tilde{h}_{\mathrm{D},j}\right\}^{\mathrm{t}} + \mathcal{O}(\delta^3)\,. \tag{10.78}$$

Here, we insert the background identities from Eq. (3.88), as

$$8\pi G\left[(\bar{\varphi}')^2 + 2a^2\bar{p}\right] \overset{(3.88)}{=} -2(2\mathcal{H}' + \mathcal{H}^2)\,, \tag{10.79}$$

$$8\pi G\left[(\bar{\varphi}')^2 + a^2(\bar{e} + \bar{p})\right] \overset{(3.88)}{=} 2(\mathcal{H}^2 - \mathcal{H}')\,. \tag{10.80}$$

For the third line of Eq. (10.78), we need the first-order equation from Eq. (3.93),

$$\mathcal{H}\tilde{h}_0 + \tilde{h}'_{\mathrm{D}} \overset{(3.93)}{\underset{(10.55)}{=}} \overbrace{4\pi G\left[a^2(\bar{e} + \bar{p})(\tilde{v} - \tilde{h}) + \bar{\varphi}'\delta\tilde{\varphi}\right]}^{\text{insert }(10.74),\,(10.75)}$$

$$= -\frac{4\pi G}{\mathcal{H}}\left\{(\bar{\varphi}')^2\mathcal{R}_\varphi + a^2(\bar{e} + \bar{p})\mathcal{R}_v + [\underbrace{(\bar{\varphi}')^2 + a^2(\bar{e} + \bar{p})}_{\text{via (10.80): }(\mathcal{H}^2-\mathcal{H}')/(4\pi G)}]\tilde{h}_{\mathrm{D}}\right\}\,, \tag{10.81}$$

which implies

$$-2\left[\tilde{h}_0 + \frac{\tilde{h}'_{\mathrm{D}}}{\mathcal{H}} + \left(1 - \frac{\mathcal{H}'}{\mathcal{H}^2}\right)\tilde{h}_{\mathrm{D}}\right] \overset{(10.80)}{\underset{(10.81)}{=}} \frac{8\pi G}{\mathcal{H}^2}\left[(\bar{\varphi}')^2\mathcal{R}_\varphi + a^2(\bar{e} + \bar{p})\mathcal{R}_v\right]\,. \tag{10.82}$$

With these substitutions, Eq. (10.78) becomes

$$
8\pi G\, T^{\mathrm{t}}_{ij} \overset{(10.78)}{\underset{(10.82)}{=}} \overbrace{-2(2\mathcal{H}' + \mathcal{H}^2)}^{\text{via (10.79)}} \vartheta^{\mathrm{t}}_{ij} + 8\pi G\, a^2 \Pi^{\mathrm{t}}_{ij}
$$

$$
+ \left\{ \frac{8\pi G}{\mathcal{H}^2} \left[(\bar{\varphi}')^2 \mathcal{R}_{\varphi,i}\mathcal{R}_{\varphi,j} + a^2(\bar{e} + \bar{p})\mathcal{R}_{v,i}\mathcal{R}_{v,j} \right] \right.
$$

$$
- 2\left(\tilde{h}_{0,i}\tilde{h}_{\mathrm{D},j} + \tilde{h}_{0,j}\tilde{h}_{\mathrm{D},i} \right) - \frac{2}{\mathcal{H}}\left(\tilde{h}_{\mathrm{D},i}\tilde{h}'_{\mathrm{D},j} + \tilde{h}_{\mathrm{D},j}\tilde{h}'_{\mathrm{D},i} \right)
$$

$$
\left. + \underbrace{(-4+2)}_{\text{3rd and 4th line of (10.78)}} \left(1 - \frac{\mathcal{H}'}{\mathcal{H}^2}\right) \tilde{h}_{\mathrm{D},i}\,\tilde{h}_{\mathrm{D},j} \right\}^{\mathrm{t}} + \mathcal{O}(\delta^3) . \tag{10.83}
$$

Finally, we express $\tilde{h}_0$ and $\tilde{h}_{\mathrm{D}}$ via ϕ and ψ, through Eqs. (10.63) and (10.64), and make use of IBP relations, from Eq. (10.67). Thereby Eq. (10.83) turns into

$$
8\pi G\, T^{\mathrm{t}}_{ij} \overset{(10.83),\,(10.67)}{\underset{(10.63),\,(10.64)}{=}} -2(2\mathcal{H}' + \mathcal{H}^2)\,\vartheta^{\mathrm{t}}_{ij} + 8\pi G\, a^2 \Pi^{\mathrm{t}}_{ij}
$$

$$
+ \left\{ \frac{8\pi G}{\mathcal{H}^2} \left[(\bar{\varphi}')^2 \mathcal{R}_{\varphi,i}\mathcal{R}_{\varphi,j} + a^2(\bar{e} + \bar{p})\mathcal{R}_{v,i}\mathcal{R}_{v,j} \right] \right.
$$

$$
- 2\left(\phi_{,i}\psi_{,j} + \phi_{,j}\psi_{,i} \right) + 2\left(\frac{\mathcal{H}'}{\mathcal{H}^2} - 1 \right)\psi_{,i}\psi_{,j} - \frac{2}{\mathcal{H}}\left(\psi_{,i}\psi_{,j} \right)'
$$

$$
\left. + 4(\psi' + \mathcal{H}\phi)\,\tilde{h}_{,ij} + 2(\mathcal{H}^2 - \mathcal{H}')\,\tilde{h}_{,i}\tilde{h}_{,j} \right\}^{\mathrm{t}} + \mathcal{O}(\delta^3) . \tag{10.84}
$$

The remaining step is to equate Eqs. (10.73) and (10.84). Transporting the source terms to the right-hand side, we obtain

$$
\boxed{
\begin{aligned}
(\partial_\tau^2 + 2\mathcal{H}\partial_\tau - \nabla^2)\vartheta^{\mathrm{t}}_{ij} &\overset{(10.73)}{\underset{(10.84)}{=}} 8\pi G\, a^2 \Pi^{\mathrm{t}}_{ij} \\
&+ \left\{ \frac{8\pi G}{\mathcal{H}^2} \left[(\bar{\varphi}')^2 \mathcal{R}_{\varphi,i}\mathcal{R}_{\varphi,j} + a^2(\bar{e} + \bar{p})\mathcal{R}_{v,i}\mathcal{R}_{v,j} \right] \right. \\
&\quad + \phi_{,i}\phi_{,j} - (\phi_{,i}\psi_{,j} + \phi_{,j}\psi_{,i}) + \left(\frac{2\mathcal{H}'}{\mathcal{H}^2} - 3 \right)\psi_{,i}\psi_{,j} - \frac{2}{\mathcal{H}}(\psi_{,i}\psi_{,j})' \\
&\quad \left. + (\psi - \phi)'\,\tilde{h}_{,ij} \right\}^{\mathrm{t}} + \mathcal{O}(\delta^3) .
\end{aligned}
}
$$

$$
\tag{10.85}
$$

A significant, but nevertheless incomplete, cancellation took place in the contributions involving $\tilde{h}$ (cf. Eq. (10.56)). In order to obtain physical predictions from Eq. (10.85), usually simplified versions thereof are solved with the Green's function method that we will illustrate in Sect. 10.5. Let us elaborate on the key points related to Eq. (10.85):

(i) Gauge dependence. In much of the literature, a gauge has been fixed ($\tilde{h} = 0$). If this is done from the outset, then one misses the information that the right-hand side of Eq. (10.85) is actually gauge dependent [9]. Gauge dependence disappears, and the result is unambiguous, only if $\psi = \phi$, which according to Eq. (7.98) requires absence of the scalar part of anisotropic stress, Π. The absence of Π alleviates the concerns about double counting that we express at the beginning of this section. If $\Pi = 0$, then Eq. (10.85) can be written in terms of gauge-invariant variables,

$$(\partial_\tau^2 + 2\mathcal{H}\partial_\tau - \nabla^2)\vartheta_{ij}^{\rm t} \overset{(10.85),\,(7.98)}{\underset{\Pi\,=\,0}{=}} 8\pi G\, a^2 \Pi_{ij}^{\rm t} \tag{10.86}$$

$$+\left\{ \frac{8\pi G}{\mathcal{H}^2}\left[(\bar{\varphi}')^2 \mathcal{R}_{\varphi,i}\mathcal{R}_{\varphi,j} + a^2(\bar{e} + \bar{p})\mathcal{R}_{v,i}\mathcal{R}_{v,j} \right] + 2\left(\frac{\mathcal{H}'}{\mathcal{H}^2} - 2 - \frac{\partial_\tau}{\mathcal{H}} \right)\psi_{,i}\,\psi_{,j} \right\}^{\rm t} + \mathcal{O}(\delta^3)\,.$$

(ii) Index placement. Often, instead of G_{ij}, the component $G^i{}_j$ of the Einstein tensor is considered (cf., e.g., Ref. [10]). At the second order, given the metric in Eq. (10.56), the relation to our expression reads

$$G_{ij}^{(2)} = [g_{i\mu}G^\mu{}_j]^{(2)} = g_{i\mu}^{(0)}[G^\mu{}_j]^{(2)} + g_{i\mu}^{(1)}[G^\mu{}_j]^{(1)} + \overbrace{g_{i\mu}^{(2)}}^{\text{via }(10.56)\,\to\,0}[G^\mu{}_j]^{(0)}$$

$$\overset{(10.56)}{=} a^2\left\{ [G^i{}_j]^{(2)} - 2\tilde{h}_{\rm D}[G^i{}_j]^{(1)} + \tilde{h}_{,i}[G^0{}_j]^{(1)} \right\}\,. \tag{10.87}$$

Here, in turn, we can write

$$a^2[G^i{}_j]^{(1)} = a^2[g^{i\mu}G_{\mu j}]^{(1)} = a^2\overbrace{[g^{i\mu}]^{(0)}}^{\text{via }(10.59):\,\delta_{i\mu}/a^2}G_{\mu j}^{(1)} + a^2[g^{i\mu}]^{(1)}\overbrace{G_{\mu j}^{(0)}}^{\text{via }(3.58):\,-(2\mathcal{H}'+\mathcal{H}^2)\delta_{\mu j}}$$

$$= \overbrace{G_{ij}^{(1)}}^{\text{insert }(3.91),\,(10.55)} - (2\mathcal{H}' + \mathcal{H}^2)\overbrace{a^2[g^{ij}]^{(1)}}^{\text{via }(10.59):\,2\tilde{h}_{\rm D}\delta_{ij}}$$

$$\overset{i\neq j}{\supset} (\tilde{h}_{\rm D} - \tilde{h}_0 - \tilde{h}' - 2\mathcal{H}\tilde{h})_{,ij}$$

$$\overset{(10.63)}{\underset{(10.64)}{=}} (\psi - \phi)_{,ij}\,, \tag{10.88}$$

$$a^2[G^0{}_j]^{(1)} = a^2[g^{0\mu}G_{\mu j}]^{(1)} = a^2\overbrace{[g^{0\mu}]^{(0)}}^{\text{via }(10.59):\,-\delta_{0\mu}/a^2}G_{\mu j}^{(1)} + a^2[g^{0\mu}]^{(1)}\overbrace{G_{\mu j}^{(0)}}^{\text{via }(3.58):\,-(2\mathcal{H}'+\mathcal{H}^2)\delta^{\mu j}}$$

$$= \overbrace{-G_{0j}^{(1)}}^{\text{insert }(3.90),\,(10.55)} - (2\mathcal{H}' + \mathcal{H}^2)\overbrace{a^2[g^{0j}]^{(1)}}^{\text{via }(10.59):\,\tilde{h}_{,j}}$$

$$= -2(\tilde{h}'_{\mathrm{D}} + \mathcal{H}\tilde{h}_0)_{,j} + \overrightarrow{(2\mathcal{H}' + \mathcal{H}^2)\tilde{h}_{,j}} - \overrightarrow{(2\mathcal{H}' + \mathcal{H}^2)\tilde{h}_{,j}}$$

$$\overset{(10.63)}{\underset{(10.64)}{=}} -2[\,\psi' + \mathcal{H}\phi + (\mathcal{H}' - \mathcal{H}^2)\tilde{h}\,]_{,j} \, . \tag{10.89}$$

Substituting Eqs. (10.88) and (10.89) in Eq. (10.87), where we write $\tilde{h}_{\mathrm{D}} = \psi + \mathcal{H}\tilde{h}$ from Eq. (10.64), we get

$$a^2 [G^i{}_j]^{(2)} \overset{(10.87)}{\underset{i \neq j}{=}} G^{(2)}_{ij} + 2 \overbrace{(\psi + \mathcal{H}\tilde{h})}^{\text{from (10.64)}} \overbrace{(\psi - \phi)_{,ij}}^{\text{from (10.88)}} + 2\tilde{h}_{,i} \overbrace{[\,\psi' + \mathcal{H}\phi + (\mathcal{H}' - \mathcal{H}^2)\tilde{h}\,]_{,j}}^{\text{from (10.89)}}$$

$$= G^{(2)}_{ij} + 2\psi(\psi - \phi)_{,ij}$$

$$+ 2\mathcal{H}\tilde{h}\,(\psi - \phi)_{,ij} + 2\tilde{h}_{,i}(\psi'_{,j} + \mathcal{H}\phi_{,j}) + 2(\mathcal{H}' - \mathcal{H}^2)\tilde{h}_{,i}\tilde{h}_{,j} \, . \tag{10.90}$$

Adding together the terms on the first line of Eq. (10.90) and on the second line of Eq. (10.65), the result agrees with the corresponding terms given in Eq. (4) of Ref. [11], where the gauge $\tilde{h} = 0$ is chosen, so that the second line of Eq. (10.90) is absent. We remark that if we want to write down the Einstein equations with this index placement, the energy-momentum tensor $a^2 [T^i{}_j]^{(2)}$ needs to be similarly worked out, however the final equation is equivalent to Eq. (10.85), if the same gauge choice is employed.

(iii) Simplifications if one energy component dominates. Depending on the physical setting, the background enthalpy density might be dominated either by the inflaton field, i.e. $(\bar{\varphi}')^2 \gg a^2(\bar{e} + \bar{p})$, or the plasma, i.e. $(\bar{\varphi}')^2 \ll a^2(\bar{e} + \bar{p})$. In these cases, we can express the corresponding curvature perturbation, $\mathcal{R}_\varphi$ or $\mathcal{R}_v$, in terms of the Bardeen potentials, through Eqs. (10.82), (10.63), and (10.64). Then only Bardeen potentials (and gauge artifacts) appear on the right-hand side of Eq. (10.85).

(iv) Remarks on the anisotropic stress, Π^{t}_{ij}. At the beginning of this section, we expressed concerns about anisotropic stress, however in the literature it is sometimes kept and re-expressed in a peculiar way. Suppose that we write $\Pi_{ij} \equiv p\,\pi_{ij}$, and consider a scalar contribution to the latter part, $\pi^{\mathrm{s}}_{ij} \equiv \pi_{,ij} - \delta_{ij}\nabla^2\pi/3$. Since π starts at first order, we may write $\Pi^{(2)}_{ij} \supset (\bar{p} + \delta p)\,\pi^{(1)}_{,ij}$. The term $\bar{p}\,\pi^{(1)}$ is identified as the scalar anisotropic stress Π in Eq. (7.98), whereas the pressure perturbation is written as $\delta p \simeq c_s^2 \,\delta e$, where c_s^2 is the speed of sound squared. Then, formally, $\{\delta e\,\pi^{(1)}_{,ij}\}^{\mathrm{t}} \neq 0$. Concretely, if δe is solved for from Eq. (3.92) (in the absence of φ), and $\bar{p}\,\pi^{(1)} = \Pi$ is solved from Eq. (7.98), this yields yet another structure expressed in terms of Bardeen potentials. However, $\Pi \neq 0$ implies that the result is gauge dependent (see point (i)). It is also unclear to what extent this second-order effect is independent of the other second-order terms in Eq. (10.85).

(v) Example of how double counting can be avoided. If the inflaton field thermalizes to a temperature T, then at small $k/a \ll T$, the contribution from $\delta\varphi_{,i}\delta\varphi_{,j}$ in Eq. (10.51), which has turned into $(\bar{\varphi}')^2\mathcal{R}_{\varphi,i}\mathcal{R}_{\varphi,j}$ in Eq. (10.85), generates a shear viscosity, η. The shear viscosity parametrizes anisotropic stress (cf. Sect. 3.7), and contributes to the gravitational-wave power spectrum according to Eq. (10.48). However, as discussed at the end of Sect. 7.6, the typical fluctuations of a thermal inflaton are so 'hard', with physical momenta $p \sim T$ (or $p \sim \sqrt{2mT}$ if the inflaton has a mass $m \gg T$), that they cannot be described by classical field theory, because it suffers from a Rayleigh-Jeans divergence. To compute the contribution of $\delta\varphi_{,i}\delta\varphi_{,j}$ to η requires tools from thermal field theory [12] (the problem is even more delicate if φ does not interact with a separate plasma, but the plasma is rather generated by its self-interactions [13]). Once this task has been accomplished, we can address the issue of double counting: it would *not* be consistent to include both Π^{t}_{ij} with η (cf. Eqs. (3.134) and (3.141)), and the *full* second-order contribution from Eq. (10.51), in a single computation. Rather, the fluctuations in $\delta\varphi$ should be divided into soft and hard momentum ranges, like in effective field theories. The contribution from hard momenta could be incorporated in the shear viscosity, whereas in Eq. (10.51) (and subsequently in Eq. (10.85)), we should include only momenta that are soft enough to be treated classically.

To summarize, scalar perturbations induce tensor perturbations at second order, as shown by Eq. (10.85). However, we should not double-count contributions from first-order anisotropic stress and explicit second-order terms. This is underlined by the fact that Eq. (10.85) is gauge dependent if $\psi \neq \phi$, which according to Eq. (7.98) happens if $\Pi \neq 0$.

10.5   From Anisotropic Stress to Gravitational-Wave Energy Density

The gravitational-wave source that we have obtained in Eq. (10.85) does not rely on the presence of an equilibrated plasma (even though it permits for this possibility), and could be active during different epochs, such as inflation, the early matter-domination era that may follow it, a reheating stage, or the subsequent radiation-dominated expansion. In the present section, we specialize on *phenomena which take place after reheating*, when the universe may be radiation-dominated. Furthermore, the processes are assumed to be localized within the Hubble horizon ($k \gg \mathcal{H}$), which according to Fig. 1.1 in Sect. 1.4 is true after some time. Local out-of-equilibrium physics could be involved, such as that induced by a first-order phase transition. Under these circumstances, we can take further steps with an expression like Eq. (10.85), converting an 'abstract' wave equation into a concrete formula for the energy density that gravitational waves carry.

As a starting point, we consider the second term of Eq. (10.42). The anisotropic stress Π^{t}_{ij} there can be interpreted as any source term, whereby the result can also be used in the context of Eq. (10.85). In Eq. (10.42), the Green's function G_{R}, defined in Eq. (10.37), determines how gravitational waves propagate from the source until observation, and we need to discuss how G_{R} behaves. Specifically, Eq. (10.37) is the harmonic oscillator equation of motion, with a time-dependent damping coefficient. When $k \gg \mathcal{H}$, Eq. (10.37) has to be solved in the *underdamped regime*.

To appreciate the challenge of determining G_{R}, let us first take the ansatz

$$G_{\mathrm{R}}(\tau, \tau_i, k) \overset{?}{\sim} A_{\mathrm{R}}(\tau, \tau_i, k) \equiv \alpha \sin\big(k(\tau - \tau_i)\big) . \tag{10.91}$$

With this ansatz, $(\partial_\tau^2 + k^2) A_{\mathrm{R}} = 0$, so we are correctly accounting for the largest terms in Eq. (10.37). But if we want to know G_{R} for a long time interval, functional dependences like $f(\mathcal{H}\tau)$ could emerge. They lead to terms like $\partial_\tau^2 f = \mathcal{H}' f' + \mathcal{H}^2 f''$. In comparison, the damping from Eq. (10.37) yields $\mathcal{H} \partial_\tau A_{\mathrm{R}} \sim k \mathcal{H} A_{\mathrm{R}} \gg \mathcal{H}^2 A_{\mathrm{R}}$. So, if $|f''| \sim |f'| \sim |f|$, the damping $\mathcal{H} \partial_\tau A_{\mathrm{R}}$ is more important than $\partial_\tau^2 f$. However, Eq. (10.91) omits the damping.

Fortunately, a better approximation can be found. Consider the ansatz

$$G_{\mathrm{R}}(\tau, \tau_i, k) \overset{?}{\sim} B_{\mathrm{R}}(\tau, \tau_i, k) \equiv \frac{\alpha\, a(\tau_i)}{a(\tau)} \sin\big(k(\tau - \tau_i)\big) . \tag{10.92}$$

It follows that

$$\partial_\tau B_{\mathrm{R}} \overset{(10.92)}{=} \frac{\alpha\, a(\tau_i)}{a(\tau)} \left\{ -\frac{a'}{a} \sin\big(k(\tau - \tau_i)\big) + k\, \cos\big(k(\tau - \tau_i)\big) \right\} , \tag{10.93}$$

$$\partial_\tau^2 B_{\mathrm{R}} \overset{(10.93)}{=} \frac{\alpha\, a(\tau_i)}{a(\tau)} \left\{ \left[-\frac{a''}{a} + \frac{2(a')^2}{a^2} - k^2 \right] \sin\big(k(\tau - \tau_i)\big) - \frac{2a'k}{a} \cos\big(k(\tau - \tau_i)\big) \right\}$$

$$\Rightarrow \big(\partial_\tau^2 + 2\mathcal{H}\partial_\tau + k^2\big) B_{\mathrm{R}} = -\frac{a''}{a} B_{\mathrm{R}} \overset{(1.4)}{=} -\big(\mathcal{H}' + \mathcal{H}^2\big) B_{\mathrm{R}} \overset{?}{\sim} 0 . \tag{10.94}$$

Now the remainder is smaller than with A_{R}, because it contains no term of $\mathcal{O}(k\mathcal{H})$. Furthermore, for $(\bar{\varphi}')^2 \ll a^2(\bar{e} - 3\bar{p})$, Eqs. (10.79) and (10.80) imply that

$$\frac{a''}{a} \overset{(1.4)}{=} \mathcal{H}' + \mathcal{H}^2 \underset{(\bar{\varphi}')^2 \ll a^2(\bar{e}-3\bar{p})}{\overset{(10.79),\,(10.80)}{\approx}} \frac{4\pi G a^2}{3}\big(\bar{e} - 3\bar{p}\big) . \tag{10.95}$$

The left-hand side of Eq. (10.95) is proportional to the Ricci scalar (cf. Eq. (1.28)), and is often referred to as *curvature* in general relativity. The combination $\bar{e} - 3\bar{p}$ appearing on the right-hand side of Eq. (10.95) is known as the *trace anomaly* in thermodynamics, with the 'trace' referring to that of the energy-momentum tensor. To give the trace anomaly a concrete meaning, we note that with the parametrization from Eq. (7.55),

$$e_r - 3p_r \overset{(7.55)}{=} \frac{2(g_* - h_*)\pi^2 T^4}{15} . \tag{10.96}$$

This vanishes in the massless non-interacting limit, where $g_* = h_*$, perhaps justifying calling it an 'anomaly'. Exceptions to its smallness are special points in the phase diagram such as phase transitions, where interactions are strong (cf., e.g., Ref. [14]), or temperatures at which some thermalized particle species 'crosses a mass threshold', i.e. has a mass of the order of the temperature. However, the latter happens for one particle species at a time, so it is usually a relatively small effect.

Given Eqs. (10.94)–(10.96), we conclude that we can normally adopt B_{R} as a good approximation for G_{R} from Eq. (10.37). There are some exceptions, and we return to them in the last bullet point below Eq. (10.152). Fixing the integration constant, α, through Eqs. (10.38) and (10.39), we thus find

$$G_{\mathrm{R}}(\tau, \tau_i, k) \underset{\substack{\approx \\ k \gg \mathcal{H}}}{\overset{\tau \geq \tau_i}{}} \frac{a(\tau_i)\,\sin(k(\tau - \tau_i))}{a(\tau)\,k} \,, \quad i \overset{(10.42)}{\in} \{1, 2\}\,. \tag{10.97}$$

Making use of $\sin c_1 \sin c_2 = \tfrac{1}{2}[\cos(c_1 - c_2) - \cos(c_1 + c_2)]$, this yields

$$G_{\mathrm{R}}(\tau, \tau_1, k)\, G_{\mathrm{R}}(\tau, \tau_2, k) \underset{\substack{\approx \\ k \gg \mathcal{H}}}{\overset{(10.97)}{}} \frac{a(\tau_1)\,a(\tau_2)}{2k^2 a^2(\tau)}\big[\cos(k(\tau_2 - \tau_1)) - \cos(k(2\tau - \tau_1 - \tau_2))\big]\,. \tag{10.98}$$

If we instead consider time derivatives of h^{t}_{ij}, which play a role in the energy density (cf. Eq. (10.24)), and insert $\cos c_1 \cos c_2 = \tfrac{1}{2}[\cos(c_1 - c_2) + \cos(c_1 + c_2)]$, then we find

$$G_{\mathrm{R}}'(\tau, \tau_1, k)\, G_{\mathrm{R}}'(\tau, \tau_2, k) \underset{\substack{\approx \\ k \gg \mathcal{H}}}{\overset{(10.97)}{}} \frac{a(\tau_1)\,a(\tau_2)}{2a^2(\tau)}\big[\cos(k(\tau_2 - \tau_1)) + \cos(k(2\tau - \tau_1 - \tau_2))\big]\,. \tag{10.99}$$

Now, we may recall from electrodynamics that when a source emits radiation, there is a 'near zone' in which the exact solution is not yet wavelike, and a 'far zone' in which it settles to a propagating wave. Similarly, for gravitational waves, their physical characteristics can be unambiguously defined only once the wave motion has had time to take shape. Concretely, this requires that $k(\tau - \tau_i) \gg 1$. In this domain, the second terms in Eqs. (10.98) and (10.99) are rapidly oscillating, and average to zero, implying that

$$\langle G_{\mathrm{R}}\, G_{\mathrm{R}} \rangle \underset{\substack{\approx \\ k(\tau - \tau_i) \gg 1}}{\overset{(10.98)}{}} \frac{a(\tau_1)a(\tau_2)\cos\big(k(\tau_2 - \tau_1)\big)}{2k^2 a^2(\tau)} \underset{\substack{\approx \\ k(\tau - \tau_i) \gg 1}}{\overset{(10.99)}{}} \frac{\langle G_{\mathrm{R}}'\, G_{\mathrm{R}}' \rangle}{k^2}\,. \tag{10.100}$$

On the other hand, the gravitational energy density from the first variant of Eq. (10.24) is proportional to the quadratic expression $G_{\mathrm{R}}' G_{\mathrm{R}}' + k^2 G_{\mathrm{R}} G_{\mathrm{R}}$. With this form, fast oscillations cancel between Eqs. (10.98) and (10.99) at *any time*, and no averaging is required.

Inserting $\langle G_{\mathrm{R}} G_{\mathrm{R}} \rangle$ from Eq. (10.100) into (10.42), which means that we are assuming $k(\tau - \tau_i) \gg 1$ and therefore averaging over oscillations, we find

$$\mathcal{P}_{\mathrm{t}}(\tau, k) \underset{(10.100)}{\overset{(10.42)}{\supset}} \frac{64k}{m_{\mathrm{pl}}^4 a^2(\tau)} \int_{-\infty}^{\tau} d\tau_1 \int_{-\infty}^{\tau} d\tau_2\, \cos(k(\tau_2 - \tau_1)) a^3(\tau_1) a^3(\tau_2) P_{\Pi}(\tau_1, \tau_2, \mathbf{k})$$

$$\underset{(10.102)}{\overset{\tau_1 \leftrightarrow \tau_2}{=}} \frac{128k}{m_{\mathrm{pl}}^4 a^2(\tau)} \int_{-\infty}^{\tau} d\tau_1 \int_{-\infty}^{\tau_1} d\tau_2\, \cos(k(\tau_2 - \tau_1)) a^3(\tau_1) a^3(\tau_2) S_{\Pi}(\tau_1, \tau_2, \mathbf{k})\,. \tag{10.101}$$

For a source active only for a finite period of time, until a moment τ_f, we may replace $\tau \to \tau_f$ in the upper bounds of the τ_1 and τ_2-integrals, and the residual time dependence of the spectrum lies in the prefactor $1/a^2(\tau)$. In the second step, we have divided the integration into the domains $\tau_1 > \tau_2$ and $\tau_1 < \tau_2$; renamed $\tau_1 \leftrightarrow \tau_2$ in the latter; and defined a time-symmetrized correlator as

$$S_\Pi(\tau_1, \tau_2, \mathbf{k}) \equiv \frac{1}{2}\left[P_\Pi(\tau_1, \tau_2, \mathbf{k}) + P_\Pi(\tau_2, \tau_1, \mathbf{k}) \right] . \tag{10.102}$$

In the classical limit, P_Π is automatically time-symmetric (cf. Eq. (10.45)), however as this is not the case in quantum mechanics, it is instructive to avoid the assumption. Writing the cosine in terms of exponentials then yields

$$\mathcal{P}_t(\tau, k) \overset{(10.101)}{\supset} \frac{128k}{m_{\rm pl}^4 a^2(\tau)} \int_{-\infty}^{\tau} d\tau_1 \int_{-\infty}^{\tau_1} d\tau_2 \, \frac{e^{ik(\tau_1-\tau_2)} + e^{ik(\tau_2-\tau_1)}}{2} \, a^3(\tau_1)a^3(\tau_2) S_\Pi(\tau_1, \tau_2, \mathbf{k})$$

$$\overset{\tau_2 = \tau_1 + \tilde{\tau}}{=} \frac{64k}{m_{\rm pl}^4 a^2(\tau)} \int_{-\infty}^{\tau} d\tau_1 \int_{-\infty}^{0} d\tilde{\tau} \left(\underbrace{e^{-ik\tilde{\tau}}}_{\tilde{\tau} \to -\tilde{\tau}} + e^{ik\tilde{\tau}} \right) a^3(\tau_1)a^3(\tau_1 + \tilde{\tau}) S_\Pi(\tau_1, \tau_1 + \tilde{\tau}, \mathbf{k})$$

$$= \frac{64k}{m_{\rm pl}^4 a^2(\tau)} \int_{-\infty}^{\tau} d\tau_1 \, a^3(\tau_1) \int_{-\infty}^{\infty} d\tilde{\tau} \, e^{ik\tilde{\tau}} a^3(\tau_1 - |\tilde{\tau}|) S_\Pi(\tau_1, \tau_1 - |\tilde{\tau}|, \mathbf{k}) . \tag{10.103}$$

Before taking the final step, let us relate $\mathcal{P}_t$ to the gravitational energy density, as it appears in Eq. (2.18). The tensor power spectrum can be expressed as

$$\langle h_{ij}^t(\tau, \mathbf{x}) h_{ij}^t(\tau, \mathbf{x}) \rangle \overset{(2.51)}{=} \int_{-\infty}^{\infty} d\ln k \sum_{i,j} \mathcal{P}_{h_{ij}^t}(\tau, k) \overset{(10.28)}{\underset{(10.29)}{=}} \int_{-\infty}^{\infty} d\ln k \, \mathcal{P}_t(\tau, k) . \tag{10.104}$$

From Eq. (10.24), adopting now the second variant, the energy density reads

$$e_{\rm gw}(\tau) \overset{(10.24)}{\simeq} \frac{1}{32\pi G a^2(\tau)} \langle h_{ij}^{t\prime}(\tau, \mathbf{x}) h_{ij}^{t\prime}(\tau, \mathbf{x}) \rangle \overset{(10.100)}{\underset{(10.104)}{\approx}} \frac{1}{32\pi G a^2(\tau)} \int_{-\infty}^{\infty} d\ln k \, k^2 \, \mathcal{P}_t(\tau, k) . \tag{10.105}$$

This yields the differential energy density

$$\frac{de_{\rm gw}(\tau, k)}{d\ln k} \overset{(10.105)}{\underset{G = m_{\rm pl}^{-2}}{\simeq}} \frac{m_{\rm pl}^2 k^2}{32\pi a^2(\tau)} \, \mathcal{P}_t(\tau, k)$$

$$\overset{(10.103)}{\supset} \frac{2k^3}{\pi m_{\rm pl}^2 a^4(\tau)} \int_{-\infty}^{\tau} d\tau_1 \, a^3(\tau_1) \int_{-\infty}^{\infty} d\tilde{\tau} \, e^{ik\tilde{\tau}} a^3(\tau_1 - |\tilde{\tau}|) S_\Pi(\tau_1, \tau_1 - |\tilde{\tau}|, \mathbf{k}) . \tag{10.106}$$

We remark that Eq. (10.106) can also be obtained *without* an oscillation average, if we do not make use of $\mathcal{P}_t$, but take the first variant of Eq. (10.24) as a starting point.

Let us then turn to the production process and inspect the evolution *rate* of the gravitational-wave energy density. There are two different physical phenomena present in Eq. (10.106): overall redshift, represented by the prefactor $1/a^4(\tau)$, and sourced production, represented by the integral over τ_1. It is convenient to handle both at the same time, as is also suggested by our notation in which τ and τ_f are not distinguished. Taking a derivative with respect to this common value, we get

$$(\partial_\tau + 4\mathcal{H})\frac{\mathrm{d}e_{\mathrm{gw}}(\tau, k)}{\mathrm{d}\ln k} \underset{k \gg \mathcal{H}}{\overset{(10.106)}{\supset}} \frac{2k^3}{\pi m_{\mathrm{pl}}^2 a(\tau)} \int_{-\infty}^{\infty} \mathrm{d}\tilde{\tau}\, e^{ik\tilde{\tau}}\, a^3(\tau - |\tilde{\tau}|) S_\Pi(\tau, \tau - |\tilde{\tau}|, \mathbf{k})\,.$$

$$(10.107)$$

Let us anticipate that below Eq. (10.153), we show how an equation of the form in Eq. (10.107) can be integrated in τ, if the right-hand side is a known function. We also remark that computing rates like in Eq. (10.107) is often safer than computing time-integrated densities, because so-called *secular terms* (which grow rapidly with time and spoil perturbative treatments) are avoided.

We can make further use of the assumption $k \gg \mathcal{H}$. The Fourier transform in Eq. (10.107) probes variations of the energy-momentum tensor, with the typical contribution given by $\tilde{\tau} \sim 1/k \ll \mathcal{H}^{-1}$. On such time scales, $a^3(\tau \pm \tilde{\tau}) \approx a^3(\tau)$. Furthermore, at short time intervals, we may assume physics to be locally time-translationally invariant. Then we find

$$S_\Pi(\tau, \tau - |\tilde{\tau}|, \mathbf{k}) = \theta(-\tilde{\tau}) \overbrace{S_\Pi(\tau, \tau + \tilde{\tau}, \mathbf{k})}^{\text{symmetry } \tau \leftrightarrow \tau + \tilde{\tau}} + \theta(\tilde{\tau}) \overbrace{S_\Pi(\tau, \tau - \tilde{\tau}, \mathbf{k})}^{\text{translation } \tau \to \tau + \tilde{\tau}}$$

$$\approx \theta(-\tilde{\tau})\, S_\Pi(\tau + \tilde{\tau}, \tau, \mathbf{k}) + \theta(\tilde{\tau})\, S_\Pi(\tau + \tilde{\tau}, \tau, \mathbf{k}) = S_\Pi(\tau + \tilde{\tau}, \tau, \mathbf{k})$$

$$\overset{(10.43)}{\underset{(10.102)}{=}} \int \mathrm{d}^3 \mathbf{x}\, e^{-i\mathbf{k}\cdot\mathbf{x}} \frac{1}{2}\big\langle \Pi^{\mathrm{t}}_{ij}(\tau + \tilde{\tau}, \mathbf{x})\, \Pi^{\mathrm{t}}_{ij}(\tau, \mathbf{0}) + \Pi^{\mathrm{t}}_{ij}(\tau, \mathbf{x})\, \Pi^{\mathrm{t}}_{ij}(\tau + \tilde{\tau}, \mathbf{0}) \big\rangle$$

$$\overset{(10.44)}{=} \int \mathrm{d}^3 \mathbf{x}\, e^{-i\mathbf{k}\cdot\mathbf{x}} \frac{1}{2}\big\langle \Pi^{\mathrm{t}}_{ij}(\tau + \tilde{\tau}, \mathbf{x})\, \Pi^{\mathrm{t}}_{ij}(\tau, \mathbf{0}) + \Pi^{\mathrm{t}}_{ij}(\tau, \mathbf{0})\, \Pi^{\mathrm{t}}_{ij}(\tau + \tilde{\tau}, \mathbf{x}) \big\rangle\,.$$

$$(10.108)$$

The parity invariance that was employed as a part of Eq. (10.44) guarantees that the value of this integral is real. Given that our derivation relied on classical field theory, we can now also invoke the corresponding assumption from Eq. (10.45), according to which operator ordering plays no role. Simplifying the notation as $\mathcal{X} \equiv (\tilde{\tau}, \mathbf{x})$ and $\mathcal{K} \cdot \mathcal{X} \equiv -k\tilde{\tau} + \mathbf{k} \cdot \mathbf{x}$, and abbreviating $\tau \equiv (\tau, \mathbf{0})$ as a space-time location, we finally get

$$(\partial_\tau + 4\mathcal{H})\frac{\mathrm{d}e_{\mathrm{gw}}(\tau, k)}{\mathrm{d}\ln k} \underset{k \gg \mathcal{H}}{\overset{(10.107),\,(10.108),\,(10.45)}{\supset}} \frac{2k^3 a^2(\tau)}{\pi m_{\mathrm{pl}}^2} \int_{\mathcal{X}} e^{-i\mathcal{K}\cdot\mathcal{X}} \big\langle \Pi^{\mathrm{t}}_{ij}(\tau)\, \Pi^{\mathrm{t}}_{ij}(\tau + \mathcal{X}) \big\rangle\,.$$

$$(10.109)$$

To summarize, the main result of this section is Eq. (10.109). It shows that the production rate of the energy density carried by gravitational waves (after oscillation-averaging, if we adopt the second variant of Eq. (10.24) as a definition), gets a contribution from the Fourier transform of the 2-point correlation function of anisotropic stress, assuming that this originates from subhorizon physics, with $k \gg \mathcal{H}$.

10.6   Graviton Production from Particle Decays and Scatterings

As we increase the momentum k of the gravitational waves considered, we see from the coefficient of the right-hand side of Eq. (10.109) that the energy density production rate increases. Simultaneously, a larger k implies that we are probing the properties of the energy-momentum tensor with an increasing resolution. At sufficiently large k, we cannot describe the plasma part of Π^{t}_{ij} with hydrodynamic degrees of freedom any more, but instead need to use elementary quantum fields in Π^{t}_{ij}, as was already done with φ in Eq. (10.85). Then, the processes responsible for generating gravitational waves are particle decays and scatterings. This is interesting at the level of fundamental physics, as it may offer a window to particles and interactions beyond the Standard Model.

Like in Sect. 10.5, we consider here an epoch late enough that a given mode is inside the Hubble horizon, $k \gg \mathcal{H}$. In terms of a physical momentum, $p \equiv k/a$, this corresponds to $p \gg H$. We take the statistical average of the 2-point correlator of Π^{t}_{ij}, assuming that the universe has reheated. Then the result also depends on the temperature T. Given that $H \sim T^2/m_{\mathrm{pl}}$ (cf. Eq. (2.36)) and that we normally have $T \ll m_{\mathrm{pl}}$, we expect $H \ll T$. We can work in a local Minkowskian frame, and there are two regimes in which p can lie, $H \ll p < T$ and $H \ll T < p$, separated by what we call the *thermal energy scale*, at $p \sim T$ (for simplicity, we focus here on ultrarelativistic particles, with masses $m \ll T$). We will see in this section that the gravitational-wave energy spectrum originating from particle scatterings *peaks at the thermal energy scale*.

As discussed in Sect. 2.4, the temperature evolves as $T \sim T_{\mathrm{reheat}}(a_{\mathrm{reheat}}/a)$ after reheating, which has the same functional form as $p = k/a$ with respect to a. So, the relation of p to T does not change much in the history of the universe. In terms of today's frequency, the thermal energy scale corresponds to $f_0|_{p \sim T} \sim 10^{11}\,\mathrm{Hz}$ (cf. Eq. (2.39)), similar to the peak frequency of the CMB Planck spectrum.

Let us compare the frequency $f_0 \sim 10^{11}\,\mathrm{Hz}$ with the interferometric measurements mentioned in Sect. 2.3. Most of them probe frequencies $f_0 \ll 10^{11}\,\mathrm{Hz}$, and therefore momenta $p \ll T$. In ultrarelativistic plasma physics, we identify the domain $p \ll T$ as a *hydrodynamic regime*, given that the corresponding wavelength is much larger than the mean free path of particle scatterings. However, the so-called ultra-high-frequency (UHF) detector concepts, mentioned at the end of Sect. 2.3, may offer for an opportunity to observe the physics of momenta $p \sim T$, where particle-gravity interactions play a role.

When we consider large momenta, we ultimately enter a domain in which the classical approximation breaks down. For thermal systems, this happens when $p \sim$

T. For non-thermal systems, the breakdown happens when the states of the Fock space that contribute to the correlation function have occupancies of order unity (rather than being $\gg 1$, which would justify the classical approximation). As an example, this implies that fermions, which obey the Pauli principle, can be treated classically only in the hydrodynamic domain, $p \ll T$, where their contribution is fully captured by their overall effect on the energy density, pressure, and viscosities.

In quantum mechanics and quantum field theory, *operator ordering* plays a role. For instance, standard scattering computations make use of time-ordered perturbation theory (reflected by the $i\epsilon$ prescription of a Feynman propagator). If time ordering is important, we cannot arbitrarily change it, as we did when going from Eq. (10.108) to Eq. (10.109). This implies that the production rate of the energy density carried by gravitational waves needs to be re-derived, by treating the perturbed Einstein-Hilbert action as a quantum field theory. In this case it is natural to refer to gravitational waves as *gravitons*.

We now explain three ways of obtaining the correct quantum-mechanical generalization of Eq. (10.109). The methods offer complementary benefits, namely that:

(i) *Boltzmann equations* are a practical tool for leading-order computations in particle cosmology (such as those related to dark matter production, or neutrino or photon decoupling), and can be adapted to graviton production as well.

(ii) *Thermal particle production* refers to a way to derive the graviton production rate directly from quantum field theory, and can thus be used for establishing the proper operator ordering of the 2-point correlator of the energy-momentum tensor.

(iii) *Linear response theory* sets graviton production in a broader context, relating it to the equilibration or damping phenomena that we met in Sect. 7.2, and providing for a physical explanation of an important Bose distribution, n_{B}, in Eq. (10.114).

To keep the expressions simple, we work with local Minkowskian coordinates in the following ($a = $ constant), denoting $\mathbf{p} \equiv \mathbf{k}/a$, $\mathbf{r} \equiv a\mathbf{x}$, and $\mathcal{R} = a\mathcal{X}$.

(i) Boltzmann equations. From the practical point of view, the most straightforward avenue to a quantum treatment of gravitons might be to abandon classical fields altogether, and to re-start from scratch, from a particle picture. The tool to be used is that of Boltzmann equations, and the variable of interest is the graviton phase-space distribution, f_{gw}. For simplicity, we assume that both polarization states are equally populated, and that f_{gw} denotes the common value. The differential energy density in gravitons reads

$$d e_{\mathrm{gw}} \supset 2p \, f_{\mathrm{gw}} \, \frac{d^3\mathbf{p}}{(2\pi)^3} \,, \tag{10.110}$$

where 2 counts the polarization states and p the energy that they carry. Assuming that the production is isotropic, so that $d^3\mathbf{p} = 4\pi p^2 dp = 4\pi p^3 \, d\ln p$, the production

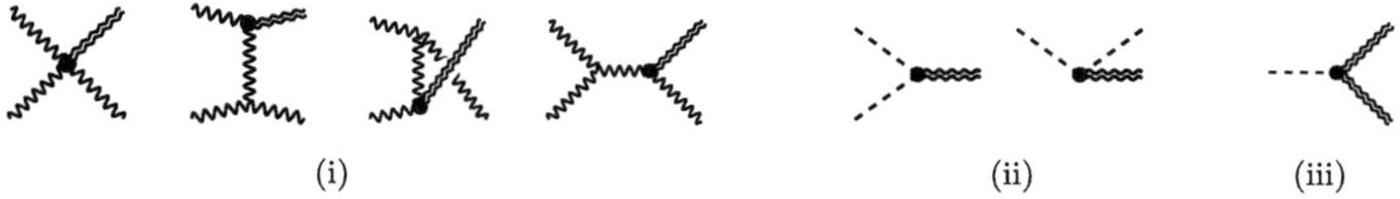

(i)   (ii)   (iii)

Fig. 10.1 Examples of *Feynman diagrams for graviton production*. Gravitons are denoted by doubled curly lines. (i) $2 \to 2$ scatterings, as described by Eq. (10.112). The wiggly lines denote gauge bosons, for instance gluons, which give the dominant contribution in the Standard Model at high temperatures, due to their large multiplicity as well as a logarithmic enhancement of the so-called t-channel process [15]. The filled circle denotes the anisotropic stress tensor, to which gravitons couple. (ii) $2 \to 1$ and $1 \to 2$ processes, for which we can draw simple diagrams, however these processes are *kinematically forbidden* in local Minkowskian space-time, due to energy conservation. The dashed line denotes a scalar particle, for instance the inflaton. (iii) $1 \to 2$ decays. The vertex needed for this process is not present for a minimally coupled scalar field, like in the Einstein-Hilbert action (1.45), however it can be induced by non-minimal couplings. Processes with two gravitons also originate at second order in anisotropic stress [16, 17]

rate then becomes

$$\frac{\mathrm{d}e_{\mathrm{gw}}}{\mathrm{d}t \, \mathrm{d}\ln p} \overset{(10.110)}{\underset{\mathrm{isotropy}}{\supset}} \frac{p^4 \dot{f}_{\mathrm{gw}}}{\pi^2} \, . \tag{10.111}$$

The time derivative of the phase-space distribution in Eq. (10.111) is given by what is referred to as a *collision term* of a Boltzmann equation. In our case it consists of 'inelastic' processes, with at least one graviton in the final state. Given that gravitons do not thermalize, only 'gain' terms need to be included. Under these conditions, a typical gain process (examples are shown in Fig. 10.1(i)) has the structure

$$\dot{f}_{\mathrm{gw}} \supset \frac{\mathbb{C}}{2p} \int \frac{\mathrm{d}^3 \mathbf{p}_{i_1}}{(2\pi)^3 \, 2\epsilon_{i_1}} \frac{\mathrm{d}^3 \mathbf{p}_{i_2}}{(2\pi)^3 \, 2\epsilon_{i_2}} \frac{\mathrm{d}^3 \mathbf{p}_{e_1}}{(2\pi)^3 \, 2\epsilon_{e_1}} (2\pi)^4 \delta^{(4)}(\mathcal{P}_{i_1} + \mathcal{P}_{i_2} - \mathcal{P}_{e_1} - \mathcal{P})$$

$$\times f_{i_1} f_{i_2} (1 \pm f_{e_1}) \sum |\mathcal{M}|^2 \, . \tag{10.112}$$

In the sum over matrix elements squared, $\sum |\mathcal{M}|^2$, it is convenient to include both graviton polarization states, as well as both particle and antiparticle states of *each* momentum. The combinatorial factor, $\mathbb{C} \equiv 1/4$, is chosen to cancel overcountings from the sums and integrals: from polarization states (factor 1/2); from non-identical initial-state particles (factor 1/2, because we should have assigned them to separate momenta); or from identical initial-state particles (factor 1/2, because we should have integrated only over half of the momentum space). The functions f_{i_n} are the phase-space distributions of initial-state particles, f_{e_1} is that of an end-state particle produced in association with the graviton, and $\pm$ refer to bosons and fermions, respectively. For the matrix elements, we need a vertex with which a graviton interacts with matter fields, and Eq. (10.112) assumes that for its determination the graviton field has been canonically normalized, like in Eq. (10.22).

Equation (10.112) corresponds to a $2 \to 2$ scattering, but in addition gravitons can be produced from $1 \to 3$ decays of a massive particle, such as the inflaton (cf., e.g., Refs. [18–21]). We remark in passing that with so-called minimal couplings, gravitons attach to the energy-momentum tensors of matter fields, which are quadratic in the fields. Therefore, we can also draw diagrams for $2 \to 1$ or $1 \to 2$ processes (cf. Fig. 10.1(ii)). However, they have two lines of equal mass and one massless line, and are thus not kinematically allowed in locally Minkowskian spacetime. If there are non-minimal vertices, which lead to kinematically allowed $2 \to 1$ or $1 \to 2$ processes (e.g. one massive particle decaying into two massless ones, cf. Fig. 10.1(iii)), they should be included (cf., e.g., Refs. [22,23]).

One strength of Boltzmann equations is that they can also be used for non-equilibrium situations. In other words, the distribution functions f_{i_n} and f_{e_1} do not necessarily need to be equilibrium ones. If the particles are in equilibrium, we should use the Bose and Fermi distributions for bosons and fermions, respectively.

Even if it is not easy to show this on an abstract level, an explicit computation demonstrates that, at leading order in coupling constants [15], Eqs. (10.111) and (10.112) yield the same result as a perturbative evaluation of the quantum-field-theory formula in Eq. (10.114), to which we now turn.

(ii) Thermal particle production. A second possibility for a quantum treatment of graviton production is to adopt the framework of *thermal particle production*. Via this formalism, two statements can be proven. First, the correct time ordering happens to be precisely the one shown in Eq. (10.109), where the space-time coordinates $\mathcal{X}$ appear on the right in the Fourier transform. This time ordering is known as one of the *Wightman correlators*, and in physical momentum space, it is denoted by $G^t_<(\mathcal{P})$. Second, in thermal equilibrium, the Wightman correlator can be related to the corresponding *retarded Green's function* (cf. Sect. 7.2), through

$$G^t_<(\mathcal{P}) \;=\; 2n_{\rm B}(p){\rm Im}G^t_{\rm R}(\mathcal{P}) \,, \quad \mathcal{P} \equiv (p, \mathbf{p}) \,. \tag{10.113}$$

Therefore, employing $\partial_\tau = a\partial_t$, $k = pa$, and $\int_{\mathcal{X}} a^4 = \int_{\mathcal{R}}$, Eq. (10.109) becomes

$$\boxed{\; \frac{{\rm d}e_{\rm gw}}{{\rm d}t\,{\rm d}\ln p} \underset{p \sim T \gg H}{\overset{(10.109),\,(10.113)}{\supset}} \frac{4p^3}{\pi m_{\rm pl}^2}\, n_{\rm B}(p)\,{\rm Im}G^t_{\rm R}(\mathcal{P}) \;} \tag{10.114}$$

We now rederive Eq. (10.114) from quantum field theory (after re-interpreting $e_{\rm gw}$ as an expectation value of an operator, i.e. $e_{\rm gw} \to \langle \hat{e}_{\rm gw} \rangle$), verifying the time ordering. The proof of Eq. (10.113), via a straightforward evaluation of the 2-point correlators in the energy eigenbasis, can be found in thermal field theory text books (cf., e.g., Ref. [24, Sect. 8.1]).

To get started we note that, after the rescaling in Eq. (10.23), and going over to local Minkowskian coordinates, we can set Eq. (10.36) in a form similar to Eq. (7.5),

$$\ddot{h}^t_{ij} - \nabla^2_{\mathbf{r}} h^t_{ij} \underset{\rm Minkowskian}{\overset{(10.36),\,(10.23)}{=}} \frac{16\pi G}{\underbrace{\sqrt{32\pi G}}_{\sqrt{8\pi}/m_{\rm pl}}} \Pi^t_{ij} \,. \tag{10.115}$$

In order to streamline the notation, it is helpful to go over to the polarization basis, with $\lambda \in \{+, \times\}$, in accordance with Eqs. (10.26)–(10.28), and omit the superscript $(...)^{\mathrm{t}}$ in $\mathbb{h}$ and Π. Furthermore, we denote

$$\frac{\sqrt{8\pi}\,\Pi_\lambda}{m_{\mathrm{pl}}} \equiv -J_\lambda\,, \tag{10.116}$$

where the (inconsequential) minus sign has been inserted in order to maintain an analogy with Eqs. (7.5) and (7.6).

Next, we quantize the graviton field and the operator to which it couples, $\mathbb{h}_\lambda \to \hat{\mathbb{h}}_\lambda$ and $J_\lambda \to \hat{J}_\lambda$. The observable that we are interested in is related to the graviton part of the Hamiltonian, normalized by a spatial volume, L^3, which will drop out in the final result,

$$\hat{e}_{\mathrm{gw}}(t) \overset{(10.24)}{\equiv} \frac{1}{L^3} \int_{\mathbf{r}} \sum_\lambda \frac{1}{2}\Big(\dot{\hat{\mathbb{h}}}_\lambda\, \dot{\hat{\mathbb{h}}}_\lambda + \nabla_{\mathbf{r}}\,\hat{\mathbb{h}}_\lambda \cdot \nabla_{\mathbf{r}}\,\hat{\mathbb{h}}_\lambda \Big)\,. \tag{10.117}$$

We represent the graviton field in terms of a mode expansion, similarly to Eqs. (5.31)–(5.33) but now in local Minkowskian space-time. Formally, this means that we go over to an *interaction picture*, in which the time evolution of the operators is determined by the free Hamiltonian (cf. Eq. (10.117)), whereas that of states originates from a perturbation (cf. Eq. (10.121)). Being pedantic about which objects are operators, we write the mode expansion as

$$\hat{\mathbb{h}}_\lambda(\mathcal{R}) = \int \frac{\mathrm{d}^3\mathbf{p}}{\sqrt{(2\pi)^3 2p}} \Big(\hat{w}_{\mathbf{p}}^\lambda\, e^{i\mathcal{P}\cdot\mathcal{R}} + \hat{w}_{\mathbf{p}}^{\lambda\dagger}\, e^{-i\mathcal{P}\cdot\mathcal{R}} \Big)\,, \tag{10.118}$$

where the free field equation requires that the energy is on-shell, $\mathcal{P} = (p, \mathbf{p})$, and we are using the signature $\mathcal{P} \cdot \mathcal{R} = -pt + \mathbf{p} \cdot \mathbf{r}$. Strictly speaking, if we go to a finite volume, like in Eq. (10.117), momentum integrals should be replaced with momentum sums. However, the finite volume is only an intermediate regulator, and in the end we take the limit $L \to \infty$. We anticipate this by maintaining the integral notation.

Inserting Eq. (10.118) in Eq. (10.117), integrating over $\mathbf{r}$, and making use of the commutators from Eq. (5.33) to 'normal-order' the appearance of the creation and annihilation operators (which means that the vacuum contribution from Sect. 10.2 is omitted), we find that our observable contains the part

$$\hat{e}_{\mathrm{gw}}(t) \overset{(10.117)}{\underset{(10.118)}{\supset}} \frac{1}{L^3} \int \mathrm{d}^3\mathbf{p}\,|\mathbf{p}| \sum_\lambda \hat{w}_{\mathbf{p}}^{\lambda\dagger} \hat{w}_{\mathbf{p}}^\lambda\,. \tag{10.119}$$

We need to compute the expectation value of this operator. If we assume that originally there are no gravitons (or, more physically, that the phase-space density of gravitons of momentum $\mathbf{p}$ is far below equilibrium), then we start from the vacuum state, $|0\rangle$. In the interaction picture, states evolve with a *time-evolution operator*,

$$\hat{U}_I(t; -\infty) \;=\; \hat{T} \exp\left[-i \int_{-\infty}^{t} \mathrm{d}t' \, \hat{H}_I(t') \right], \qquad (10.120)$$

where $\hat{T}$ is the time-ordering operator, and $\hat{H}_I$ denotes the interaction Hamiltonian (cf. Eq. (7.6)), taken in the interaction picture (cf. Eq. (7.9)),

$$\hat{H}_I(t) \overset{(7.6)}{\underset{(7.9)}{=}} \int_{\mathbf{r}} \sum_{\lambda} \hat{\mathbb{h}}_{\lambda}(\mathcal{R}) \, \hat{J}_{\lambda}(\mathcal{R}) , \quad \mathcal{R} \equiv (t, \mathbf{r}) . \qquad (10.121)$$

Thus the initial vacuum state evolves into

$$|0(t)\rangle \;\equiv\; \hat{U}_I(t; -\infty)|0\rangle . \qquad (10.122)$$

In this state, the expectation value of Eq. (10.119) is non-zero,

$$\langle 0(t) | \hat{w}_{\mathbf{p}}^{\lambda\,\dagger} \hat{w}_{\mathbf{p}}^{\lambda} | 0(t) \rangle \overset{(10.122)}{=} \langle 0 | \hat{U}_I^{\dagger}(t; -\infty) \, \hat{w}_{\mathbf{p}}^{\lambda\,\dagger} \hat{w}_{\mathbf{p}}^{\lambda} \, \hat{U}_I(t; -\infty) | 0 \rangle$$

$$\overset{(10.120)}{\underset{\mathcal{O}(\hat{H}_I)}{\supset}} \langle 0 | \int_{-\infty}^{t} \mathrm{d}t' \, \hat{H}_I^{\dagger}(t') \, \hat{w}_{\mathbf{p}}^{\lambda\,\dagger} \underbrace{\mathbb{1}}_{\sum_n |n\rangle\langle n|} \hat{w}_{\mathbf{p}}^{\lambda} \underbrace{\int_{-\infty}^{t} \mathrm{d}t' \, \hat{H}_I(t')}_{\text{linear in } \hat{w}_{\mathbf{p}}^{\lambda}, \, \hat{w}_{\mathbf{p}}^{\lambda\,\dagger}} | 0 \rangle$$

$$= \langle 0 | \int_{-\infty}^{t} \mathrm{d}t' \, \hat{H}_I^{\dagger}(t') \, |\mathbf{p}; \lambda\rangle \, \langle \mathbf{p}; \lambda| \int_{-\infty}^{t} \mathrm{d}t' \, \hat{H}_I(t') | 0 \rangle , \qquad (10.123)$$

where in the last step we have realized that only the vacuum state contributes in the sum $\sum_n |n\rangle\langle n|$, and subsequently denoted $|\mathbf{p}; \lambda\rangle \equiv \hat{w}_{\mathbf{p}}^{\lambda\,\dagger}|0\rangle$. We remark that expectation values like in Eq. (10.123) are referred to as the *'in-in' formalism*, because constraints are imposed only on states at the initial time (final states originate from the unit operator).

To proceed, we need to include the heat bath in the computation. We thus consider an initial and final state of the type

$$|I\rangle \equiv |i\rangle \otimes |0\rangle , \quad |F\rangle \equiv |f\rangle \otimes |\mathbf{p}; \lambda\rangle , \qquad (10.124)$$

where $|i\rangle$ and $|f\rangle$ are states in the Fock space of the heat bath. Generalizing on Eq. (10.123), the *transition matrix element* and its absolute value squared read

$$T_{FI}(\boldsymbol{p};\lambda) \equiv \langle F \,| \int_{-\infty}^{t} \mathrm{d}t' \, \hat{H}_I(t') \,|I\rangle$$

$$\overset{(10.124)}{\underset{(10.121),\,(10.118)}{=}} \int_{\mathcal{R}'} \frac{e^{-i\mathcal{P}\cdot\mathcal{R}'}}{\sqrt{(2\pi)^3 2p}} \, \langle f \,|\hat{J}_\lambda(\mathcal{R}')|i\,\rangle \,, \tag{10.125}$$

$$|T_{FI}|^2 \overset{(10.125)}{=} \frac{1}{(2\pi)^3 2p} \int_{\mathcal{R}',\mathcal{R}''} e^{i\mathcal{P}\cdot(\mathcal{R}''-\mathcal{R}')} \langle i \,|\hat{J}_\lambda^\dagger(\mathcal{R}'')|f\,\rangle\langle f \,|\hat{J}_\lambda(\mathcal{R}')|i\,\rangle \,,$$

$$\tag{10.126}$$

where $\mathcal{R}' \equiv (t', \mathbf{r}')$. Like in Eq. (10.123), the final states, $|f\,\rangle$, are a representation of the unit operator, $\sum_f |f\,\rangle\langle f\,| = \mathbb{1}$, and can be removed. The initial states of the heat bath are summed over, with the assumption that they are thermally distributed, with the weights $e^{-E_i/T}/\mathcal{Z}_{\mathrm{bath}}$, where $\mathcal{Z}_{\mathrm{bath}} \equiv \sum_i e^{-E_i/T}$ is the *canonical partition function*. A quantum-statistical average is defined as

$$\langle\ldots\rangle \equiv \frac{1}{\mathcal{Z}_{\mathrm{bath}}} \sum_i e^{-E_i/T} \langle i\,|\ldots|i\,\rangle \,. \tag{10.127}$$

Then, from Eqs. (10.119), (10.123), (10.126), and (10.127), the thermally averaged production rate of the gravitational-wave energy density becomes

$$\frac{\mathrm{d}\langle\hat{e}_{\mathrm{gw}}\rangle}{\mathrm{d}t\,\mathrm{d}^3\mathbf{p}} \overset{(10.119)}{\underset{(10.123)}{\supset}} \lim_{L\to\infty} \frac{|\mathbf{p}|}{L^3} \sum_{\lambda,f,i} \frac{\mathrm{d}|T_{FI}|^2}{\mathrm{d}t} \frac{e^{-E_i/T}}{\mathcal{Z}_{\mathrm{bath}}}$$

$$\overset{(10.126)}{\underset{(10.127)}{=}} \frac{1}{16\pi^3} \lim_{L\to\infty} \frac{1}{L^3} \sum_\lambda \frac{\mathrm{d}}{\mathrm{d}t} \int_{\mathcal{R}',\mathcal{R}''} e^{i\mathcal{P}\cdot(\mathcal{R}''-\mathcal{R}')} \langle\, \hat{J}_\lambda^\dagger(\mathcal{R}'')\hat{J}_\lambda(\mathcal{R}')\,\rangle. \tag{10.128}$$

The time dependence in Eq. (10.128) follows the pattern

$$\frac{\mathrm{d}}{\mathrm{d}t} \int_{-\infty}^{t} \mathrm{d}t' \int_{-\infty}^{t} \mathrm{d}t'' \, e^{ip(t'-t'')} \, f(t'',t')$$

$$= \int_{-\infty}^{t} \mathrm{d}t'' \, e^{ip(t-t'')} \underbrace{f(t'',t)}_{t''=t-\tilde{t}} + \int_{-\infty}^{t} \mathrm{d}t' \, e^{ip(t'-t)} \underbrace{f(t,t')}_{t'=t+\tilde{t}}$$

$$= \int_{0}^{\infty} \mathrm{d}\tilde{t}\, e^{ip\tilde{t}} \underbrace{f(t-\tilde{t},t)}_{\approx\,\mathrm{time-translation\ invariant}} + \int_{-\infty}^{0} \mathrm{d}\tilde{t}\, e^{ip\tilde{t}} \, f(t,t+\tilde{t})$$

$$\approx \int_{-\infty}^{\infty} \mathrm{d}\tilde{t}\, e^{ip\tilde{t}} \, f(t,t+\tilde{t}) \,. \tag{10.129}$$

For the spatial integrals in Eq. (10.128), we analogously get

$$\lim_{L\to\infty} \frac{1}{L^3} \int_{\mathbf{r}',\mathbf{r}''} e^{i\mathbf{p}\cdot(\mathbf{r}''-\mathbf{r}')} \underbrace{g(\mathbf{r}'',\mathbf{r}')}_{\mathbf{r}'=\mathbf{r}''+\tilde{\mathbf{r}}}$$

$$= \lim_{L\to\infty} \frac{1}{L^3} \int_{\tilde{\mathbf{r}},\mathbf{r}''} e^{-i\mathbf{p}\cdot\tilde{\mathbf{r}}} \underbrace{g(\mathbf{r}'',\mathbf{r}''+\tilde{\mathbf{r}})}_{\text{spatial}-\text{translation invariant}}$$

$$= \int_{\tilde{\mathbf{r}}} e^{-i\mathbf{p}\cdot\tilde{\mathbf{r}}}\, g(\mathbf{0},\tilde{\mathbf{r}}) \,. \tag{10.130}$$

The final steps are to write $\mathrm{d}^3\mathbf{p} = 4\pi p^3 \,\mathrm{d}\ln p$; insert the Hermitean interaction term $\hat{J}_\lambda = -\sqrt{8\pi}\,\hat{\Pi}_\lambda/m_{\text{pl}}$ from Eq. (10.116); and define

$$G^{\text{t}}_<(\mathcal{P}) \equiv \sum_\lambda \int_{\tilde{\mathcal{R}}} e^{-i\mathcal{P}\cdot\tilde{\mathcal{R}}} \big\langle \hat{\Pi}^{\text{t}}_\lambda(t)\,\hat{\Pi}^{\text{t}}_\lambda(t+\tilde{\mathcal{R}})\big\rangle \,, \tag{10.131}$$

where now $t \equiv (t,\mathbf{0})$, and the integral over $\tilde{\mathcal{R}} \equiv (\tilde{t},\tilde{\mathbf{r}})$ runs over the full space-time according to Eqs. (10.129) and (10.130). Thereby we finally get

$$\frac{\mathrm{d}\langle\hat{e}_{\text{gw}}\rangle}{\mathrm{d}t\,\mathrm{d}\ln p} \overset{(10.128)-(10.131)}{\supset} \frac{2p^3}{\pi m_{\text{pl}}^2}\, G^{\text{t}}_<(\mathcal{P}) \,. \tag{10.132}$$

This is also valid for non-thermal systems, provided that we adjust the expectation value in Eq. (10.127) accordingly. Inserting $G^{\text{t}}_<(\mathcal{P}) = 2n_{\text{B}}(p)\,\mathrm{Im}G^{\text{t}}_{\text{R}}(\mathcal{P})$ from Eq. (10.113), which applies to an equilibrated plasma (n_{B} contains T), Eq. (10.132) verifies Eq. (10.114).

(iii) Linear response argument. The third possibility for treating graviton production at the thermal scale $p \sim T$, or above it, is that we adapt our quantum-mechanical linear-response treatment of the friction coefficient Υ, from Sect. 7.2, to the present context. Combining with a *detailed balance* argument, we can again obtain the production rate. We should stress that this last logic is conceptually non-trivial, because physically speaking gravitons are not close to equilibrium, which is often considered a necessary requirement for the validity of linear-response theory. The argument turns out to be viable for a different reason, namely that gravitons are weakly coupled to the equilibrium plasma.

Let us start again from Eq. (10.115), but now in momentum and helicity space,

$$\ddot{\mathrm{h}}_\lambda + p^2 \mathrm{h}_\lambda \overset{(10.115)}{=} \frac{\sqrt{8\pi}}{m_{\text{pl}}}\, \Pi_\lambda \overset{(10.116)}{=} -J_\lambda \,. \tag{10.133}$$

From Eq. (7.27), we get a friction coefficient associated with the gravitational field,

$$\Upsilon_\lambda(p) \overset{(7.27)}{=} \frac{\mathrm{Im}G_{\text{R}}[J_\lambda]}{p} \overset{(10.133)}{=} \frac{8\pi}{m_{\text{pl}}^2}\frac{\mathrm{Im}G_{\text{R}}[\Pi_\lambda]}{p} \,. \tag{10.134}$$

With the friction coefficient known, we need to review its role in the evolution equations. The principle of detailed balance asserts that every interacting particle species in principle approaches equilibrium sooner or later, implying that

$$\dot{f}_\lambda = -\Upsilon_\lambda(p)\big(f_\lambda - n_{\mathrm{B}}(p)\big) + \mathcal{O}\!\left(\frac{1}{m_{\mathrm{pl}}^4}\right) . \tag{10.135}$$

Close to equilibrium, this functional form could be postulated as a leading term in an expansion in $f_\lambda - n_{\mathrm{B}}(p)$. For us, more important is that the same form also applies far from equilibrium, at leading order in the interaction strength between the graviton and the plasma, i.e. $1/m_{\mathrm{pl}}^2$, as can be established through a non-trivial analysis [25].

In the early universe, gravitons are far from reaching equilibrium, so that $f_\lambda \ll n_{\mathrm{B}}(p)$ in most momentum bins (there could be exceptions at specific momenta, if gravitational waves are produced by non-equilibrium processes such as phase transitions). Then Eq. (10.135) implies $\dot{f}_\lambda \approx \Upsilon_\lambda(p)\, n_{\mathrm{B}}(p)$. For single polarization modes, Eq. (10.110) becomes

$$\mathrm{d}e_\lambda \overset{(10.110)}{\supset} p\, f_\lambda\, \frac{\mathrm{d}^3\mathbf{p}}{(2\pi)^3} \overset{\text{isotropy}}{=} f_\lambda\, \frac{p^4}{2\pi^2}\, \mathrm{d}\ln p , \tag{10.136}$$

which then yields

$$\frac{\mathrm{d}e_{\mathrm{gw}}}{\mathrm{d}t\, \mathrm{d}\ln p} \overset{(10.136)}{\supset} \frac{p^4}{2\pi^2} \sum_\lambda \dot{f}_\lambda \overset{(10.135)}{\underset{f_\lambda \ll n_{\mathrm{B}}}{\approx}} \frac{p^4}{2\pi^2} \sum_\lambda \Upsilon_\lambda(p)\, n_{\mathrm{B}}(p)$$

$$\overset{(10.134)}{=} \frac{4p^3}{\pi m_{\mathrm{pl}}^2}\, n_{\mathrm{B}}(p)\, \mathrm{Im}\, G_{\mathrm{R}}^{\mathrm{t}}(\mathcal{P}) , \quad \mathcal{P} = (p, \mathbf{p}) . \tag{10.137}$$

This once again agrees with Eq. (10.114).

Let us summarize the findings of this section. If we consider physical momenta $p > T$, the growth of the gravitational-wave spectrum that we had observed at smaller momenta (cf. Eq. (10.42)), finally turns around into an exponential falloff, because of the Bose factor, $n_{\mathrm{B}}(p)$, in Eq. (10.137). The physical reason is that, to produce a graviton of momentum p, the energy needs to be extracted from fluctuations, which in a thermal plasma are Boltzmann-suppressed at high energies. Non-trivial contributions at energies higher than $p \sim T$ can only originate from non-equilibrium processes, for instance if the gravitons are produced from the decays of heavy inflaton particles, whose abundance can be much above the equilibrium value during an early matter-dominated epoch (cf. Sect. 1.7).

10.7   Transfer Function for Tensor Modes

The generic result for the tensor power spectrum (cf. Eq. (10.42)), which in turn determines the gravitational-wave energy density (cf. Eq. (10.105)), originates from two kinds of sources. First of all there are the perturbations that were produced by vacuum fluctuations and then extended beyond the Hubble horizon by a period of inflationary expansion. This component has the special property that the corresponding power spectrum 'froze out' for a long period of time, while the modes were outside of the Hubble horizon. And second, there are the gravitational waves that are continuously produced by physical phenomena within the Hubble horizon, notably during the period of reheating and afterwards. Gravitational waves can also get damped by the plasma that they traverse, via the anisotropic stress induced by the shear viscosity (cf. Eq. (10.35)) or, more generally, by a damping coefficient that can be computed from linear-response theory (cf. Eq. (10.135)). The purpose of the present section is to explain how the primordial power spectra get converted to the gravitational-wave energy density that is measurable today.

Given that a distinction is made between modes that did or did not exit the Hubble horizon, it is important to know which today's frequency, f_0, is the maximal one that did so, $f_0^{(\mathrm{max})}$. The answer is not unique, but depends on the energy scale of inflation, and also on the expansion history that followed inflation (for instance, on whether there was an extended matter-domination period after inflation, cf. Sect. 1.7). As can be deduced from Fig. 1.2 in Sect. 1.7, modes crossed outside of the Hubble horizon until very briefly after inflation ended. If we *assume* that the universe is radiation-dominated immediately when the modes re-enter inside the Hubble horizon, then the subsequent scaling is as discussed in Sect. 2.4. For instance, if $T_{\mathrm{re\text{-}entry}}^{(\mathrm{max})} = 10^{15}\,\mathrm{GeV}$, then $f_0^{(\mathrm{max})} = 2.6 \times 10^7\,\mathrm{Hz}$ (cf. Eq. (2.40)). If, instead, there is a matter-dominated period after inflation, then the very same initial momentum mode, $(k/a_i)^{(\mathrm{max})}$, corresponds to a smaller wave vector and frequency today (cf. Fig. 7.2 (right) in Sect. 7.7).

We now turn to a concrete evaluation of the transfer function for the two cases mentioned above. The discussion applies to any realization of the post-inflationary history, however if a perturbation re-enters inside the Hubble horizon during radiation domination, then the final result can be simplified quite a bit (cf. the discussion around Eq. (10.151)).

(i) Vacuum-generated tensor perturbations.

The present-day $(\tau \to \tau_0, a \to a_0)$ gravitational-wave energy density from Eq. (10.24), now written in terms of helicity states, reads

$$e_{\mathrm{gw},0} \overset{(10.24)}{\simeq} \frac{1}{32\pi G a_0^2} \sum_{\lambda} \left\langle h_\lambda'(\tau_0, \mathbf{x})\, h_\lambda'(\tau_0, \mathbf{x}) \right\rangle , \qquad (10.138)$$

where the left-hand side does not depend on $\mathbf{x}$, because of translational invariance (cf. Eq. (2.51)), and the angular brackets denote an average over an oscillation period and a wavelength. Recalling that $h_\lambda(\tau_{\mathrm{out}}, k)$ is constant if τ_{out} is a moment at which the mode is outside of the Hubble horizon, we express the subsequent evolution in comoving momentum space as a functional of this initial value,

$$h_\lambda(\tau_0, k) \equiv X_{\mathrm{t}}(\tau_0, \tau_{\mathrm{out}}, k)\, h_\lambda(\tau_{\mathrm{out}}, k) , \qquad (10.139)$$

$$X_{\mathrm{t}}(\tau_{\mathrm{out}}, \tau_{\mathrm{out}}, k) \equiv 1 , \qquad \partial_\tau X_{\mathrm{t}}(\tau, \tau_{\mathrm{out}}, k) \overset{\tau = \tau_{\mathrm{out}}}{=} 0 . \qquad (10.140)$$

Then, from Eqs. (10.3), (10.138), (10.139), and (2.51),

$$\frac{d\Omega_{\text{gw},0}}{d\ln k} \overset{(10.3)}{=} \frac{8\pi}{3m_{\text{pl}}^2 H_0^2} \frac{de_{\text{gw},0}}{d\ln k} \overset{(10.138)}{\underset{(10.139)}{\approx}} \underbrace{\frac{[\partial_{\tau_0} X_{\text{t}}(\tau_0, \tau_{\text{out}}, k)]^2}{12 a_0^2 H_0^2}}_{\equiv\, \mathcal{T}_{\text{t}}(k)} \underbrace{\mathcal{P}_{\text{t}}(\tau_{\text{out}}, k)}_{\text{like (2.51)}} \,. \qquad (10.141)$$

Here $\mathcal{T}_{\text{t}}$ denotes the *transfer function* in the tensor channel.

The remaining task is to determine the time evolution of X_{t} from Eq. (10.139). It satisfies the same equation as the tensor perturbations,

$$(\partial_\tau^2 + 2\mathcal{H}\partial_\tau + k^2)h_\lambda \overset{(10.36)}{=} 16\pi G a^2 \Pi_\lambda \,. \qquad (10.142)$$

As we have accounted for the right-hand side via the second term in Eq. (10.42), it will be omitted from the transfer function of the vacuum contribution, and we therefore consider

$$(\partial_\tau^2 + 2\mathcal{H}\partial_\tau + k^2)\, X_{\text{t}} \overset{\text{for assumptions}}{\underset{\text{see below}}{\approx}} 0 \,. \qquad (10.143)$$

As for the assumptions behind this approximation, we recall that, apart from being produced, tensor modes can also be damped by the anisotropic stress (cf. Eq. (10.35)). Indeed it has been realized that decoupled particle species, which are free-streaming in a late universe, can have an influence on the transfer function of the tensor modes that re-enter inside the Hubble horizon at late times [26]. For instance, *neutrinos decouple* from the electromagnetic plasma at $T \sim 2\,\text{MeV}$. So, according to Eq. (2.41), we expect them to have an influence at frequencies $f_0 < 4 \times 10^{-11}\,\text{Hz}$. A concrete computation shows an effect at $f_0 < 10^{-9}/(2\pi)\,\text{Hz}$ [27]. If we have in mind an interferometric observation of gravitational waves at larger frequencies, then this effect can probably be ignored, and we will do this in the following. Explorations of the influence of viscous damping at larger frequencies can be found, e.g., in Refs. [28, Fig. 1] and [29, Fig. 1].

Despite its simple structure, the solution of Eq. (10.143), with initial conditions from Eq. (10.140), is quite non-trivial. In fact, the solution is not unlike that shown in Fig. 9.1 in Sect. 9.3: initially, it is constant, and then it starts oscillating fast.

The strategy that is normally adopted for the solution is as follows [30]. In general, we may integrate the equation numerically, similarly to what was done in Sect. 9.7, until a time when the oscillations become fast. For the latter domain, it is helpful to work out an analytical approximation, because integrating over many fast oscillations is numerically expensive. The numerical and analytical solutions are matched onto each other in a regime where both are valid; let us call this moment τ_{match}. Subsequently, the analytic approximation is used for extrapolating to present day.

As far as the analytic approximation goes, it is discussed around Eqs. (10.92)–(10.95). Taking over the form from Eq. (10.92), we find that

$$X_{\text{t}}(\tau, \tau_{\text{out}}, k) \overset{\tau \gg \tau_{\text{out}}}{\underset{(10.92)}{\approx}} \frac{\alpha\, a(\tau_{\text{out}})}{a(\tau)} \sin\big(k(\tau - \tau_{\text{out}}) + \beta\big) \,, \qquad (10.144)$$

where α and β are two integration constants. The integration constants can be fixed by matching the numerical solution and its time derivative to Eq. (10.144),

$$\overbrace{X_{\rm t}(\tau_{\rm match}, \tau_{\rm out}, k)}^{\text{determined numerically}} \overset{(10.144)}{=} \frac{\alpha\, a(\tau_{\rm out})}{a(\tau_{\rm match})} \sin\big(k(\tau_{\rm match} - \tau_{\rm out}) + \beta\big) \,, \qquad (10.145)$$

$$\underbrace{(\partial_\tau + \mathcal{H})X_{\rm t}(\tau_{\rm match}, \tau_{\rm out}, k)}_{\text{determined numerically}} \overset{(10.144)}{=} \frac{\alpha\, k\, a(\tau_{\rm out})}{a(\tau_{\rm match})} \cos\big(k(\tau_{\rm match} - \tau_{\rm out}) + \beta\big) \,. \qquad (10.146)$$

Subsequently, making use of the hierarchy $k \gg \mathcal{H}$ valid at $\tau \geq \tau_{\rm match}$, the transfer function from Eq. (10.141) can be approximated as

$$\boxed{\; \mathcal{T}_{\rm t}(k) \overset{(10.141)}{\underset{(10.144)}{\approx}} \frac{\alpha^2}{12} \frac{k^2 a^2(\tau_{\rm out})}{a_0^4\, H_0^2} \cos^2\big(k(\tau_0 - \tau_{\rm out}) + \beta\big) \,, \;} \qquad (10.147)$$

where α and β are fixed through Eqs. (10.145) and (10.146). We remark that, as discussed below Eq. (10.106), we could eliminate the rapid oscillations from Eq. (10.147) by taking the first variant of Eq. (10.24) as the definition of the gravitational-wave energy density or, alternatively, we could replace the oscillations by their average, 1/2.

There is now a remarkable point, following from Eqs. (10.94) and (10.95). Namely, if $\bar{e} - 3\bar{p} = 0$ during the time at which modes re-enter inside the Hubble horizon, then Eq. (10.144) is an *exact* solution of Eq. (10.143), even when $k \sim \mathcal{H}$ or $k \ll \mathcal{H}$. In this case, we just need to fix the integration constants. Imposing the boundary conditions of Eq. (10.140) at the time $\tau = \tau_{\rm out}$ yields

$$1 \overset{(10.140)}{\underset{(10.145)}{=}} \alpha \sin \beta \,, \qquad (10.148)$$

$$0 \overset{(10.140)}{\underset{(10.146)}{=}} -\alpha \mathcal{H}_{\rm out} \sin \beta + \alpha k \cos \beta \,, \qquad (10.149)$$

and consequently we find that

$$\tan \beta \overset{(10.149)}{=} \frac{k}{\mathcal{H}_{\rm out}} \overset{k \ll \mathcal{H}_{\rm out}}{\Longrightarrow} \beta \approx \frac{k}{\mathcal{H}_{\rm out}} \overset{(10.148)}{\Longrightarrow} \alpha \approx \frac{\mathcal{H}_{\rm out}}{k} \overset{(1.6)}{=} \frac{a_{\rm out} H_{\rm out}}{k} \,. \qquad (10.150)$$

In Eq. (10.147), the k-dependence $\alpha^2 \sim 1/k^2$ cancels against the explicit factor k^2, so that

$$\mathcal{T}_{\rm t}(k) \overset{\bar{e} - 3\bar{p} \approx 0}{\underset{(10.147),\,(10.150)}{\approx}} \frac{1}{12} \left(\frac{a_{\rm out}}{a_0}\right)^4 \left(\frac{H_{\rm out}}{H_0}\right)^2 \cos^2\big(k(\tau_0 - \tau_{\rm out})\big) \,. \qquad (10.151)$$

The same comments as below Eq. (10.147) can be made about the oscillations.

Let us summarize how the transfer function from Eq. (10.147), or its approximate form from Eq. (10.151), affects the spectral shape and amplitude of the energy density carried by vacuum-generated gravitational waves:

- As for the shape, Eq. (10.151) is to a good approximation independent of k, just like the original power spectrum from Eq. (10.31) (the latter corresponds to the tensor spectral tilt being very small, cf. Eq. (10.34)). Therefore, for frequencies small enough to cross outside of the Hubble horizon during inflation (cf. the discussion in the paragraph above Eq. (10.138)), and to re-enter it during an epoch where the trace anomaly is small, the current energy density is almost independent of the frequency.

- As for the overall amplitude of the gravitational-wave energy density, its order of magnitude can be approximated from Eq. (10.151). Inserting the evolution of the scale factor (cf. Eq. (2.34)) and the Hubble rate (cf. Eq. (2.36)) as well as the critical energy density (cf. Eq. (10.1)), we obtain

$$\left(\frac{a_{\mathrm{out}}}{a_0}\right)^4 \left(\frac{H_{\mathrm{out}}}{H_0}\right)^2 \underset{(10.1)}{\overset{(2.36),(2.34)}{=}} \left(\frac{\bar{s}_0/T_0^3}{\bar{s}_{\mathrm{out}}/T_{\mathrm{out}}^3}\right)^{4/3} \left(\frac{\bar{e}_{\mathrm{out}}/T_{\mathrm{out}}^4}{\bar{e}_{\gamma,0}/T_0^4}\right) \overbrace{\left(\frac{\bar{e}_{\gamma,0}}{e_{\mathrm{crit}}}\right)}^{\equiv\,\Omega_{\gamma,0}}, \qquad (10.152)$$

where $\bar{e}_{\gamma,0}$ is the current energy density of CMB photons. The last factor in Eq. (10.152) is the numerically most significant one: even if thermal photons carry a large fraction of the current entropy density, they contribute very little to the current energy density, $\Omega_{\gamma,0} \sim 5 \times 10^{-5}$, which is instead dominated by dark energy and dark matter. As concerns the other two factors, they do differ from unity, because the effective numbers of light degrees of freedom, as defined in Eq. (7.55), change with the temperature. However, this dependence is moderate, and the two factors also partly compensate against each other. As a consequence, the choice of τ_{out} has little influence. Including the factor $1/12$, the numerical magnitude of the transfer function at $f_0 > 10^{-8}$ Hz is then $h^2 \mathcal{T}_{\mathrm{t}} \sim 10^{-6}$.

- We arrived at Eq. (10.151) by assuming the absence of a trace anomaly, i.e. that $\bar{e} - 3\bar{p} = 0$. However, certain frequencies re-enter inside the Hubble horizon when the trace anomaly is substantial [14], for instance during the QCD crossover at $T \sim 150$ MeV. For these frequencies, $\mathcal{T}_{\mathrm{t}}$ has to be determined numerically, as specified by Eqs. (10.145) and (10.146). This produces small but non-trivial 'features' in the current gravitational-wave energy density, interpolating between the flat parts existing at other frequencies. We illustrate these patterns in Fig. 10.2. If a sufficient level of precision can be reached one day, such features offer for a direct way to see the thermal history of the early universe imprinted on the gravitational-wave background [32].

(ii) Tensor perturbations from reheating and post-inflationary dynamics.

Between Eqs. (10.141) and (10.152), we have determined the transfer function for tensor perturbations produced by vacuum fluctuations that were extended beyond the Hubble horizon during inflationary expansion. However, as discussed in Sects. 10.3–10.6, later moments in the history of the universe can produce additional contributions to the gravitational-wave background. Here we discuss how these get redshifted to the current day.

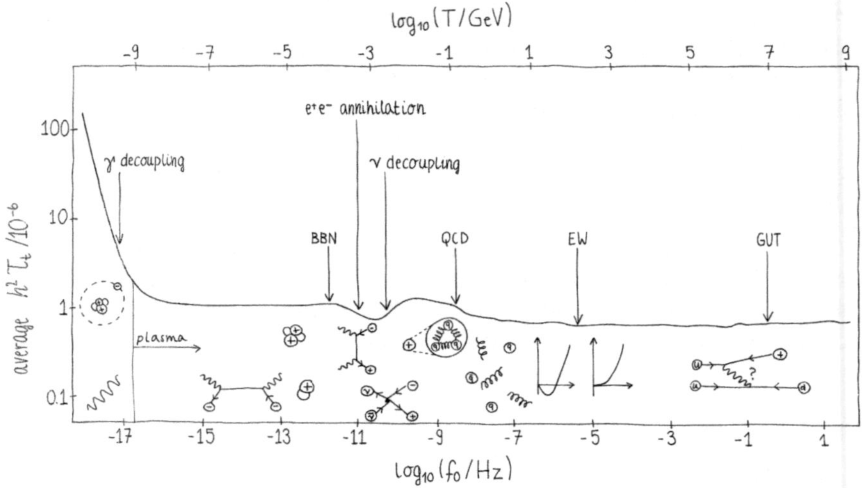

Fig. 10.2 A sketch of the transfer function for the tensor channel, $h^2 T_t$, from Eq. (10.147). The lower horizontal axis displays the current frequency, f_0/Hz, and the upper one the temperature at which the corresponding mode re-enters the Hubble horizon, cf. Eqs. (2.40) and (2.41). The small features correspond to mass thresholds, or the QCD and electroweak (or higher-scale) phase transitions, which induce a trace anomaly in the thermodynamics of the radiation plasma (cf., e.g., Ref. [14]). Quantitative evaluations of $h^2 T_t$ can be found, e.g., in Refs. [30,31], with the former displaying also the rapid oscillations in Eq. (10.147). In the sketch above, we only show the physically meaningful average, in line with the discussion below Eq. (10.147)

As a starting point, we re-express the gravitational energy density production rate from Eq. (10.109) with microscopic information according to Eq. (10.114),

$$\left(\partial_\tau + 4\mathcal{H}\right) \frac{\mathrm{d}e_{\mathrm{gw}}(\tau, k)}{\mathrm{d}\ln k} \overset{\substack{(10.109),\,(10.114)\\ k/a,\,T \gg H}}{\approx} \frac{4k^3}{\pi m_{\mathrm{pl}}^2 a^2} n_{\mathrm{B}}\!\left(\frac{k}{a}\right) \mathrm{Im}\, G_{\mathrm{R}}^{\mathrm{t}}\!\left(\frac{\mathcal{K}}{a}\right) . \tag{10.153}$$

Here $G_{\mathrm{R}}^{\mathrm{t}}$ is the retarded 2-point correlator associated with anisotropic stress, and the Fourier transform is now taken in local Minkowskian coordinates ($\int_{\mathcal{R}} e^{-i\mathcal{P}\cdot\mathcal{R}} = a^4 \int_{\mathcal{X}} e^{-i\mathcal{K}\cdot\mathcal{X}}$). We transform Eq. (10.153) into physical time, which according to Eqs. (1.6) and (1.7) goes via $\partial_\tau + 4\mathcal{H} = a(\partial_t + 4H)$. Then we define

$$\mathcal{E}(t, k) \equiv \frac{a^4}{a_0^4} \frac{\mathrm{d}e_{\mathrm{gw}}(t, k)}{\mathrm{d}\ln k} \quad \Rightarrow \quad \frac{\mathrm{d}e_{\mathrm{gw}}(t, k)}{\mathrm{d}\ln k} = \frac{a_0^4}{a^4} \mathcal{E}(t, k) . \tag{10.154}$$

It follows that

$$\frac{4k^3}{\pi m_{\mathrm{pl}}^2 a^3} n_{\mathrm{B}}\!\left(\frac{k}{a}\right) \mathrm{Im}\, G_{\mathrm{R}}^{\mathrm{t}}\!\left(\frac{\mathcal{K}}{a}\right) \overset{\substack{(10.153)\\(10.154)}}{=} (\partial_t + 4H)\left[\frac{a_0^4}{a^4} \mathcal{E}(t, k)\right] = \frac{a_0^4}{a^4} \partial_t \mathcal{E}(t, k) ,$$

$$\tag{10.155}$$

which can be integrated into

$$\mathcal{E}(t_0, k) - \mathcal{E}(t_e, k) = \int_{t_e}^{t_0} dt \, \frac{4k^3 a}{\pi m_{\text{pl}}^2 a_0^4} \, n_{\text{B}}\!\left(\tfrac{k}{a}\right) \text{Im} G_{\text{R}}^{\text{t}}\!\left(\tfrac{\mathcal{K}}{a}\right)$$

$$\overset{(10.154)}{\Leftrightarrow} \quad \frac{de_{\text{gw}}(t_0, k)}{d \ln k} = \frac{a_e^4}{a_0^4} \frac{de_{\text{gw}}(t_e, k)}{d \ln k} \tag{10.156}$$

$$+ \int_{t_e}^{t_0} dt \, \frac{a^4}{a_0^4} \frac{4}{\pi m_{\text{pl}}^2} \left(\frac{k}{a}\right)^3 n_{\text{B}}\!\left(\tfrac{k}{a}\right) \text{Im} G_{\text{R}}^{\text{t}}\!\left(\tfrac{\mathcal{K}}{a}\right) .$$

Here the initial moment, t_e, where the subscript refers to the beginning of emission (which may coincide with the end of inflation), should be chosen late enough that the production is well back within the Hubble horizon (otherwise we need to return to a transfer function like in Eq. (10.147)).

The physical interpretation of Eq. (10.156) can be as follows. The first term on the right-hand side indicates that, if there is an initial inside-horizon spectrum at time t_e, it redshifts until today with the factor a_e^4/a_0^4. The second term tells that, in addition, we need to integrate over the production taking place between t_e and t_0, redshifting the momenta at which the production happens (k/a), so that we consider a fixed frequency today. Moreover, we need to redshift the energy densities produced at a given time by a^4/a_0^4.

Let us now illustrate the use of Eq. (10.156) with a practical example. In order to have a simple right-hand side, we consider the signal sourced by *hydrodynamic fluctuations*, previously considered in Eq. (10.48). Given that hydrodynamic fluctuations appear in the classical domain, $H \ll k/a \ll T$, we rewrite the retarded Green's function in Eq. (10.156) in terms of the Wightman function from Eq. (10.113), which has a direct classical limit,

$$\frac{de_{\text{gw}}(t_0, k)}{d \ln k} \overset{(10.156)}{\underset{(10.113)}{\supset}} \int_{t_e}^{t_0} dt \, \frac{a^4}{a_0^4} \frac{2}{\pi m_{\text{pl}}^2} \left(\frac{k}{a}\right)^3 \int_{\mathcal{R}} e^{-i\frac{\mathcal{K}}{a}\cdot\mathcal{R}} \langle \Pi_{ij}^{\text{t}}(t) \, \Pi_{ij}^{\text{t}}(t + \mathcal{R}) \rangle .$$

$$\tag{10.157}$$

The 2-point correlator of the anisotropic stress can be determined with Eqs. (3.24), (3.127) and (3.141). Replacing Π_{ij} by its noise part S_{ij} according to Eq. (3.127), the correlator can be expressed in comoving momentum space as

$$\int_{\mathcal{R}} e^{-i\frac{\mathcal{K}}{a}\cdot\mathcal{R}} \langle \Pi_{ij}^{\text{t}}(t) \, \Pi_{ij}^{\text{t}}(t + \mathcal{R}) \rangle \overset{\mathcal{R}=a\mathcal{X}}{\underset{a \approx \text{constant}}{\supset}} a^4 \int_{\mathcal{X}} e^{-i\mathcal{K}\cdot\mathcal{X}} \langle S_{ij}^{\text{t}}(\tau) \, S_{ij}^{\text{t}}(\tau + \mathcal{X}) \rangle \tag{10.158}$$

$$\overset{(8)}{\underset{(9)}{=}} a^4 \int_{\tau'} e^{ik\tau'} \underbrace{\int_{\mathbf{q}} \langle S_{ij}^{\text{t}}(\tau, \mathbf{q}) \, S_{ij}^{\text{t}}(\tau + \tau', \mathbf{k}) \rangle}_{\text{insert } (3.141)} .$$

The traceless and transverse index contraction ($S_{ij}^{\text{t}} S_{ij}^{\text{t}} = \mathbb{T}_{ij}^{kl} S_{ij} S_{kl}$) yields

$$\overbrace{\frac{1}{2}\left(\mathbb{P}_i^k \mathbb{P}_j^l + \mathbb{P}_i^l \mathbb{P}_j^k - \mathbb{P}_{ij}\mathbb{P}^{kl}\right)}^{\mathbb{T}_{ij}^{kl}\ \text{from (3.24)}} \overbrace{\left[\eta\left(\delta_{ik}\delta_{jl} + \delta_{il}\delta_{jk}\right) + \left(\zeta - \frac{2\eta}{3}\right)\delta_{ij}\delta_{kl}\right]}^{\text{part of } \langle S_{ij} S_{kl}\rangle \text{ from (3.141)}} \tag{10.159}$$

$$= \frac{\eta}{2}\left(2\mathbb{P}_i^k \mathbb{P}_j^j + 2\mathbb{P}_i^j \mathbb{P}_j^i - 2\mathbb{P}_i^j \mathbb{P}_j^i\right) + \frac{1}{2}\left(\zeta - \frac{2\eta}{3}\right)\left(2\mathbb{P}_i^j \mathbb{P}_j^i - \mathbb{P}_i^i \mathbb{P}_j^j\right) = 4\eta\,.$$

Putting everything together, we find

$$\int_{\mathcal{R}} e^{-i\frac{\mathcal{K}}{a}\cdot\mathcal{R}} \langle \Pi_{ij}^{\mathrm{t}}(t)\, \Pi_{ij}^{\mathrm{t}}(t+\mathcal{R})\rangle \underset{(3.141)}{\overset{(10.158),(10.159)}{\supset}} 8\,T\eta = 8\,T^4\hat{\eta}\,, \tag{10.160}$$

where the rescaled $\hat{\eta} \equiv \eta/T^3$ is dimensionless. We plug this into Eq. (10.157), and convert the time integral to a temperature one with the help of Eq. (1.89),

$$\frac{d e_{\mathrm{gw}}(t_0, k)}{d\ln k} \underset{(10.160)}{\overset{(10.157)}{\supset}} \int_{t_e}^{t_0} dt\, \frac{a^4}{a_0^4}\frac{2}{\pi m_{\mathrm{pl}}^2}\left(\frac{k}{a}\right)^3 8\,T^4\hat{\eta}$$

$$\overset{(1.89)}{\approx} \frac{3\sqrt{5}}{2\sqrt{\pi^3}}\frac{16}{\pi m_{\mathrm{pl}}}\left(\frac{k}{a_0}\right)^3 \int_{T_0}^{T_e} dT\, \frac{\hat{\eta}}{\sqrt{g_*}}\frac{aT}{a_0}\,. \tag{10.161}$$

The integrand is slowly varying, so for $T_e \gg T_0$ the value of the integral is dominated by the region close to the upper boundary, and the result can be approximated as

$$\int_{T_0}^{T_e} dT\, \frac{\hat{\eta}}{\sqrt{g_*}}\frac{aT}{a_0} \overset{T_e \gg T_0}{\approx} T_e\frac{\hat{\eta}_e}{\sqrt{g_{*,e}}}\frac{a_e T_e}{a_0} \overset{(2.34)}{\underset{(7.55)}{=}} T_e T_0\frac{\hat{\eta}_e}{\sqrt{g_{*,e}}}\left(\frac{h_{*,0}}{h_{*,e}}\right)^{1/3}\,. \tag{10.162}$$

Furthermore, we write $k = a_0 p_0$ (cf. Eq. (2.37)) and $e_{\gamma,0} = g_{\gamma,0}\pi^2 T_0^4/30$ (cf. Eq. (7.55)). With these, the current spectrum takes the form

$$\frac{d e_{\mathrm{gw}}(t_0, k)}{d\ln k} \underset{(10.162)}{\overset{(10.161)}{\supset}} \frac{3\sqrt{5}}{2\sqrt{\pi^3}}\frac{16\,T_e}{\pi m_{\mathrm{pl}}}\left(\frac{p_0}{T_0}\right)^3 \frac{\hat{\eta}_e}{\sqrt{g_{*,e}}}\left(\frac{h_{*,0}}{h_{*,e}}\right)^{1/3}\overbrace{\frac{30\,e_{\gamma,0}}{\pi^2 g_{\gamma,0}}}^{T_0^4}\,. \tag{10.163}$$

The final step is to normalize the result to the critical energy density (cf. Eq. (10.3)),

$$\frac{h^2 d\Omega_{\mathrm{gw},0}}{d\ln f_0} \underset{(10.3)}{\overset{(10.163)}{\supset}} \frac{720\sqrt{5}}{\pi^3\sqrt{\pi^3}}\overbrace{\left(\frac{p_0}{T_0}\right)^3}^{(2.39)}\overbrace{\frac{1}{g_{\gamma,0}\sqrt{g_{*,e}}}}^{\approx \frac{1}{2.0\sqrt{106.75}}}\overbrace{\left(\frac{h_{*,0}}{h_{*,e}}\right)^{1/3}}^{\approx \frac{3.93}{106.75}}\overbrace{\frac{h^2 e_{\gamma,0}}{e_{\mathrm{crit}}}}^{\approx 2.47\times 10^{-5}}\frac{\hat{\eta}_e T_e}{m_{\mathrm{pl}}}$$

$$\approx 2.02\times 10^{-29}\frac{\hat{\eta}_e T_e}{m_{\mathrm{pl}}}\left(\frac{f_0}{\mathrm{kHz}}\right)^3\,, \qquad f_0 \ll 10^{11}\,\mathrm{Hz}\,. \tag{10.164}$$

This shows that the maximal temperature (and shear viscosity) of the radiation-dominated epoch is in principle visible, as the coefficient of a f_0^3 growth of the gravitational energy density in the frequency domain of the planned ET and CE interferometers (cf. Sect. 2.3 and Fig. 10.3). However, an exquisite resolution might be needed for observing the signal.

10.8   Overview of Frequency Domains

Let us summarize the results of this chapter. Tensor perturbations, or gravitational waves, carry energy density (cf. Sect. 10.1), and can be produced from vacuum fluctuations by the same mechanism as curvature perturbations (cf. Sect. 10.2). After these modes re-enter inside the Hubble horizon, their frequency redshifts, however the spectrum remains almost flat, with small features superimposed on it by special epochs during the subsequent expansion (cf. Sect. 10.7).

There are also new sources for gravitational waves. A model-independent contribution originates from hydrodynamic fluctuations of the reheated plasma that fills the universe (cf. Sect. 10.3 and Eq. (10.164)). Moreover, the curvature perturbations that we have discussed in previous chapters, produce a gravitational-wave background at second order (cf. Sect. 10.4). In contrast to vacuum fluctuations, the later sources generate a spectrum which shows overall growth with frequency, modified by the 2-point correlation function of anisotropic stress (cf. Sect. 10.5). Ultimately, going to very high frequencies, gravitational waves are produced by quantum-mechanical elementary particle decays and scatterings (cf. Sect. 10.6). However, the energy carried by them must come out of somewhere. Above the scale of thermal motion, $p \sim T$, there are no guaranteed sources present, and the amplitude of the primordial spectrum will likely turn down.

As has been reviewed in Sect. 2.3, these primordial sources can conceivably be probed with future observations, in particular by combing interferometric searches sensitive to different frequency domains. Of course, the sensitivity of each individual experiment is limited by various 'backgrounds', ranging from instrumental noise to astrophysical foregrounds. At the time of writing, the conceptual and technical development of the instruments is work in progress, and it is difficult to judge what their sensitivities will be. However, even if the observed gravitational-wave background will be dominated by astrophysical sources, it is clear that we are facing the advent of a new probe for the physics of the early universe.

Complementary to interferometers, there is an important indirect constraint on gravitational waves in any frequency range, and that is their contribution to the *effective number of neutrinos*, N_{eff}, as explained around Eq. (2.24). Traditionally, this is referred to as a BBN constraint, because big bang nucleosynthesis works successfully only if the overall expansion rate, and therefore the energy density at that time ($T_{\mathrm{BBN}} \sim 0.1\,\mathrm{MeV}$), is very close to that predicted by the Standard Model. In addition, the overall energy density is constrained by the physics of photon decoupling ($T_{\mathrm{dec}} \sim 0.3\,\mathrm{eV}$). Even though these two constraints, $[\Delta N_{\mathrm{eff}}]_{\mathrm{BBN}}$ and $[\Delta N_{\mathrm{eff}}]_{\mathrm{dec}}$, are independent, their numerical relationship to today's $\Omega_{\mathrm{gw},0}$ is the same, once we note that the BBN constraint is by convention evaluated at $T \ll T_{\mathrm{BBN}}$, so that the electron mass can be approximated as heavy, $\exp(-m_e/T) \ll 1$. In phenomenological determinations, the BBN and CMB data sets are often combined, in order to obtain a more precise estimate of the observed value of N_{eff}. Therefore, the additional subscript from ΔN_{eff} is usually dropped. Below we illustrate how the bound is imposed, focussing on the CMB case for concreteness (cf. Eq. (10.167)).

Following Eq. (2.24), we write the gravitational energy density as an undetermined radiation energy contribution allowed by observational uncertainties around CMB decoupling,

$$e_{\mathrm{gw,dec}} \equiv \Delta N_{\mathrm{eff}} \frac{7}{8}\left(\frac{4}{11}\right)^{4/3} \bar{e}_{\gamma,\mathrm{dec}} . \tag{10.165}$$

This can be redshifted to today according to the first term on the right-hand side of Eq. (10.156), and subsequently normalized to the critical energy density (cf. Eq. (10.3)),

$$\Omega_{\mathrm{gw,0}} \underset{(10.156)}{\overset{(10.3)}{=}} \frac{e_{\mathrm{gw,dec}}}{e_{\mathrm{crit}}} \frac{a_{\mathrm{dec}}^4}{a_0^4} . \tag{10.166}$$

Then we insert the evolution of the scale factor from Eq. (2.34), and also divide and multiply with the energy density in CMB photons, like in Eq. (10.152),

$$\Omega_{\mathrm{gw,0}} \underset{(10.165),(10.166)}{\overset{(2.34)}{=}} \Delta N_{\mathrm{eff}} \frac{7}{8}\left(\frac{4}{11}\right)^{4/3} \left(\frac{\bar{s}_0/T_0^3}{\bar{s}_{\mathrm{dec}}/T_{\mathrm{dec}}^3}\right)^{4/3} \left(\frac{\bar{e}_{\gamma,\mathrm{dec}}/T_{\mathrm{dec}}^4}{\bar{e}_{\gamma,0}/T_0^4}\right) \frac{\bar{e}_{\gamma,0}}{e_{\mathrm{crit}}}$$

$$\overset{(7.55)}{=} \Delta N_{\mathrm{eff}} \frac{7}{8}\left(\frac{4}{11}\right)^{4/3} \underbrace{\left(\frac{h_{*,0}}{h_{*,\mathrm{dec}}}\right)^{4/3} \left(\frac{g_{\gamma,\mathrm{dec}}}{g_{\gamma,0}}\right) \overbrace{\frac{\bar{e}_{\gamma,0}}{e_{\mathrm{crit}}}}^{\approx 2.47\times 10^{-5}/h^2}}_{\approx 5.62\times 10^{-6}/h^2} . \tag{10.167}$$

We recall from the discussion below Eq. (2.34) that, by convention, the current entropy density encompasses the effect of the decoupled neutrinos, whereas in the energy density, only the thermalized electromagnetic plasma is included. For the numerical estimates, the photon energy densities have been approximated via $g_{\gamma,\mathrm{dec}} \approx g_{\gamma,0} \approx 2.0$, and we have also set $h_{*,\mathrm{dec}} \approx h_{*,0}$, given that neutrino decoupling is completed well before we reach T_{dec}. From the observational upper bound on ΔN_{eff} from BBN and CMB physics, we therefore obtain an empirical upper bound on gravitational-wave energy density,

$$h^2\Omega_{\mathrm{gw,0}} \overset{(10.167)}{<} 5.62 \times 10^{-6}\, \Delta N_{\mathrm{eff}}^{\mathrm{obs}} , \tag{10.168}$$

where $\Delta N_{\mathrm{eff}}^{\mathrm{obs}} \simeq 0.15 \pm 0.11$ according to Eqs. (2.23) and (2.25).

It should be stressed that Eq. (10.168) represents a constraint on the *integral* over the gravitational-wave energy density spectrum,

$$h^2 \Omega_{\mathrm{gw},0} = \int_{-\infty}^{\infty} \mathrm{d} \ln f_0 \, \frac{h^2 \mathrm{d} \Omega_{\mathrm{gw},0}}{\mathrm{d} \ln f_0} \, . \tag{10.169}$$

Quite often in the literature, the result from Eq. (10.168) is displayed as a frequency-independent upper bound, and we repeat this in Fig. 10.3. Unfortunately, there is also a widespread convention of denoting the *differential* spectrum by $\Omega_{\mathrm{gw},0}$, having no separate notation for the integrated gravitational-wave energy density, which increases the likelihood of misunderstanding. In any case, if there is a peak in the spectrum, the maximal amplitude of $h^2 \mathrm{d} \Omega_{\mathrm{gw},0}/\mathrm{d} \ln f_0$ could well violate Eq. (10.168), as long as the peak is narrow, so that the overall gravitational-wave energy density satisfies it. Conversely, if the spectrum is nearly flat, the would-be constraint is stronger than Eq. (10.168) — for example, a constant of *any* magnitude is excluded, as the integral in Eq. (10.169) would diverge.

Let us also briefly quantify the constraint on $h^2 \mathrm{d} \Omega_{\mathrm{gw},0}/\mathrm{d} \ln f_0$ from CMB polarization, discussed around Eq. (2.17). Adapting Eq. (10.141), we can express it as

$$\frac{h^2 \mathrm{d} \Omega_{\mathrm{gw},0}}{\mathrm{d} \ln f_0} \overset{(10.141)}{\simeq} \underbrace{h^2 \mathcal{T}_{\mathrm{t}}(k)}_{\text{Fig. 10.2: } \sim 10^{-6}} \underbrace{\frac{\mathcal{P}_{\mathrm{t}}(\tau_{\mathrm{out}}, k)}{\mathcal{P}_{\mathrm{s}}(\tau_{\mathrm{out}}, k)}}_{(2.17):\, <\, 4\times 10^{-2}} \underbrace{\mathcal{P}_{\mathrm{s}}(\tau_{\mathrm{out}}, k)}_{(2.8):\, \sim\, 2\times 10^{-9}} \lesssim 10^{-16} \, . \tag{10.170}$$

This is much stronger than Eq. (10.168), however only applicable at very small frequencies: the momentum scale $k/a_0 = 0.002\,\mathrm{Mpc}^{-1} = 2\pi/\lambda_0$ from Eq. (2.17) corresponds to $f_0 \approx 3 \times 10^{-18}\,\mathrm{Hz}$ according to Eq. (2.45).

Finally, we would like to explain the relation of the gravitational-wave energy density to another quantity relevant for the observational side. It is meant as an instrumental tool, as it gives a direct feeling about the relative magnitude of a space-time distortion that is caused by a gravitational wave entering an interferometer.

As a starting point, let us inspect Eq. (10.24), normalized to the critical energy density like in Eq. (10.3), but now in physical time,

$$\Omega_{\mathrm{gw},0} \overset{(10.3)}{\underset{(10.24)}{\simeq}} \frac{1}{12 H_0^2} \sum_{ij} \langle (\dot{h}_{ij}^{\mathrm{t}})^2 \rangle \, . \tag{10.171}$$

The idea is to re-parametrize the differential spectrum, $\mathrm{d} \Omega_{\mathrm{gw},0}/\mathrm{d} \ln f_0$, through a new quantity, $\mathrm{h}_{\mathrm{t}}(f_0)$, called the *gravitational strain*. A recipe for this is to replace h_{ij}^{t} in Eq. (10.171) by a plane wave of angular frequency $\omega_0 = 2\pi f_0$, which is 'unpolarized', so that the sum $\sum_{ij}$ is replaced by a factor $\Sigma \equiv 4$. In addition, as explained below Eq. (10.147), we should take an oscillation average, multiplying the result by $1/2$, even if this is not always implemented in the literature; we indicate this with $\langle \Sigma \rangle$.

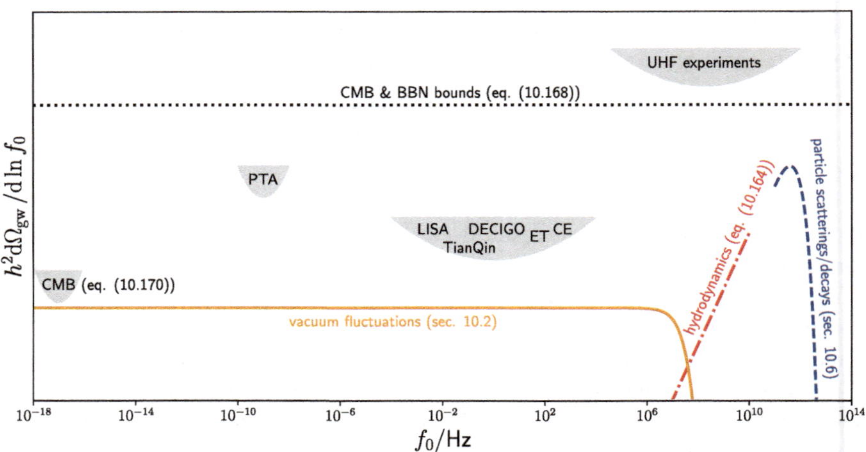

Fig. 10.3 An illustration of the contributions of inflation and post-reheating physics to the current gravitational-wave energy density. Only features that are guaranteed to be present are displayed, even though their amplitudes are still unknown (they depend, respectively, on the energy scale of inflation, and on the maximal temperature after reheating). The curves that we show are oversimplified; an example of a more accurate determination of the vacuum part, from Eqs. (10.31) and (10.141), can be found in Fig. 8 of Ref. [30], whereas model realizations of the high-frequency spectrum can be found, e.g., in Refs. [33–35]. The 'CMB & BBN bounds' from Eq. (10.168), describing the gravitational-wave contribution to the effective number of neutrinos (N_{eff}), applies to the integral over the spectrum rather than differentially. The planned interferometers, described in Sect. 2.3, are only indicated schematically, given that their ultimate resolutions are difficult to anticipate, due to astrophysical foregrounds and instrumental challenges. We remark that in the literature, the y-axis of this type of plots is frequently labelled by $h^2\Omega_{\mathrm{gw},0}$, even though what is meant is the differential spectrum, $h^2\mathrm{d}\Omega_{\mathrm{gw},0}/\mathrm{d}\ln f_0$

After these steps, we write the differential spectrum as

$$\frac{\mathrm{d}\Omega_{\mathrm{gw},0}}{\mathrm{d}\ln f_0} \equiv \frac{\langle\Sigma\rangle\,\pi^2 f_0^2 \mathrm{h}_\mathrm{t}^2(f_0)}{3H_0^2} \equiv \frac{\langle\Sigma\rangle\,\pi^2 f_0^3}{3H_0^2}\,S_{\mathrm{h}_\mathrm{t}}(f_0)\,, \tag{10.172}$$

where $S_{\mathrm{h}_\mathrm{t}} \equiv \mathrm{h}_\mathrm{t}^2(f_0)/f_0$ is the *strain sensitivity* (often $S_{\mathrm{h}_\mathrm{t}}^{1/2}[\mathrm{Hz}^{-1/2}]$ is plotted). The factor f_0^3 in Eq. (10.172) matches the k^3 in Eq. (10.109). Assuming that technology allows to probe similar strain sensitivities over many frequencies, this suggests opportunities for mapping the physics induced by anisotropic stress across a broad spectrum of phenomena.

A schematic summary of various sources and detection opportunities is shown in Fig. 10.3. Here it is appropriate to repeat the warning from below Eq. (10.169): in the literature, the y-axis of this type of plots is frequently labelled by $h^2\Omega_{\mathrm{gw},0}$, even though what is meant is the differential spectrum, $h^2\mathrm{d}\Omega_{\mathrm{gw},0}/\mathrm{d}\ln f_0$.

10.9   Computer Algebra for Einstein Tensor at Second Order

In the script below, we carry out the main computations that were reported in Sect. 10.4. First, Eq. (10.60) is checked. Then its scalar-induced part is moved to the right-hand side of the Einstein equation, where it is combined with the relevant parts of Eq. (10.78). The variables are expressed in terms of Bardeen potentials via Eqs. (10.63) and (10.64). What is *not* implemented are the IBP relations from Eqs. (10.67), (10.70), and (10.71). Therefore, the final result is a bit more complicated than Eq. (10.85), however the expressions are equivalent if IBP relations are subsequently employed.

```python
# computer-algebraic determination of the source for scalar-induced GWs [symbolic_sigw.py]
#
# basic coordinates and functions appearing
from sympy import *
i, j, k, l = symbols('i j k l',cls=Idx); x = IndexedBase('x'); eps = symbols('eps')
a = Function('a')(x[0]); H = Function('H')(x[0])
h = Function('h')(x[0],x[1],x[2],x[3])
h0 = Function('h0')(x[0],x[1],x[2],x[3]); hD = Function('hD')(x[0],x[1],x[2],x[3])
phi = Function('phi')(x[0],x[1],x[2],x[3]); psi = Function('psi')(x[0],x[1],x[2],x[3])

# partial derivative with respect to x[i]
def d(h,i): return diff(h,x[i],evaluate=True)
def dd(h,i,j): return diff(h,x[i],x[j],evaluate=True)          # doesn't symmetrize as should :(
def dds(h,i,j): return (dd(h,i,j)+dd(h,j,i))/2                 # symmetrized
def ddd(h,i,j,k): return diff(h,x[i],x[j],x[k],evaluate=True)  # doesn't symmetrize as should :(
def ddds(h,i,j,k): return (ddd(h,i,j,k)+ddd(h,i,k,j))/2        # symmetrized in last two args

# restrict to eps^order for possibly more efficient execution
def trunc2(h): return (h.expand()).subs([[(eps**5,0),(eps**4,0),(eps**3,0)]])
def trunc1(h): return (h.expand()).subs([[(eps**2,0)]])
order = 2

# input metric with both indices down
a00 = a**2*(-1-2*eps*h0)
a01 = a**2*(eps*d(h,1));   a02 = a**2*(eps*d(h,2)); a03 = a**2*(eps*d(h,3))
a10 = a01; a20 = a02; a30 = a03
a11 = a**2*(1-2*eps*hD); a22=a11; a33=a11
a12 = 0; a13 = 0; a23 = 0; a21 = a12; a31 = a13; a32 = a23
gdown=Matrix([[a00,a01,a02,a03],[a10,a11,a12,a13],[a20,a21,a22,a23],[a30,a31,a32,a33]])

# input metric with both indices up
b00 = a**(-2)*(-1+2*eps*h0 + eps**2*( -4*h0**2 + d(h,1)**2 + d(h,2)**2 + d(h,3)**2 ))
b01 = a**(-2)*d(h,1)*(eps +  eps**2*(  2*hD - 2*h0  ))
b02 = a**(-2)*d(h,2)*(eps +  eps**2*(  2*hD - 2*h0  ))
b03 = a**(-2)*d(h,3)*(eps +  eps**2*(  2*hD - 2*h0  ))
b10 = b01; b20 = b02; b30 = b03
b11 = a**(-2)*(1 + 2*eps*hD + eps**2*( 4*hD**2 - d(h,1)**2 ))
b22 = a**(-2)*(1 + 2*eps*hD + eps**2*( 4*hD**2 - d(h,2)**2 ))
b33 = a**(-2)*(1 + 2*eps*hD + eps**2*( 4*hD**2 - d(h,3)**2 ))
b12 = a**(-2)*(0 + eps**2*( - d(h,1)*d(h,2) ))
b13 = a**(-2)*(0 + eps**2*( - d(h,1)*d(h,3) ))
b23 = a**(-2)*(0 + eps**2*( - d(h,2)*d(h,3) ))
b21 = b12; b31 = b13; b32 = b23
gup=Matrix([[b00,b01,b02,b03],[b10,b11,b12,b13],[b20,b21,b22,b23],[b30,b31,b32,b33]])
if order==1: gup = trunc1(gup)

# check inversion
test=(gdown*gup)
if order==1: print(trunc1(test))
else: print(trunc2(test))
print("inversion checked")

# christoffel symbols
def dgdown(m,n,i): return d(gdown[m,n],i)
def gamma(r,m,n):
```

```
        return Rational(1/2)*Sum( gup[r,i]*(dgdown(i,m,n)+dgdown(i,n,m)-dgdown(m,n,i)),(i,0,3)).doit()

# ricci tensor with both indices down
def prericcidown(m,n):
    return ( Sum( d(gamma(k,m,n),k) - d(gamma(k,m,k),n),(k,0,3) ).doit() +
        Sum( gamma(l,m,n)*gamma(k,l,k) - gamma(l,m,k)*gamma(k,l,n),(k,0,3),(l,0,3) ).doit() )
riccidown=trunc2(Matrix(4,4,prericcidown))
if order==1: riccidown=trunc1(riccidown)

# ricci tensor with indices up and down
ricciupdown = trunc2(gup*riccidown)
if order==1: ricciupdown=trunc1(ricciupdown)

# ricci scalar
ricciscalar = ricciupdown.trace()

# einstein tensor with both indices down
einstein = trunc2(riccidown - ricciscalar*gdown/2)
if order==1: einstein=trunc1(einstein)

# replace derivatives of the scale factor by the hubble rate
einstein = einstein.subs([(Derivative(a,x[0]),a*H),(Derivative(a,x[0],2),a*(d(H,0)+H**2))])

# adjust expression until the difference subtracts to zero
testeinstein12 = (0 + eps*( - dds(h0,1,2) + dds(hD,1,2) - ddds(h,0,1,2) - 2*H*dds(h,1,2) )
                    + eps**2*(
                        + d(h0,1)*d(h0,2) + 3*d(hD,1)*d(hD,2)
                        + 2*( h0*dds(h0,1,2) + hD*dds(hD,1,2) )
                        - d(h0,1)*d(hD,2) - d(hD,1)*d(h0,2)
                        + ( d(h0,0) + d(hD,0) )*dds(h,1,2)
                        + 4*H*h0*dds(h,0,1,2) + 2*h0*ddds(h,0,1,2)
                        - d(hD,1)*dd(h,0,2) - d(hD,2)*dd(h,0,1)
                        - dd(hD,0,1)*d(h,2) - dd(hD,0,2)*d(h,1)
                        - 2*H*( d(hD,1)*d(h,2) + d(hD,2)*d(h,1) )
                        + ( dd(h,1,1) + dd(h,2,2) + dd(h,3,3) )*dds(h,1,2)
                        - ( dd(h,1,1)*dds(h,1,2) + dds(h,2,1)*dd(h,2,2) + dd(h,3,1)*dd(h,3,2) )
        #### remnant of non-symmetrized derivatives
                        + (dd(h,1,1)-dd(h,2,2))*(dd(h,2,1)-dd(h,1,2))/2    ) )
if order==1: testeinstein12=trunc1(testeinstein12)
print(expand((einstein[1,2]+einstein[2,1])/2 - testeinstein12))
print("einstein 12 checked to 2nd order")

# construct effective energy-momentum tensor by moving metric perturbations to right-hand side
rhs12=-testeinstein12 + eps**2*(
    -2*( d(h0,1)*d(hD,2) + d(hD,1)*d(h0,2) )
    -2/H*( dd(hD,0,1)*d(hD,2) + dd(hD,0,2)*d(hD,1) )
    +2*(d(H,0)/H**2 - 1)*d(hD,1)*d(hD,2) )
if order==1: rhs12=trunc1(rhs12)

# substitute bardeen potentials
rhs12=rhs12.subs([(h0,phi-d(h,0)-H*h),(hD,psi+H*h)]).doit()
if order==1: rhs12=trunc1(rhs12)

# adjust expressions until the difference subtracts to zero
testrhs12 = (0 + eps*( dds(phi,1,2)-dds(psi,1,2) )
                + eps**2*(
                    -d(phi,1)*d(phi,2) - 2*phi*dds(phi,1,2)
                    -( d(phi,1)*d(psi,2) + d(phi,2)*d(psi,1) ) - 2*psi*dds(psi,1,2)
                    +( 2*d(H,0)/H**2 - 5 )*d(psi,1)*d(psi,2)
                    -2/H*( dd(psi,0,1)*d(psi,2) + dd(psi,0,2)*d(psi,1) )
                    +2*H*( h*dds(phi,1,2) - phi*dds(h,1,2) )
                    -d(phi,0)*dds(h,1,2) + 2*dds(phi,1,2)*d(h,0) + d(phi,1)*dd(h,0,2) + d(phi,2)*dd(h,0,1)
                    -2*H*( d(psi,1)*d(h,2) + d(psi,2)*d(h,1) + psi*dds(h,1,2) + h*dds(psi,1,2) )
                    -d(psi,0)*dds(h,1,2) - dd(psi,0,1)*d(h,2) - dd(psi,0,2)*d(h,1)
                    +dd(h,1,1)*dds(h,1,2) + dds(h,2,1)*dd(h,2,2) + dd(h,3,1)*dd(h,3,2)
                    -( dd(h,1,1) + dd(h,2,2) + dd(h,3,3) )*dds(h,1,2)
                    +dd(h,0,0)*dds(h,1,2) - dd(h,0,1)*dd(h,0,2) + 2*H*d(h,0)*dds(h,1,2)
        #### remnant of non-symmetrized derivatives
                    - (dd(h,1,1)-dd(h,2,2))*(dd(h,2,1)-dd(h,1,2))/2    ) )
if order==1: testrhs12=trunc1(testrhs12)
print(expand(rhs12-testrhs12))
print("sigw 12 checked to 2nd order")
```

References

1. Isaacson, R.A.: Gravitational Radiation in the Limit of High Frequency. I. The Linear Approximation and Geometrical Optics. Phys. Rev. **166**, 1263 (1968). https://doi.org/10.1103/PhysRev.166.1263

2. Choi, S.Y., Shim, J.S., Song, H.S.: Factorization and polarization in linearized gravity. Phys. Rev. D **51**, 2751 (1995). https://doi.org/10.1103/PhysRevD.51.2751, hep-th/9411092

3. Grishchuk, L.P.: Amplification of gravitational waves in an isotropic universe. Sov. Phys. JETP **40**, 409 (1975) [Zh. Eksp. Teor. Fiz. 67 (1974) 825]

4. Starobinsky, A.A.: Spectrum of relict gravitational radiation and the early state of the universe. JETP Lett. **30**, 682 (1979) [Pisma Zh. Eksp. Teor. Fiz. 30 (1979) 719]

5. Rubakov, V.A., Sazhin, M.V., Veryaskin, A.V.: Graviton creation in the inflationary universe and the grand unification scale. Phys. Lett. B **115**, 189 (1982). https://doi.org/10.1016/0370-2693(82)90641-4

6. Abbott, L.F., Wise, M.B.: Constraints on generalized inflationary cosmologies. Nucl. Phys. B **244**, 541 (1984). https://doi.org/10.1016/0550-3213(84)90329-8

7. Domcke, V.: Discovery Opportunities with Gravitational Waves, Contribution to TASI (2024). 2409.08956

8. Laine, M., Procacci, S.: General SIGW source for reheating dynamics. 2512.04482

9. Hwang, J.-c., Jeong, D., Noh, H.: Gauge Dependence of Gravitational Waves Generated from Scalar Perturbations. Astrophys. J. **842**, 46 (2017). https://doi.org/10.3847/1538-4357/aa74be, 1704.03500

10. Acquaviva, V., Bartolo, N., Matarrese, S., Riotto, A.: Gauge-invariant second-order perturbations and non-Gaussianity from inflation. Nucl. Phys. B **667**, 119 (2003). https://doi.org/10.1016/S0550-3213(03)00550-9, astro-ph/0209156

11. Baumann, D., Steinhardt, P.J., Takahashi, K., Ichiki, K.: Gravitational wave spectrum induced by primordial scalar perturbations. Phys. Rev. D **76**, 084019 (2007). https://doi.org/10.1103/PhysRevD.76.084019, hep-th/0703290

12. Klose, P., Laine, M., Procacci, S.: Gravitational wave background from non-Abelian reheating after axion-like inflation. JCAP **05**, 021 (2022). https://doi.org/10.1088/1475-7516/2022/05/021, 2201.02317

13. Jeon, S.: Hydrodynamic transport coefficients in relativistic scalar field theory. Phys. Rev. D **52**, 3591 (1995). https://doi.org/10.1103/PhysRevD.52.3591, hep-ph/9409250

14. Caldwell, R.R., Gubser, S..S: Brief history of curvature. Phys. Rev. D **87**, 063523 (2013). https://doi.org/10.1103/PhysRevD.87.063523, 1302.1201

15. Ghiglieri, J., Jackson, G., Laine, M., Zhu, Y.: Gravitational wave background from Standard Model physics: complete leading order. JHEP **07**, 092 (2020). https://doi.org/10.1007/JHEP07(2020)092, 2004.11392

16. Ghiglieri, J., Schütte-Engel, J., Speranza, E.: Freezing-in gravitational waves. Phys. Rev. D **109**, 023538 (2024). https://doi.org/10.1103/PhysRevD.109.023538, 2211.16513

17. Ghiglieri, J., Laine, M., Schütte-Engel, J., Speranza, E.: Double-graviton production from Standard Model plasma. JCAP **04**, 062 (2024). https://doi.org/10.1088/1475-7516/2024/04/062, 2401.08766

18. Nakayama, K., Tang, Y.: Stochastic gravitational waves from particle origin. Phys. Lett. B **788**, 341 (2019); *ibid.* **839**, 137787 (2023) (erratum). https://doi.org/10.1016/j.physletb.2018.11.023, 1810.04975

19. Barman, B., Bernal, N., Xu, Y., Zapata, Ó.: Gravitational wave from graviton Bremsstrahlung during reheating. JCAP **05**, 019 (2023). https://doi.org/10.1088/1475-7516/2023/05/019, 2301.11345

20. Kanemura, S., Kaneta, K.: Gravitational waves from particle decays during reheating. Phys. Lett. B **855**, 138807 (2024). https://doi.org/10.1016/j.physletb.2024.138807, 2310.12023

21. Tokareva, A.: Gravitational waves from inflaton decay and bremsstrahlung. Phys. Lett. B **853**, 138695 (2024). https://doi.org/10.1016/j.physletb.2024.138695, 2312.16691

22. Alonzo-Artiles, A., Avilez-López, A., Díaz-Cruz, J.L., Larios-López, B.O.: The Higgs-Graviton Couplings: from Amplitudes to the Action. 2105.11684
23. Ema, Y., Mukaida, K., Nakayama, K.: Scalar field couplings to quadratic curvature and decay into gravitons. JHEP **05**, 087 (2022). https://doi.org/10.1007/JHEP05(2022)087, 2112.12774
24. Laine, M., Vuorinen, A.: Basics of Thermal Field Theory. Lect. Notes Phys. **925**, 1 (2016). https://doi.org/10.1007/978-3-319-31933-9, 1701.01554
25. Bödeker, D., Sangel, M., Wörmann, M.: Equilibration, particle production, and self-energy. Phys. Rev. D **93**, 045028 (2016). https://doi.org/10.1103/PhysRevD.93.045028, 1510.06742
26. Weinberg, S.: Damping of tensor modes in cosmology. Phys. Rev. D **69**, 023503 (2004). https://doi.org/10.1103/PhysRevD.69.023503, astro-ph/0306304
27. Watanabe, Y., Komatsu, E.: Improved calculation of the primordial gravitational wave spectrum in the standard model. Phys. Rev. D **73**, 123515 (2006). https://doi.org/10.1103/PhysRevD.73.123515, astro-ph/0604176
28. Mirón-Granese, N.: Relativistic viscous effects on the primordial gravitational waves spectrum. JCAP **06**, 008 (2021). https://doi.org/10.1088/1475-7516/2021/06/008, 2012.11422
29. Stoetzel, T., Fechtner, R., Floerchinger, S.: Energy-momentum response to metric perturbations in the fluid dynamic regime. 2508.16562
30. Saikawa, K., Shirai, S.: Primordial gravitational waves, precisely: the role of thermodynamics in the Standard Model. JCAP **05**, 035 (2018). https://doi.org/10.1088/1475-7516/2018/05/035, 1803.01038
31. Klose, P., Laine, M., Procacci, S.: Gravitational wave background from vacuum and thermal fluctuations during axion-like inflation. JCAP **12**, 020 (2022). https://doi.org/10.1088/1475-7516/2022/12/020, 2210.11710
32. Schwarz, D.J.: Evolution of gravitational waves through cosmological transitions. Mod. Phys. Lett. A **13**, 2771 (1998). https://doi.org/10.1142/S0217732398002941, gr-qc/9709027
33. Ghiglieri, J., Laine, M.: Gravitational wave background from Standard Model physics: qualitative features. JCAP **07**, 022 (2015). https://doi.org/10.1088/1475-7516/2015/07/022, 1504.02569
34. Ringwald, A., Tamarit, C.: Revealing the cosmic history with gravitational waves. Phys. Rev. D **106**, 063027 (2022). https://doi.org/10.1103/PhysRevD.106.063027, 2203.00621
35. Xu, X.-J., Xu, Y., Yin, Q., Zhu, J.: Full-Spectrum Analysis of Gravitational Wave Production from Inflation to Reheating. 2505.08868

Index

M. Laine and S. Procacci, *From Inflation to Hot Big Bang*, Lecture Notes in Physics 1047, https://doi.org/10.1007/978-3-032-09893-1